Lecture Notes in Management and Industrial Engineering

Series Editor

This book series provides a means for the dissemination of current theoretical and applied research in the areas of Industrial Engineering and Engineering Management. The latest methodological and computational advances that can be widely applied by both researchers and practitioners to solve new and classical problems in industries and organizations contribute to a growing source of publications written for and by our readership.

The aim of this book series is to facilitate the dissemination of current research in the following topics:

- Strategy and Entrepreneurship
- Operations Research, Modelling and Simulation
- Logistics, Production and Information Systems
- Quality Management
- Product Management
- Sustainability and Ecoefficiency
- Industrial Marketing and Consumer Behavior
- Knowledge and Project Management
- Risk Management
- Service Systems
- Healthcare Management
- Human Factors and Ergonomics
- Emergencies and Disaster Management
- Education

More information about this series at https://link.springer.com/bookseries/11786

Carmen Avilés-Palacios · Miguel Gutierrez
Editors

Ensuring Sustainability

New Challenges for Organizational Engineering

Springer

Editors
Carmen Avilés-Palacios
E.T.S.I. de Montes, Forestal y del Medio Natural
Universidad Politécnica de Madrid
Madrid, Spain

Miguel Gutierrez
E.T.S.I. Industriales
Universidad Politécnica de Madrid
Madrid, Spain

ISSN 2198-0772 ISSN 2198-0780 (electronic)
Lecture Notes in Management and Industrial Engineering
ISBN 978-3-030-95966-1 ISBN 978-3-030-95967-8 (eBook)
https://doi.org/10.1007/978-3-030-95967-8

This Springer imprint is published by the registered company Springer Nature Switzerland AG
The registered company address is: Gewerbestrasse 11, 6330 Cham, Switzerland

Preface

The present volume compiles revised and extended versions of a selection of papers presented at the 14th International Conference on Industrial Engineering and Industrial Management (CIO 2020) which took place from 9 to 10 July 2020. The conference was promoted by Association for the Development of Organizational Engineering (ADINGOR) and European Academy for Industrial Management (AIM), and it was organized by the Universidad Politécnica de Madrid (UPM) and Universidad Carlos III de Madrid (uc3m).

The conference constituted a forum to disseminate, to all branches of academy and industry, information on the most recent and relevant research, theoretical proposals, and practice experiences in industrial engineering and operations management. The conference motto of CIO2020 was: "Ensuring sustainability. New challenges for organizational engineering." The interdisciplinary character of the papers presented at the conference reflected in the topics covered: Sustainability; Eco-efficiency and Quality Management; Strategy, Innovation, Networks and Entrepreneurship; Operations Research, Modelling and Simulation; Supply Chain Management and Logistics; Production Planning and Control; Management Information Systems and Knowledge Management; Project and Process Management; Service Systems; Human Resources and Organizational Design; Product Design, Industrial Marketing and Consumer Behaviour; Education in Organizational Engineering.

After a double-blind review process carried out by three reviewers per each submitted paper, 99 communications were selected to be presented at the conference as oral presentations or posters (the abstracts were published in "14th ICIEIM and CIO2020 Book of Abstracts" ISBN 978-84-96442-94-8). A total of 251 authors from 13 countries contributed to the works. This special volume includes 30 selected communications that, based on the reviewers' evaluation, were invited to submit revised and extended versions. The contents have been arranged in seven sections; yet, we notice the intrinsic linkage among them with the discipline of organizational engineering as the unifying thread:

- Sustainability and Circular Economy
- Supply Chain Management
- Production Planning and Control
- Operations Research
- Project and Process Management
- Business Management, Entrepreneurship and Innovation
- Organizational Engineering Education

We would like to express our gratitude to all the authors for their quality contributions and members of the Program Committee for their valuable aid throughout the review process.

Madrid, Spain
March 2021

Carmen Avilés-Palacios
Miguel Gutierrez

Contents

Part I
Sustainability and Circular Economy

Chapter 1
Users or Taxpayers? Drafting a Pay-As-You-Throw Programme for Madrid's Districts

R. del-Amo and R. Carrasco-Gallego

Abstract Most Spanish municipalities, including Madrid, cover the costs of their waste management programmes through fixed fees or directly from taxes, regardless of the amount of waste generated in each household. In this study, we investigate the features of variable fee schemes for municipal solid waste services and we identify two successful European case studies of pay-as-you-throw systems. Next, we set the foundations for the design of a variable rate pricing system in Madrid, complemented with a survey capturing some key aspects required to gain citizenship acceptance. The results show that approaching waste services design from an usage perspective, just like other utilities, such as electricity or gas, provides economic incentives for waste prevention and recycling increased rates, and it is aligned with the European objectives of diverting waste from landfills.

Keywords Municipal solid waste (MSW) · Circular economy · SDG 11 cities

1.1 Introduction

The rise of Municipal Solid Waste (MSW) in large cities such as Madrid leads to more complex management and treatment system. The increase in MSW generation is mainly due to overpopulation society, economic growth and higher consumption rates [5]. The latest report on MSW generation by the Spanish Ministry of Ecological Transition (MITECO) [8] shows that each citizen produced in average 385 kg of waste that year, which 65% ended in landfills, 18% in energy valorization treatment plant and only 13% was recycled in 2016. The actual solid waste management in Madrid is not only unsustainable but also it has led to many issues between municipalities within the region.

R. del-Amo (✉) · R. Carrasco-Gallego
Department of Industrial Management Engineering, Técnica Superior de Ingenieros Industriales, Universidad Politécnica de Madrid, C/José Gutiérrez Abascal 2, Madrid, Spain

C. Avilés-Palacios and M. Gutierrez (eds.), *Ensuring Sustainability*, Lecture Notes in Management and Industrial Engineering,
https://doi.org/10.1007/978-3-030-95967-8_1

In this context, it is of pivotal importance to move from a linear to a circular economy with the aim of understanding waste as a resource, encouraging its prevention and source separation activities. The European Union has recently issued a number of new ambitious goals dealing with MSW management:

1. "Member States shall take the necessary measures to ensure that by 2035 the amount of municipal waste landfilled is reduced to 10% or less of the total amount of municipal waste generated (by weight)". (Directive 2018/850 on the landfill of waste)
2. Increasing recycling rates for 2025 by weight for specific materials in packaging waste; 50% of plastic; 70% of glass; 75% of paper and cardboard; 50% of aluminium; 70% of ferrous metals. (Directive 2018/852 on packaging and packaging waste)
3. Member States must take measures to prevent waste generation such as promote sustainable production and consumption models, re-use of products or encourage the reduction of food waste in households. (Directive 2018/851 on waste and Directive 2018/852 on packaging and packaging waste).

In order to achieve these goals, member states are allowed and encouraged to use economic instruments, for instance landfill and incineration fees or Pay-As-You-Throw (PAYT) schemes.

A research carried out by the European Environment Agency in 2016 shows that PAYT programmes help to reach a recycling rate up to 45% of the total MSW generated [4]. In PAYT systems, citizens are charged a variable fee for the amount of residual waste, either on the volume or on the weight, thrown away. These programmes create a direct economic incentive to prevent waste, to foster re-use, refurbishment and recycling and to apply the "polluter pay" principle [13]. The most common charge method for financing the waste management system in Spain is through a flat rate scheme in which the fee is applied independently of the amount of waste generated, so it does not provide any incentive to reduce the waste or recycle [9].

The structure of the paper is as follows. Section 1.2 describes the objectives and methods applied to conduct the research. Section 1.3 carries out a literature review to describe the state of the art on PAYT programmes. Section 1.4 proposes an approach for implementing a PAYT system in Madrid, while Sect. 1.5 describes the results of a survey on citizenship willingness to accept a variable charge for MSW management. Finally, the most relevant conclusions of this research are provided in Sect. 1.6.

1.2 Background and Methodology

Researchers have found that MSW generation is one of the most challenging issues that urban areas must face due to the large and non-stop increasing amounts of household waste [1]. This research studies an economic incentive known as pay-as-you-throw, designed to encourage citizens' awareness on the amount of waste they generate and to promote a more sustainable use of resources.

The methodology used in this research is based on two main approaches. First, a literature review followed by a comparative case study research in which we study the best practices found after implementing the PAYT system. Secondly, a survey has been designed and delivered to Madrid citizens to quantify the environment awareness among the population as well as the PAYT programme acceptance might have, once it is implemented.

1.3 Literature Review and Benchmarking

This chapter further analyses the actual context and legislation regarding MSW in Europe through a literature review approach and a case study research in which we describe the best practices and results found after applying the PAYT system (Fig. 1.1).

The figure above describes the amount of MSW generated (kg per inhabitant) in EU-27 and in Spain in 2018 (Eurostat). Even if the total amount of waste produced was similar, the proportion that ends up in landfills differs significantly (23.6% in the 27 European countries while 50.1% in Spain). Madrid's city performance is below average in the Spanish context: the latest available data show that in 2016 landfills received the amount of 2.5 millions of tonnes of household waste (65% of the total MSW that was produced in Madrid city centre) equivalent to 342 times Eiffel Tower's weight.

The European Union has approved the restriction of waste landfilled up to a 10% of the total amount generated for all member states by 2035. This new legislation aims

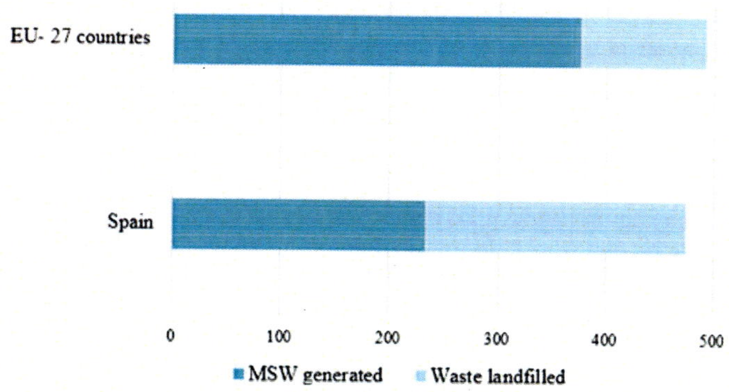

Fig. 1.1 MSW generation and disposal in landfills (kg per inhabitant) in European member states and Spain in 2018 (own development from Eurostat database)

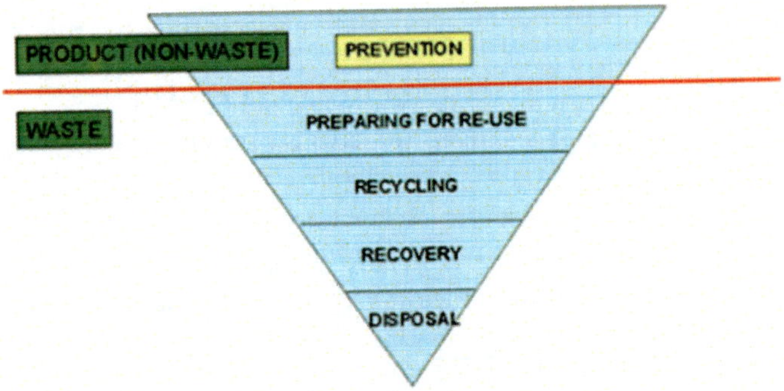

Fig. 1.2 Waste management hierarchy (Directive 2008/98/CE on waste)

at enforcing society to follow the waste hierarchy proposed in Directive 2008/98/CE (shown in Fig. 1.2), which has revealed an effective strategy to achieve several goals in terms of waste management.

The PAYT system is considered an economic incentive that follows the waste hierarchy in which it promotes "the best waste is no waste" through reinforcing prevention. This system lies in applying a proportional fee based on the amount of waste that a citizen generates and therefore the corresponding service obtained for its management. The most common methods to define the service fee are [3]:

1. Volume based (container volume): size of the container
2. Weight based
3. Pickup frequency based: actual number of emptying the container
4. Volume based (actual volume collected).

Once the fee has been defined, how to identify the user that throws the waste is the next step. To do so, there are many ways, for instance using a RFID access card to open the container, identifying the individual containers if the fee is defined based on the pickup frequency or its size. Another method is using standardized bags for the fraction of waste the system charges (option 4, volume-based fee) [13, 9].

The different waste fractions are paper and cupboard, glass, packaging, organic and residual. Researchers have found that charging the amount of waste that goes to residual fraction[1] and providing the others free of charge is the best way to encourage not only reduction of residual waste but also to improve the source separation activities [2].

Some undesirable practices or behaviours would come up after applying this system such as illegal dumping or disposal the waste in the wrong fraction container because is free of charge. However, surveys showed low actual incidence of illegal dumping problems [6].

[1] Residual waste is the fraction of solid waste that is non-compostable and non-recyclable. Therefore, it ends in sanitary landfill.

This research identifies, through a case study approach, two regions that have achieved the best results after applying the PAYT system. Firstly, the region of Flanders in Belgium where the recycling rate have been increased to a 51%; the landfill rate decreased to a less than 1% and prevention have been encouraged through this system due to the reduction of 10% of total waste generation in 4 years (Public Waste Agency of Flanders).

The second region that has been analysed is Aschaffenburg in Germany that achieved a 74% recycled rate and 0% landfill rate, and the 86% of the total MSW generated came from selective collection which means that the citizens in this region actively participate in recycling their waste.

Therefore, we can conclude that applying a PAYT system not only improves the recycling rate but also promotes awareness of environmental protection with the aim of reducing waste generation and its disposal in the right container.

1.4 Madrid Implementation Proposal

As the capital city of an EU-27 member state, Madrid must accomplish the ambitious goals in terms of waste diverted from landfills set up in the new directives. However, our review on the current solid waste management system shows that holistic policies encouraging the waste hierarchy are still lacking. The local government has made some improvement in the last two years regarding selective collection and encouraging prevention through communications campaigns. In addition, separate collection for organic waste has been gradually introduced in Madrid's 21 districts in order to increase the composting treatment rate, which at the same time improves the amount of MSW diverted form landfill. However, the percentage of organic waste is roughly 30% of the MSW weight [7], so additional measures will be required to reduce the amount of waste that ends in landfills even more significantly. Thus, the implementation of a PAYT scheme is proposed in this context.

The PAYT system that we proposed is based on charging a variable fee depending on the amount of waste thrown into the residual fraction container (orange bin), meanwhile the other fractions remain free of any charge (Fig. 1.3). This decision is made due to the outcomes of many researches in which the authors highlight that is

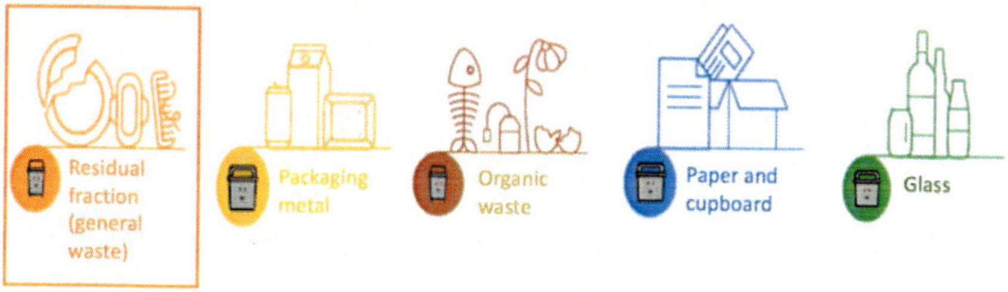

Fig. 1.3 Selective waste containers in Madrid (own development)

the best way to reduce waste generation and improve source separation activities [2] and because of our conclusion in Chap. 3 after studying the PAYT system that have been already running in different countries.

To determine the most effective way to identify the user and measure the waste, it is required to study the current waste collection system in the city. As of November 2020, there are two ways of household waste collection in Madrid either through large containers placed in the street (800, 2400 or 3400 L) or through door-to-door collection with small or individual containers (120, 240 or 320 L). Madrid's citizens are encouraged to participate in the selective waste collection through the use of selective waste (large and individual) containers shown in the Fig. 1.3.

The method proposed to identify the user in a large container is with RFID access card to open the dumpster. In order to measure the waste, the container must be equipped with a weight sensor. The process is depicted in Fig. 1.4a.

Regarding the door-to-door collection, we propose to identify the small container through a barcode tag and measure the waste by the size of the container (volume-based fee). Therefore, the charge is calculated according to the container's volume times the pickup frequency (Fig. 1.4b). For instance, in a building in Madrid City centre equipped with a 240-L container, the fee will be more expensive than in the case of a 120-L container.

In order to prevent undesirable behaviours, the following actions could be applied: firstly, providing all containers with electronic locks and weight sensors to keep track of the amount of waste that a citizen throws in each container. Doing so, we are able to calculate the average waste that is dumped in each fraction and identify the users that are throwing a significant deviate quantity from the mean. For instance, if the average amount generated is 385 kg/inhabitant per year in which 33% corresponds to food

Fig. 1.4 **a** PAYT implementation for large containers in Madrid (own development) and **b** PAYT implementation for door-to-door collection in Madrid (own development)

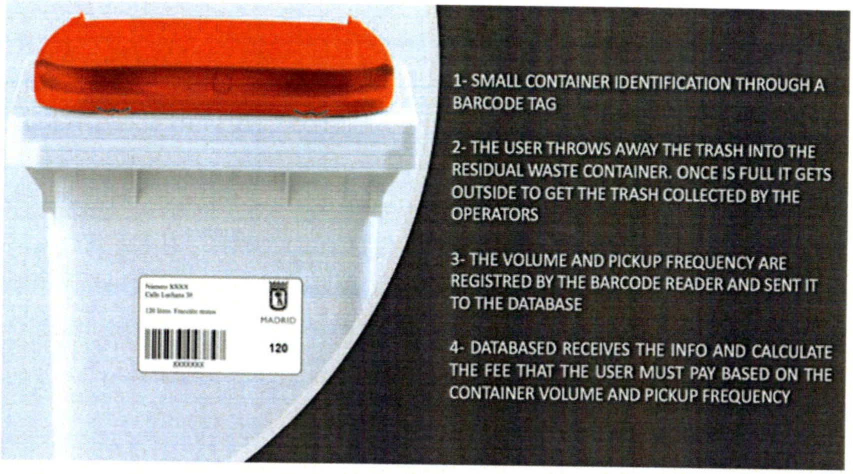

Fig. 1.4 (continued)

waste that must go to organic fraction container and if a household is throwing nearly 70% to this container, it might be the case of an undesirable practice in which the citizen throws organic and residual waste into the same container in order to avoid the charge placed for residual fraction. Secondly, encouraging waste collectors to found and notify unwelcome behaviours. Thirdly, promoting awareness campaigns about the importance of every action made.

Since the proposed PAYT system directly engages Madrid's citizens, it is reasonable that prior to the implementation, the local government should give the society the possibility of expressing themselves about the system. Hence, changes on the current regulation and logistic processes of collective MSW are needed. Researchers agreed that the steps that should be followed before implementing a PAYT are the following [10]. Firstly, consult citizens' opinion through for instance a citizen participation forum which could lead to Changes on legislation (taxes). Study and implementation of the best technical and logistic solution. After that one of the most important stages is the communication campaign. Later on, the local government should select one area or district to start implementing the PAYT with the aim of monitoring and controlling any problem that might happen. Figure 1.5 shows an estimation about the duration of each phase.

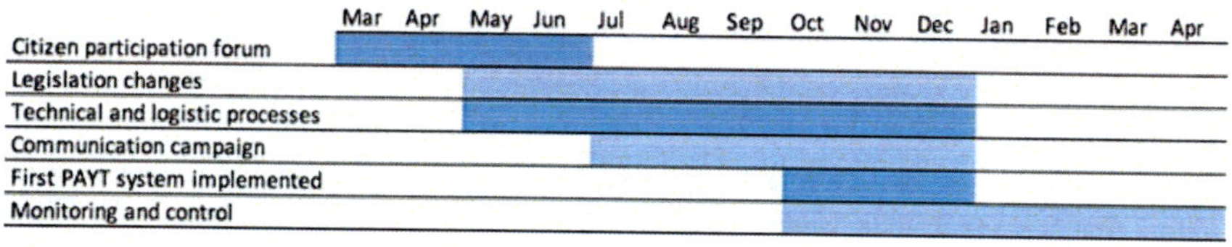

Fig. 1.5 Stages (adapted from Puig Ventosa et al. [10])

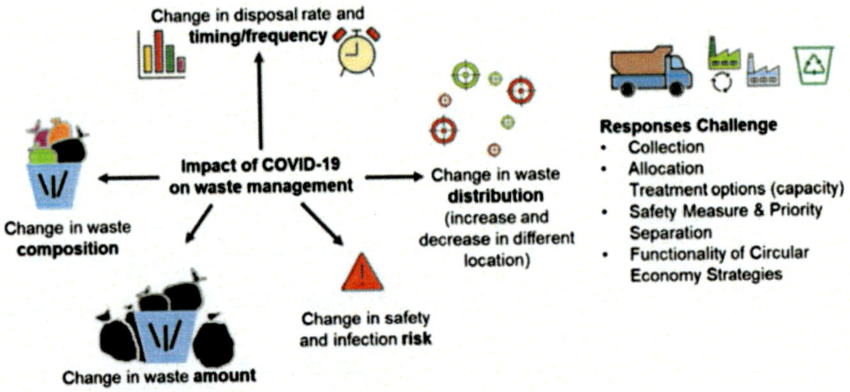

Fig. 1.6 Impacts and challenges of COVID-19 on waste management [11]

As of November 2020, there are several regions in Spain in which the PAYT system has been successfully implemented. Those are mainly in Catalonia (15 districts) and in the Balearic Islands where the PAYT has been implementing progressively through all the regions.

The COVID-19 pandemic has brought us new challenges for solid waste management such as the significantly increase in medical waste, toxic residues, plastic packaging, single used items such as masks, gloves or hand sanitizers. For instance, the medical waste in China (Hubei Province) has increased by 600% from 40 tonnes per day to 240 tonnes [12]. Researchers have found that common plastic packaging waste (PP, PET, etc.) is likely to continue increasing due to online shopping [11]. Therefore, now more than ever we do need a change to improve the efficiency of the MSW system and spread awareness about the huge impact of our waste. Figure 1.6 shows the impact and challenges that waste management systems have been facing due to the pandemic.

1.5 Survey Results

A survey has been conducted to quantify society's environmental concerns and PAYT acceptance. The survey has been delivered to Madrid's citizens mainly through social media platforms. Even though these platforms are more frequently used by millennials, we have obtained answers from different ages. Figure 1.7 shows the age ranges of the interviewees. Respondents were anonymous, so personal data were not collected during the surveying process. The purpose of the study was clearly described in the heading of the questionnaire and respondents consented the use of their responses for that purpose.

Two hundred seventy-eight answers were received in a two-week time frame. The major outcomes of the survey are: first is the rise of environmental concerns due to

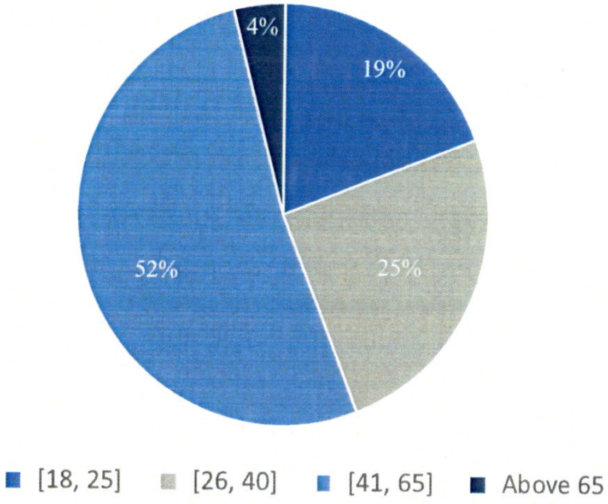

Fig. 1.7 Interviewees' age range (own development)

the high percentage obtained about "How important is environment protection for you?" with nearly the 100% (97%) answering "high relevance". Second, the 97% respondents claimed that they ignore any information regarding waste management in Madrid (recycling rate; collection and treatment costs; landfill rates). Third, the 65% replied that they consider it reasonable to apply a "pay-as-you-throw" and "polluters pay" principles. Lastly, the following question was asked "Why do you think some citizens tend to not recycle their domestic waste?" and 50% replied that the waste system is not trustful meaning that they recycle doing the source separation activities but ignore what happens with their waste if, for instance, was finally recycled or landfilled.

Survey results show that a relevant share of citizens distrusts Madrid's MSW management system, and most of them ignore the costs and results that come out from waste treatment and collection. Local authorities could possibly increase their transparency and communication efforts in this regard.

1.6 Conclusions

Through our work, we have identified a complex and unsustainable situation in Madrid concerning municipal solid waste management. PAYT system has been identified as one of the possible courses of action that, combined, could address this challenge. Variable fee schemes significantly improve recycling rates while reducing the waste dumped to landfills. Since PAYT systems price the service depending on the amount of waste thrown away, users have a clear economic incentive to prevent waste generation and engage in source separation activities. PAYT is a transparent and fair system that applies the principle "polluters pay" and transfers the costs of waste collection and treatment services from taxpayers to users. However, it is not exempt of difficulties in its implementation such as free-riding behaviours, impact on low-income households, waste management costs distribution in residents' associations and acceptance of a variable fee by citizenship in general (that currently show a very scarce knowledge on their MSW system as the survey results reveal). Those challenges will be further researched in upcoming studies.

Compliance with Ethical Standards The Ethical Committee for Research Activities at UPM ("Comité de Ética de actividades de I + D + i de la UPM", https://www.upm.es/Investigacion/soporte/ComiteEtica) approves the survey procedure conducted in this research. No personal data of the respondents of the survey were collected or treated (survey respondents were anonymous); hence, there is no need for a particular reference for the authorization for the experiment, as mentioned in the waiver (MOD.6 Identificación de Cuestiones Éticas en los Proyectos, Contratos, Subvenciones o Colaboraciones).

References

1. Al-Khatib IA, Monou et al (2010) Solid waste characterization, quantification and manage ment practices in developing countries. A case study: Nablus district–Palestine. J Environ Manag 91(5):1131–1138
2. Connett P (2013) The zerowaste solution: untrashing the planet one community at a time. Chelsea Green Publishing
3. Dohogne J-J (2014) Waste management costs & financing and options for cost recovery
4. Environment European Agency (EEA) (2016) Municipal Waste Management across European Countries
5. Guerrero LA, Maas G, Hogland W (2013) Solid waste management challenges for cities in developing countries. Waste Manage 33(1):220–232
6. Hogg D, Wilson D et al (2006) Modelling the impact of household charging for waste in England. Department for Environment Food and Rural Affairs (defra). Eunomia Research and Consulting Ltd
7. Madrid City Council (2019) Estrategia de Prevención y Gestión de Residuos
8. MITECO, Ministry for Ecological Transition and Demographic Challenge (2016) Memoria annual de generación y gestion de residuos de competencia municipal
9. Puig-Ventosa I (2008) Charging systems and PAYT experiences for waste management in Spain. Waste Manage 28(12):2767–2771
10. Puig Ventosa I, Calaf Forn M, Mestre Montserrat M (2010) Guía para la implementación de sistemas de pago por generación de residuos. Agència de Residus de Catalunya
11. Sharma HB, Vanapalli KR, Cheela VS, Ranjan VP, Jaglan AK, Dubey B, Goel S, Bhattacharya J (2020) Challenges, opportunities, and innovations for effective solid waste management during and post COVID-19 pandemic. Resour Conserv Recycl 162:105052
12. Van Fan Y, Jiang P, Hemzal M, Klemeš JJ (2020) An update of COVID-19 influence on waste management. Sci Total Environ 754:142014
13. Zeschmar-Lahl B, Schoenberger H et al. (2016) Background report on best environmental management practice in the waste management sector. Report for the European Commission's Joint Research Centre, Seville

Chapter 2
Circular Economy Analysis Applying Ellen MacArthur Model: Spanish Glass Sector Case

J. Morcillo-Bellido, E. Santos-Iscoa, and L. Isasi-Sánchez

Abstract Circular economy (CE) has reached a quite relevant position on academic research studies nowadays, and it has been considered a key driver for long-term organization supply chains sustainability and competitiveness increase. Some theoretical models have been developed trying to identify the most relevant key parameters and conditions that allow to define a specific supply chain, or a sector or organization, as "circular economy practice". Between the models, one of them was developed and established by the well-known Ellen MacArthur Foundation (EMF). This research tries to analyze the applicability and fulfillment of EMF model to the Spanish glass sector, through a deep analysis of real practices already applied in this industrial sector. After the analysis, it is possible to infer that this sector meets almost all the EMF model characteristics. Nevertheless, there is still some run for circularity improvement in the sector, since still renewable energy application is below current demanding standards.

Keywords Circular economy · Spanish glass sector · Sustainability · Circular economy case study

2.1 Introduction

Since the beginning of the First Industrial Revolution, companies have been focused on exploiting resources to create cheap and massive products, and as consequence of that has been created a society with a large capacity of waste generation. Final result has been a high consumption of new additional resources to produce massive consumer products and a large amount of waste [1, 2]. The increasing interest forced the development of the well-known Brundtland report [3].

J. Morcillo-Bellido (✉) · E. Santos-Iscoa · L. Isasi-Sánchez
Escuela Politécnica Superior, Área de Ingeniería de Organización, Universidad Carlos III de Madrid, Avenida de la Universidad nº 30, 28911 Leganés, Madrid, Spain
e-mail: morcillo@ing.uc3m.es

L. Isasi-Sánchez
e-mail: lisasi@ing.uc3m.es

C. Avilés-Palacios and M. Gutierrez (eds.), *Ensuring Sustainability*, Lecture Notes in Management and Industrial Engineering,
https://doi.org/10.1007/978-3-030-95967-8_2

This report detailed the environment deterioration and the need to take care of the ecosystem through a better use of raw material and product´s consumption, in order to achieve a sustainable economic growth [3–5].

The concept of sustainable development was defined based on the three main sustainability pillars: environmental, economic and social [3]. According to [6], a high qualified economic system is able to meet the sustainable characteristics of taking care for the environment, achieving competitive economic objectives and improving social parameters. Following [7], the sustainability and the ecosystem have become a top priority integral issue being able to combine social, environmental and economic angles. Considering the environmental angle, circular economy (CE) is based on using resources in a very efficient way, but always seeking to reduce waste and emissions [8].

Natural resources are finite, and the demand is increasing daily due to the exponential population growth (from a total population of 2.6 billion people in 1950 to more than 7 billion people in 2020). It worries many collectives, who have developed different CE concepts over last decades. A summary of its evolution is shown in Table 2.1.

Most recent studies [17, 18] on CE propose a design adaptation in the supply chain, increasing quality and speed of manufacturing using resources efficiently. This leads to higher supply chain performance in terms of reused ability maintaining its technical properties [19].

In order to promote a social, economic and environmental balance, Unites Nations (UN) sponsored The Kyoto Protocol (1997), aiming to reduce greenhouse gases, improving the worldwide quality of life and reducing environment issues [20, 21]. After some initial trials, the so-called 2030 Agenda aims to achieve sustainable planet development in a medium-long term and implementing a friendly environmental

Table 2.1 Circular economy concept evolution. Source: author´s elaboration

Industrial ecology [9]	It proposes to study the materials inflows and outflows, using a sustainable system applying renewable energy and an efficient resources use
Regenerative design [10]	The supply chain is a self-supply closed system. Balance social and environmental needs in highly promoted
Performance economy [11, 12]	It proposes an economy working in loops, with a longer useful life of the product, and a reconditioning products for its new use
Cradle to cradle [13]	It bases the supply chain on the environment, where all organisms can be connected to each other generating zero waste
Blue economy [14]	The waste of one system becomes part of the input of another, without waste generation, and always, using the cascade concept
Ellen MacArthur model [15, 16]	Expose the feasibility and profitability of applying a circular economy. Model increase its relevancy since there is a high interest on finding real application on business environment

global circular economy, where the use of renewable energy is encouraged and all products must increase their capability to be reused [22–24].

To achieve stablished objective, 169 goals were defined based on the seventeen UN Sustainable Development Goals, which are interrelated and connected to three different angles: (i) social, such as the end of poverty or hunger, (ii) environmental, like ecosystem care; and (iii) economical, where entities must generate added value through business innovation on circular products and higher renewable energy use [22, 23, 25, 26].

Ellen MacArthur Foundation (EMF) has developed a circular economy model concept embedded into UN Sustainable Development Goals. Following EMF companies and governments must jointly work to improve the transformation processes toward a more efficient system.

In this research, authors conduct a research to apply EFM to a sector practices, and glass sector has been chosen as study field. Main study objective will be related to the EMF application feasibly to this particular industry.

2.2 A Reference Circular Economy Model

The CE aims to give maximum utility and reuse to resources [1, 27] by manufacturing products able to be reused, recycled and/or repaired [15, 28]. As consequence, balance could be achieved in the system in order to mitigate the supply risk [29, 30]. To create a product that fulfills the circular characteristics, it must be considered that in the last phase of the product life circle, the raw material must be recovered as much as possible [15, 31]. To get the most out of resources, [32] created some theories in order to avoid waste in the transformation process. This also entails a lower use of energy, trying to use renewable as much as possible because of the current situation of environmental risk [16, 17].

On top of that, Ellen MacArthur proposes to interconnect different systems in such a way that waste materials from a system are useful as new raw material for others [15]. For that reason, [15] has developed during last decade a conceptual model that tries to integrate certain characteristics that should be fulfilled by any organization that it could be considered a "circular economy organization" (see Table 2.2).

This EMF circular model tries to keep inside all the raw material waste generated by different systems as a loop. This means that scraps from one of the system could be a manufacturing input for other [15]. It means less externalized traditional environmental costs and other associated risks related to waste management [8]. Organizations should achieve the highest level of resource self-recovery to reduce the generation of non-reusable waste [33].

Table 2.2 Circular economy characteristics based on Ellen MacArthur Foundation model Source: author´s elaboration based on EMF 2014

	Characteristic 1: The inner circle When the end user considers the product is a waste, product has the possibility of being reused, recycled, remanufactured or shared within its own system
	Characteristic 2: Infinite loop The product considered waste can reproduce an inner circle in an infinite way. Becoming input of the same system without losing the quality it had at the beginning
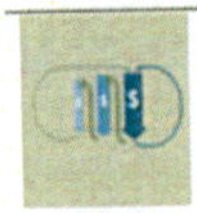	Characteristic 3: Cascade manufacturing The raw material can be part of several interconnected systems among themselves. For example, a garment can be part of its own system by reusing it, or it can be part of another subsequent system (e.g., cotton fiber)
	Characteristic 4: There are no losses An essential characteristic of the circular economy is that the raw material remains within the value chain, without the existence of toxic waste or emissions

2.3 Objective and Study Methodology

This study is part of a global research line that aims to deepen the understanding of circular economy practices and also to make specific proposals for improving management practices, potentially applicable to other sectors. The objective of this particular study was to check if EC practices at Spanish glass sector fit, or not, with a specific EC model characteristics proposed by Ellen McArthur Foundation.

In an initial step of the study, an extensive search of published data (basically through web searching in several web pages from Spanish sector entities) regarding the glass industry has been performed, and due to the specific nature and novelty of this issue, it was decided to carry out a case study.

Case study method, according to Eisenhardt [34], Rialp [35] and Voss et al. [36], is a very suitable tool for issues related to strategic management decisions. Yin [37] suggests using case study analysis when boundaries between the context and the phenomenon to observe are not obvious. Case studies have been gradually recognized, despite some criticism, as a very correct way "to address contemporary organizational problems and established credibility" [38].

2.4 Circular Economy in Spanish Glass Sector. Case Study

2.4.1 Circular Economy at the Sector

Spanish glass manufacturers and bottling companies are organized in large sectorial associations divided by products (flat glass, bottles, ...) to carry out specific technical practices. The most important associations are ANFEVI (manufacturers of glass bottles), which produces 66.4% of total Spanish glass consumption and has a turnover over 1,000 million euros; and FAVIPLA (manufacturers of flat glass) with a share of 26.4% of total Spanish glass consumption and 700 million euros in turnover [39].

Glass is a non-biodegradable material, so if it was not recycled, it would mean a serious environmental problem [40]. The history of glass recycling in Spain begins in 1980 when recycling it was applied to a production methodology to recycle the entire bottle without generating non-biodegradable waste [41]. At that time, glass recycling started to be promoted as a fundamental element to obtain energy savings in smelting furnaces, getting the possibility of raw materials recovering to reuse in new bottle manufacturing. In this regard, hollow glass companies were the first Spanish industrial sector to launch a comprehensive recycling system in collaboration with local entities and citizens [42]. Recycling in this industry achieved following indicators level by 2018:

- 230,950 glass "collection containers", which facilitates the empty bottle collection and recycling, reach 896,450 tons in 2019 [41].
- 51% of bottle manufacturing raw materials are sourced within a radius of less than 300 km [39], what optimizes supply chains resources use.
- A ton of recycled glass saves 1.2 tons of raw materials, 30% of energy and the emission of 670 kg of CO_2 [43]. In Spain, 84% of people recycle glass which means that 24.4 million MW of energy and 13.2 million tons of raw materials have been saved in the last 20 years. Also, 7 million tons of CO_2 have been reduced as pollution emissions to the atmosphere [39].
- A recycling rate of 76.5% is obtained in Spain (calculated as a ratio between recycle and sold bottles). This ratio is higher than the European Union environmental objectives within the 2030 Agenda. It implies a saving of more than 1 million tons of raw material, which means avoiding more than 500,000 tons of CO_2 into the atmosphere [40].

The Spanish legal normative process took several years. These recycling activities were regulated by EU Directive 94/62/CE [44] and its transposition into Ley Orgánica 11/1997 [45] regarding containers management and waste within their supply chain. This was intended to prevent and reduce the impact on the environment, recovering waste throughout bottle life cycle [45]. Given this regulation, glass manufacturers created an integrated management system (SIG) called ECOVIDRIO, which is a non-profit association responsible for the glass collection and recovery [41].

In 2011, a new legal regulation (Decreto ley 22/2011) [46] regarding contaminated soils and waste recovering established full responsibility on the producer. In this case, glass companies must recover empty bottles at the end of the supply chain and from soils [46]. SIG was passed to be called as SCRAP (expanded collective responsibility producer system).

Spanish companies have introduced new techniques which reduce energy consumption and therefore CO_2 emissions. For instance, they are preheating the furnace incoming air because energy saving is always a priority objective for manufacturers. More than 70% of total energy consumption corresponds to furnace heating, and they consume a high amount of energy, between 20 and 30% of the production cost [43]. As consequence of all these efforts, it has achieved a reduction of 50,000 CO_2 tons of emissions per year.

2.4.2 *Circular Economy Model (EMF) Applied to Glass Sector*

In order to understand the Spanish glass bottle supply chain, the study puts attention on the area of the SC that includes the processes from bottles manufacturing to their recovery and reuse as raw material into manufacturing process. The filled bottles follow alternative channels within the supply chain. On the one hand, the filled bottles can be distributed in retail establishments (restaurants, retail shops, etc.) which are responsible for returning the bottles already used to the bottler company (such as a returnable bottle) to be cleaned and refilled. And, on the other hand, at the consumer level, the bottle is recovered through the "igloo" containers, which are collected by the SCRAP collection systems and returned to the bottler company through the own SCRAP. Thus, the raw material always remains within the system itself, there is no residue or waste, and there is a positive impact on energy saving and materials on the whole production system. To analyze whether the glass industry contains in its supply chain practices of CE, its practices are analyzed through the glass life cycle in relation to the model developed by Ellen MacArthur Foundation [15] as described in Table 2.3.

2.5 Conclusions

From the analysis that has been managed, when the EMF circular model is applied to the Spanish glass sector, it could be possible to infer with respect to characteristics 1 (inner circle) and 2 (infinite loop) that the glass sector widely complies with both characteristics due to the fact that being a material that can practically be reincorporated into the productive process as raw material many times and in many different products. So, both characteristics are fully found and clearly verified in the practices of the Spanish glass sector.

Table 2.3 Summary of EMF model application to Spanish glass sector

Characteristic 1: The inner circle The glass container is recyclable and becomes another bottle similar to the previous one, without losing its value, since from one container there is another of identical characteristics; therefore, it can be said that it meets this requirement at 100%
Characteristic 2: Infinite loop In the case of glass containers, there are two options: the returnable bottle and the non-returnable one (one way). In the case of one way products, the packaging is recycled. In the case of refillable containers, the container becomes packaging plant to be reused for an indefinite number of cycles
Characteristic 3: Cascade production Due to its inert material capacity, it can be used in infinite applications. For example, 69% of glass wool comes from recycled glass
Characteristic 4: There are no losses All collected bottles are reused or recycled

Source: author´s own elaboration

On top of that, being the glass a material that also constitutes a relevant input for other sectors manufacturing, such as decoration and construction, it can feed other sectors supply chains as an important raw material. That fact totally fulfills the specific requirements that the used reference model (EMF) establishes as "cascade production".

Respect to the last characteristic of the EMF model described as "there are no losses". The recycling and reuse levels managed by companies operating within the Spanish glass sector have achieved a degree of efficiency that the waste in terms of materials and pollution generated by those companies is really a very low percentage of used resources. However, even though this sector nearly approaches to the total fulfillment of this characteristic in the way in which every returned glass bottle is reincorporated as a raw material to the production process, it does not fully comply with EMF model because still it does not use exclusively renewable energies along their supply chain.

From the analysis carried out, it could be possible to infer that the Spanish glass sector complies with most of the characteristics established by the circular economy EMF model, since the recycling and reuse of bottles in a continuous way have been internalized as part of its sector transformation process, without losing its properties and quality of use. Still there is a possibility of improving the sector's circularity performance by expanding the use of renewable energy in its manufacturing process and increasing the level of consumer recycling awareness.

References

1. Lieder M, Rashid A (2016) Towards circular economy implementation: a comprehensive review in context of manufacturing industry. J Clean Prod 115:36–51
2. Sariatli F (2017) Linear economy versus circular economy: a comparative and analyzer study for optimization of economy for sustainability. Visegrad J Bioecon Sustain Dev 6(1):31–34
3. WCED (1987) Report of the world commission on environment and development: our common future acronyms and note on terminology chairman's foreword. Oxford University Press, Brundtland
4. Daly HE, Von Droste B (1997) Medio ambiente y desarrollo sostenible: más allá del informe Brundtland (No. 333.7 MEDamb)
5. Rexhague J, Murphy D (2010) Sustainable development: From Bruntland to Rio 2012 (Background Paper), UN
6. Shrivastava P (1995) The role of corporations in achieving ecological sustainability. Acad Manag Rev 20(4):936–960
7. Elkington J (1997) Cannibals with Forks: the triple bottom line of 21st century. Capstone, Oxford
8. Geissdoerfer M, Savaget P, Bocken NM, Hultink EJ (2017) The circular economy—a new sustainability paradigm? J Clean Prod 143:757–768
9. Frosch RA (1992) Industrial ecology: a philosophical introduction. Proc Natl Acad Sci 89(3):800–803
10. Lyle TJ (1994) Regenerative design for sustainable development. Editorial Wiley and Sons, Nueva York. ISBN 978-04-7155-582-7
11. Stahel W (1994) The utilization-focused service economy: resource efficiency and product-life extension. The greening of industrial ecosystems, pp 178–190
12. Stahel RW (2010) The performance economy, 2nd edn. Basingstoke, Palgrave Macmillan, pp 6–7
13. McDonough W, Braungart M (2002) Cradle to cradle: remaking the way we make things. North Point Press
14. Pauli GA (2010) Blue economy
15. Ellen MacArthur Foundation (2014) Towards the circular economy: accelerating the scale-up across global supply chains, vol 3. https://ellenmacarthurfoundation.org/towards-the-circular-economy-vol-3-accelerating-the-scale-upacross-global
16. Ellen MacArthur Foundation (2015) Delivering the circular economy - A toolkit for policy-makers. https://ellenmacarthurfoundation.org/a-toolkit-for-policymakers
17. Morseletto P (2020) Targets for a circular economy. Resources, Conservation and Recycling 153:104553
18. Slorach PC, Jeswani HK, Cuéllar-Franca R, Azapagic A (2019) Environmental and economic implications of recovering resources from food waste in a circular economy. Sci Total Environ 693:133516
19. Kuo-Yi L (2018) User experience-based product design for smart production to empower industry 4.0 in the glass recycling circular economy. Comput Ind Eng 125:729–738
20. Grubb M, Vrolijk C, Brack D (2018) Routledge revivals: Kyoto protocol (1999): a guide and assessment. Routledge
21. Santilli M, Moutinho P, Schwartzman S, Nepstad D, Curran L, Nobre C (2005) Tropical deforestation and the Kyoto protocol. Clim Change 71(3):267–276
22. Colglazier W (2015) Sustainable development agenda: 2030. Science 349(6252):1048–1050
23. Schroeder P, Anggraeni K, Weber U (2019) The relevance of circular economy practices to the sustainable development goals. J Ind Ecol 23(1):77–95
24. Tsalis TA, Malamateniou KE, Koulouriotis D, Nikolaou IE (2020) New challenges for corporate sustainability reporting: United Nations' 2030 Agenda for sustainable development and the sustainable development goals. Corporate Social Responsibility and Environmental Management

25. Ministerio de Asuntos Exteriores (MAE) (2015) Agenda 2030. Gobierno de España. Retrieved form http://www.exteriores.gob.es/Portal/es/PoliticaExteriorCooperacion/Agenda2030/Paginas/Inicio.aspx. Consultado: 23/10/2019
26. Stafford-Smith M, Griggs D, Gaffney O, Ullah F, Reyers B, Kanie N, Stigson B, Shrivastava P, Leach M, O'Connell D (2017) Integration: the key to implementing the sustainable development goals. Sustain Sci 12(6):911–919
27. Sariatly F (2017) Linear economy versus circular economy: a comparative and analyzer study for optimization of the economy for sustainability. Visegrad J Bioecon Sustain Dev 1:31–34
28. Balboa CH, Somonte MD (2014) Economía circular como marco para el ecodiseño: el modelo ECO-3. Informador técnico 78(1):82–90
29. Govindan K, Hasanagic M (2018) A systematic review on drivers, barriers, and practices towards circular economy: a supply chain perspective. Int J Prod Res 56(1–2):278–311
30. Korhonen J, Honkasalo A, Seppälä J (2018) Circular economy: the concept and its limitations. Ecol Econ 143:37–46
31. Bellopede R, Zichella L, Marini P (2020) Glass Waste3: a preliminary study for a new industrial recovery processing. Sustainability 12(5):1997
32. Ellen MacArthur Foundation (2015) Towards the circular economy. Business rationale for an accelerated transition. Ellen MacArthur Foundation, Isle of Wight
33. Ghisellini P, Ripa M, Ulgiati S (2018) Exploring environmental and economic costs and benefits of a circular economy approach to the construction and demolition sector. A literature review. J Clean Prod 178:618–643
34. Eisenhardt KM (1989) Building theories from case study research. Acad Manag Rev 14(4):532–550
35. Rialp A (1998) El Método del Caso como Técnica de Investigación y Su Aplicación al Estudio de la Función Directiva. IV Taller de Metodología ACEDE, La Rioja
36. Voss C, Tsikriktsis N, Frohlich M (2002) Case research in operations management. Int J Oper Prod Manag 22(2):195–219
37. Yin RK (1994) Case study research, vol 2. Sage, Thousand Oaks
38. Ketokivi M, Choi T (2014) Renaissance of case research as a scientific method. J Oper Manage 32(5):232–240
39. Vidrio España (2017) Contribución económica, ambiental y social del sector del vidrio en España. Vidrio España y la economía circular Balance 2014/2016. Retrieved form https://www.vidrio.org/wp-content/uploads/2017/11/informe-Vidrio-Espana-Economia-Circular-Contribucion-Economica-Ambiental-Social-2014-2016.pdf
40. Ecovidrio (2018) Informe anual sobre sostenibilidad. Retrieved from https://www.ecovidrio.es/sites/default/files/2019-11/Ecovidrio_Informe-Sostenibilidad-2018_1.pdf
41. ANFEVI (2020) ANFEVI informe. Retrieved from http://www.ANFEVI.com/
42. ANFEVI (2014) La contribución medioambiental, social y económica del sector de envases de vidrio en España. Retrieved from http://www.ANFEVI.com/wp-ontent/uploads/2016/09/estudio -economia-circular-eyenvases-de-vidrio.pdf
43. Roudier S, Sancho LD, Scalet B, Muñoz MG (2013) Best Available Techniques (BAT) Reference document for the manufacture of glass: industrial emissions directive 2010/75/EU: (Integrated Pollution Prevention and Control) (No. JRC78091). Joint Research Centre (Seville site)
44. Directiva del Consejo 94/62/CE, de 31 de diciembre de 1995, relativa a envases y residuos de envases
45. Ley Orgánica 11/1997, del 24 de abril, de envases y residuos de envases. (BOE núm. 99, del 25 de abril del 1997)
46. Ley Orgánica 22/2011, del 28 de julio, de residuos y suelos contaminados. (BOE núm. 181, del 29 de julio del 2011)

Chapter 3
Innovation in a Company Committed to Sustainability Culture

M. J. Alvarez, C. Jaca, M. Ormazabal, and J. Rincon

Abstract Sustainability and the circular economy are the answer to today's challenge of resource depletion and waste generation. The transition to circular economy requires technological and process innovations: However, the capacity for innovation has often been a point of difficulty, and the problem is worse in small-sized companies because of the scarcity of resources. Our aim is to show how a company committed to sustainability overcomes the obstacles in the innovation process.

Keywords Sustainability · Circular economy · Eco-innovation · Case study · SME

3.1 Introduction

The increase in the world's population, the expansion of middle-class consumers in developing countries and the growth of urbanized areas are inflating the demand for products and services [20]. These facts make the current production system unsustainable. Therefore, sustainability can no longer be a simple matter of reputation for companies. It is not just a matter of companies. Governments and society must address the big challenge of this millennium. Circular economy (CE), inspired by natural ecosystems, is an alternative to the non-sustainable take-make-waste system that constitutes the linear economy. The need for an alternative system is patent given the fact that a significant proportion of non-renewable resources is diminishing, and natural resource volatility is increasing. In that sense, economic growth should be

M. J. Alvarez (✉) · C. Jaca · M. Ormazabal · J. Rincon
Tecnun, Universidad de Navarra, C/Manuel de Lardizabal, 13, San Sebastián, Spain
e-mail: mjalvarez@tecnun.es

C. Jaca
e-mail: cjaca@tecnun.es

M. Ormazabal
e-mail: mormazabal@tecnun.es

J. Rincon
e-mail: jrincon@tecnun.es

C. Avilés-Palacios and M. Gutierrez (eds.), *Ensuring Sustainability*,
Lecture Notes in Management and Industrial Engineering,
https://doi.org/10.1007/978-3-030-95967-8_3

decoupled from the consumption of natural resources, which is possible with an economic model based on CE principles. CE is "an economic system that represents a change of paradigm in the way that human society is interrelated with nature and aims to prevent the depletion of resources, close energy and materials loops and facilitate sustainable development through its implementation at the micro- (enterprises and consumers), meso- (economic agents integrated in symbiosis) and macro- (city, regions and governments) levels. Attaining this circular model requires cyclical and regenerative environmental innovations in the way society legislates, produces and consumes" [14].

In recent decades, governments have been implementing policies to boost sustainability. Germany led the way with the Waste Disposal Act in 1976 and Japan with the Law for Effective Utilization of Recyclables in 1991 [6]. China was a pioneer in the circular economy with the Circular Promotion Law in 2009 [16]. The European Union's 2008 Directive 2008/98/EC on waste sets the region on the road to the European Green Deal by the end of 2019. At a global level, in 2015, the United Nations proposed 17 Sustainable Development Goals (SDGs), a broad framework for more sustainable production and consumption that is to be achieved by 2030.

3.1.1 Circular Economy and Innovation

Innovation is defined as the "generation, acceptance and implementation of new ideas, processes, products or services" [17]. The term was introduced by Schumpeter in 1939 [15] "as the introduction of new goods, new methods of production, new markets, the conquest of new sources of supply and the carrying out of a new organization of any industry". Nowadays, innovation is important for any business. In fact, it is considered to be a way to obtain competitive advantage. Companies must continually adapt its activity to the needs of their customers. To be competitive, they must discover new products or services or offer them differently. If a company wants to survive, it has to continually adapt, anticipate the needs of the clients and/or discover business models. For that reason, innovation should be present in any organization.

As previously stated, nowadays, we face big challenges such as climate change, pollution and resources depletion. Eco-innovation arises to help to mitigate these problems. The notion of eco-innovation was introduced as a sort of innovation that improves environmental performance. Eco-innovation is defined as a set of technological and non-technological innovations that prevent, mitigate and allow recovery from environmental damage. Today, companies must provide their services or products in response to social demands for sustainability and the environmental challenges and climate change. This implies that the process of innovation must aim at sustainability as one of its goals. It is understandable, then, that many authors have stressed the importance of innovation in achieving the transition toward sustainability [4, 8].

3.1.2 *The Role of the Small and Medium Enterprises (SMEs)*

SMEs are vital to the challenge of achieving a more sustainable production system, as they are an important part of the business network of any society. SMEs represent 98% of the companies in Europe [11] and 99% of all businesses and between 50 and 60% of value-added across the OECD [13]. Therefore, they must take part in this global transformation toward a more sustainable productivity system. However, most of the research on the role of business in advancing the CE has focused on large companies [19]. Besides, small companies face great difficulties into innovate due to their limited resources [12]. SMEs are very limited due to barriers such as the lack of financial and specialized professionals that are needed to be able to innovate. They usually do not have a R&D department, and they lack the required knowledge to innovate successfully. Also, failure of innovation projects may be financially catastrophic for them [2]. One way to overcome these barriers is to collaborate with other companies and organizations. There are many collaboration models: partnerships, networks, strategic alliances, etc. This collaboration aims to achieve complementarity, knowledge sharing and access to financial resources [7]. Collaboration with universities and research institutes has been proved successful [18].

A special category of companies worth paying attention to is start-ups that usually arise from an innovation. Most of them are SMEs. The difficulties that they meet are of such magnitude that the "failure rate among start-ups currently stands at an alarming rate of approximately 90%" [3]. SMEs and large firms innovate differently because they possess different organizational structures [10]. Small companies have the advantage of flexibility, and they can be more dynamic. However, as it has been mentioned above, their difficulties in innovating are quite important.

3.1.3 *Sustainability Culture*

We understand organizational values as shared and accepted values within the organization [1]. This paper aims how the culture of sustainability within a SME drives the overcoming of difficulties it faces when trying to innovate.

3.2 Methodology

We adopt an explorative case study research strategy to empirically investigate how a sustainable culture influences innovation behavior in companies. More specifically, we chose a SME whose main activity is closing the loop for a material that is problematic from an environmental standpoint: plastics. It is a company with a deep-rooted culture of sustainability. The period covered by the research is from the beginning of 2016 to the end of 2018. We have used a multi-method approach to answer our research question using data collection techniques such as semi-structured interviews and desk research. A description of the methods used is described below:

1. Semi-structured interviews with the CEO and senior manager of the company. They were conducted to gain information about the experiences of individuals [5].
2. Information available from Web sites, reports and press releases or news items, to triangulate the information and to help ensure that the data the company is answering to this research is consistent with the reports [9].

In accordance with the objective of this paper, the case study aims to verify how the culture of sustainability of this company drives innovation. The results of the following section present innovation projects that have emerged as a result of having a sustainable culture.

3.3 Results

Eko-REC is a small-sized company specialized in the recycling of PET plastic obtained from the municipal garbage collection system (the yellow recycling container in Spain). The company manufactures packaging and products for the automotive and food sectors. It embraces sustainability principles and is a leader in the transition toward the CE. It is the only factory in Europe that uses all the flakes it produces in the manufacture of the final products. In addition, sustainability is a cultural value of the company that is shared by all employees, who are involved in generating strategy and making decisions.

Eko-REC was founded in 2012 as a company whose economic model was based on creating value by reusing assets from a company that went bankrupt. Eko-REC owners saw an opportunity in the ruined company's PET recycling plant. The productive assets acquired by Eko-REC were the ones that the company then used to close the loop for a particular kind of plastic, namely by transforming PET waste into flakes that are then used as raw materials in their own production process.

The idea behind Eko-REC's business came from the fact that companies were looking for ways to make products with less environmental impact, and that those companies were willing to work with a supplier who could offer them recycled and recyclable materials.

The company is experiencing steady growth. During the period under consideration, the number of employees has grown from 100 by the end of 2015 to 115 at the beginning of 2018. Nowadays, they recycle 23,000 tons of plastic per year, which means that 1 out of every 3 bottles that end up on a yellow container in Spain is recycled by Eko-REC.

3.3.1 Manufacturing Processes

There are three types of manufacturing processes in the Eko-REC facilities: PET is recycled into flakes, which are in turn transformed into one of two products, PET sheets or polyester fiber.

Recycling PET into flakes: The raw material (bales of plastic) is bought from separating companies; to start the process first, the bales go through a bale-breaker that further separates the materials, so they are easier to process in the next stages. After that, the broken bale goes through three kinds of sorters, first, a magnet extracts little pieces of metal materials, after that, the remaining unsuitable materials are removed with the help of an automatic optic selector, which also are in charge of separating the remaining PET into a transparent group and a multicolor group, and our last sorter is manual, that is, two workers are in charge of manually sorting the remaining groups with the sole purpose of correcting any mistakes the automatic processes might have made.

Afterward, the 60–70% of material left undergoes further processing: first, grinding process because there are still tags and other materials that will be latter classified as waste. Second, a flotation process, in which the grinded materials are put into a liquid that separates by flotation tags, caps and little residues; the caps are sold to clients (to manufacture other products), and the tags and residue are classified as waste. Third, after eliminating the unwanted materials, the rest undergoes a cleaning and drying process by centrifuge. Finally, an optic control process takes place, and two different products can be differentiated at the end: transparent flakes and multicolored flakes.

Manufacturing PET sheets: In the facility where this process takes place, there are two big production lines plus a smaller line for testing. The process begins with clear flakes as the raw material (obtained in the previous process); first, they are mixed with virgin and recycled flakes provided by suppliers. Second, they go through a simple extrusion process, after which the extruded material gets formed and silicone is added. Then, the material is dried, reeled, weighed, cut and shipped to clients. This process has as a result PET sheets of different colors and thickness; the final product, the reels; sometimes have different measures, because each client usually asks for certain measures to fit their specific production machines.

Manufacturing polyester fiber: In this process, the multicolored flakes obtained in the first process are mixed with other multicolored flakes purchased to suppliers and are transformed into polyester fiber of all kinds of colors. In the facility where this manufacturing process takes place, there is one large production line, a medium size one and a small one for special colors.

First, the flakes are melted in an infrared drier; after, they are mixed with colorant in order to achieve the requested colors by the clients (usually black). Then, the mix is extruded, and strained, using cylinders as strainers. After collecting the threat created by the holes, continuous and straight fiber is obtained; that is, latter passed through thermostable rollers converting it into a wavy fiber. Finally, the fiber is dried, stabilized and cut; ready to be packed in bales for clients. Unlike the final product of the PET sheet manufacturing process, the final product obtained from this process has standard measures, which means that every client receives the exact same bales in terms of height and weight, being the color the only feature that may change upon request of clients.

Another interesting process that takes place in Eko-REC's facilities is the energy conversion process; in which gas is converted into different types of energy. In the cogeneration plant, natural gas is bought and transformed into two very different things. First is transformed into electric energy that is then sold to the electric network. And, second is converted into thermal energy in the form of hot water, cold water and vapor steam that is used for Eko-REC's own processes and also sold to other companies.

3.3.1.1 Raw Materials and Manufactured Products

To sum up, here is a list of the main raw materials that Eko-REC needs and the products that are manufactured as the result of the processes mentioned above:

1. Raw materials

- PET bottles: They are the main raw material used by the company. Most are provided by Ecoembes (the organization that coordinates the recycling container network in Spain). Other providers are bottle manufacturing companies that produce defective batches.
- Virgin flakes: Food sector regulations require that a thin outer layer of the PET sheet used in packaging not be made of recycled plastic because it will come into contact with the food. Therefore, 10% of the sheets for the food sector must come from virgin flakes.
- Other raw materials: Some of the processes include other materials such as the colorants and additives needed to achieve the required properties of the final products. All of them are recyclables.

2. Manufactured products

- PET sheets: Eko-REC produces sheets of different colors and thicknesses from clear flakes. Clients (usually from the food sector) use them for packaging.
- Flakes: The PET bottles are converted into different colored flakes. These flakes are the raw material for other final products manufactured by Eko-REC.
- Polyester fiber: Eko-REC manufactures polyester fibers of diverse colors using multicolor flakes and colorants. Most of the fiber bales are destined for the automotive industry for car interior elements (carpets and upholsteries).

3.3.2 Projects

As a company, Eko-REC is involved in many different projects that show its commitment to its values and brand. The degree of evolution they have experienced in such a short time is pleasantly surprising, mainly because of the company's size.

In general, in a company's early years, it tends to be more conservative and strategic when it comes to selecting and doing projects. What can be observed in Eko-REC is a refreshing enthusiasm toward new projects. Eko-REC's CEO firmly states that their goal as a company is to have 70% of their billings, in just 10-year time, come from products that they do not currently manufacture. An ambitious goal such as this can only be achieved by a very intense and effective R&D department, whose aim is to find new ways to give waste a second life that is not currently possible, whether is because the technology is not able to accomplish it yet or those ways have not been explored as viable because of their financial risk. Currently, the R&D department is one of the foundations of the company, and it is well known in the sector for the continuous innovative projects it undertakes under its direction.

As evidence of this, we compiled their numerous projects over the last few years, along with their goals, results and partners.

Table 3.1 shows the projects in which Eko-REC has been involved internally and Table 3.2 in collaboration with other companies. We consider the number and quality of these projects to be an adequate indicator of this company's attitude toward innovation. All of them are related to sustainability. In some cases, the purpose is to take advantage of waste from internal manufacturing processes, and in other cases, it is to make better use of raw materials or even to launch a spin-off company to provide solutions to major environmental problems such as the garbage in the seas. But all in all, the goals of the projects can be categorized into 2 main orientations: The improvement of energy efficiency in production or the search for alternative uses to all the waste that comes out of their PET sheet and polyester fiber manufacturing processes. The number of projects and the ambitions of its goals are surprising considering the size of the company.

Another manifestation of the company's innovative character is the launch of a new brand in 2017 Ekomodo, which offers a new set of products: office accessories. A brand Ekomodo represents a more conscious, responsible and sustainable way of life; making products with functional and esthetic design out of plastic waste. The products and the brand follow a sustainable philosophy and target a new type of customer that feels they have a compromise with social and environmental improvement. This business is completely different from the main activity of the company, and since all products are made from recycled plastic, which is in line with their values, they maintain the culture of sustainability.

Table 3.1 Inner projects (where only Eko-REC participated)

Project	Year	Goal	Key challenges	Results
Particle board	2018	Reduce the generated waste in the processes that are carried out in Eko-REC	Find the machine that would fit the needs of this project	Through the acquisition of a new machine that manages to bind the waste particles and produce PET flakes, they eliminate 500 tons of waste per year and produce 500 tons more of flakes at the same buying cost
Black master batch	2016	Fabricate the colorant (black master batch) in the company	Expand the industrial capacities to coextrude PET pellets with carbon black and fabricate master batch	Fabricating the master batch at Eko-REC had higher costs than buying it; it had a low quality; therefore, the company did not go forward with this project
3D 100% recycled	2016	Prove and demonstrate that it is possible to produce filament with recycled PET plastic, for the 3D impression	Reduce the environmental impact of the growing additive industry by using recycled and recyclable filament	This project proved that it was possible to use a filament made out of recycled PET in the 3D printing industry. However, the quality of this product is still far from the quality of the products made with filament from non-recycled plastic. Therefore, there are no companies interested in this product

Table 3.2 Collaboration projects (where only Eko-REC joins forces with other companies)

Project	Year	Partners	Goal	Results
District heating	2017	Irizar e mobility	Use as much residual heat as possible (obtained from the industrial processes) by designing a heating circuit in order to strengthen the industrial symbiosis of the company and area	The project has become a regular activity between both companies, transporting the energy from Eko-REC's cogeneration plant to Irizar's plant in the most efficient way
Orlegi Sarea	2016	Ihobe & Sea2See	Reduce marine litter, by recycling fishing nets made out of polyamide	It was proven that it is possible to recycle net waste and give them a second life by using this waste to manufacture frames. But, the process had no economic viability due to the nets treatment
Eko-Koopera	2017	Ihobe & Koopera	Reduce textile waste, by recycling synthetic textiles made of polyester transforming them to pellets and using them for car carpets	This project proved that it was possible to transform textile waste into PET sheets. But, the project is industrially and economically unfeasible due to the tedious manufacturing process and the quality of the of the final product proven to be much worse than the one obtained using recycled bottles
Ekotex	2017	European union & other companies	Reduce textile waste by closing the cycle of waste items made out of polyester that cannot be reused or reintroduced in a manufacturing process, through a chemical recycling process for the production of monomers that are fit to use to produce new textile products	It had positive results and reached the goal of fabricating textile products with textile waste. However, Eko-REC's current plans do not include fabricating textile products as something regular. It has been proven that it is feasible. The EU is part of the project, so it will be easier to find a company that is interested in starting this kind of project

3.4 Conclusions

The company engages in innovative behavior, as evidenced by the amount and quality of projects that it has undertaken in a short time. Because most of the projects tackle challenges related to sustainability, it can be deduced that this cultural value of the company acts as a motivation for innovation. The core business of this company is to close the loop for a specific kind of plastic (one of the most polluting materials due to the length of time it takes to degrade). But, it is not only a matter of business, it is a commitment to sustainability by the company and its employees. For that reason, there is great motivation to tackle projects that provide innovative solutions. The fact that projects do not always achieve the objectives (of technical and/or economic viability) does not stop new ideas from emerging or the company from being involved in new projects. These innovation projects help the company to grow and become more sustainable and more competitive.

The company overcomes the barrier of its small size (which implies limited resources) by undertaking projects in collaboration with other companies and public institutions, which gives it access to the financial support, knowledge and experience of its partners. This is due, in part, to the company's internal motivation and commitment to its values.

The research presented here has limitations in scope because it is focused on a single company. It would be interesting to extend the study to other SMEs in different industrial sectors.

Compliance with Ethical Standards The results of this work are supported by semi-structured interviews with the CEO and senior manager of the Eko-REC, who voluntarily agreed and gave informed consent to participate in the process and has also give written consent to publish the contents of this chapter regarding the aforementioned interviews. The "Research Ethics Com-mittee of Universidad de Navarra" (Ref. 2021.161) approved the research metho-dology.

References

1. Argandoña A (2003) Fostering values in organizations. J Bus Ethics 45:15–28
2. Bos-Brouwers HEJ (2010) Corporate sustainability and innovation in SMEs: evidence of themes and activities in practice. Bus Strategy Environ 19:417–435
3. Boso N, Adeleye I, Donbesuur F, Gyensare M (2019) Do entrepreneurs always benefit from business failure experience? J Bus Res 98:370–370
4. De Jesus A, Antunes P, Santos R, Mendoca S (2018) Eco-innovation in the transition to a circular economy: an analytical literature review. J Clean Prod 172:2999–3018
5. DiCicco-Bloom B, Crabtree BF (2006) The qualitative research interview. Med Educ 40:314–321
6. Ghisellini P, Cialani C, Ulgiati S (2016) A review on circular economy: the expected transition to a balanced interplay of environmental and economic systems. J Clean Prod 114:11–32
7. Iturrioz C, Aragón C, Narvaiza L (2015) How to foster shared innovation within SMEs' networks: social capital and the role of intermediaries. Eur Management J 33(2):104–115
8. Kemp R (2010) Eco-innovation: definition, measurement, and open research issues. Economia Politica 3:397–420

9. Khagram S, Nicholas KA, Bever DM, Warren J (2010) Thinking about knowing: conceptual foundations for interdisciplinary environmental research. Environ Conserv 37:388–397
10. Klewitz J, Hansen EG (2014) Sustainability-oriented innovation of SMEs: a systematic review. J Clean Prod 65:57–75
11. Lukács E (2005). The economic role of SMEs in world economy, especially in Europe. Eur Integr Stud 4(1):3–12
12. Nieto MJ, Santamaria L (2010) Technological collaboration: bridging the innovation gap between small and large firms. J Small Bus Manage 48(1):44–69
13. Organisation for Economic Co-Operation and Development (OECD) (2019) OCDE small and medium enterprise outlook. OECD Publishing, Paris
14. Prieto-Sandoval V, Jaca C, Ormazabal M (2018) Towards a consensus on the circular economy. J Clean Prod 179:605
15. Schumpeter JA (1939) Business cycles. a theoretical, historical, and statistical analysis of the capitalist process. McGraw-Hill, New York
16. Smol M, Kulczycka J, Avdiushchenko A (2017) Circular economy indicators in relation to eco-innovation in European regions. Clean Technol Environ Policy 19:669–678
17. Tarafdar M, Gordon SR (2007) Understanding the influence of information systems competences on process innovation: a resource-based view. J Strateg Inf Syst 16(4):353–392
18. Triguero A, Moreno-Mondejar L, Davia MA (2013) Drivers of different types of eco-innovation in European SMEs. Ecol Econ 92:25–33
19. Veleva V, Bodkin G (2018) Corporate-entrepreneur collaborations to advance a circular economy. J Clean Prod 188:20–37
20. WWF (2018) Living planet report—2018: aiming higher. In: Grooten M, Almond REA (eds). WWF, Gland

Chapter 4
Supply Chain 4.0 to Enhance Circular Economy

L. Davila, J. Mula, and R. Sanchis

Abstract The main theme of this literature review is supply chain (SC) models that focus on circular economy (EC), Industry 4.0 (I4.0) and green operations. Special interest is paid to optimisation models and intelligent and digitalised supply chains, whose operations are based on circular business models. This circularity approach allows resources to be reused to strike a balance between economic growth and environmental concerns. This literature review analyses the current research state according to: objectives and the context; research methodology and methods; benefits; limitations and critical points.

Keywords Supply chain · Circular economy · Green · Industry 4.0 · Optimisation

4.1 Introduction

Genovese et al. [13] distinguish between a green and sustainable supply chain (SC) and the economy circular (CE) concept. A green SC refers to the strategy of integrating environmental issues into manufacturing organisations by reducing material flows or minimising the negative consequences of production processes [31], i.e. a linear SC. The CE concept is related to production methods that are self-sustaining and true to nature, where materials are used over and over again [26]. Here Geissdoerfer et al. [12] consider sustainable development to be a broader and more intangible concept than CE, which could be converted into a more tangible way to organise the society and economy, and to define circular SC management as the configuration and

L. Davila · J. Mula (✉) · R. Sanchis
Research Centre on Production Management and Engineering (CIGIP), Universitat Politècnica de València, Alarcón, 1, 03801 Alcoy, Alicante, Spain
e-mail: fmula@cigip.upv.es

L. Davila
e-mail: luida@epsa.upv.es

R. Sanchis
e-mail: rsanchis@cigip.upv.es

C. Avilés-Palacios and M. Gutierrez (eds.), *Ensuring Sustainability*,
Lecture Notes in Management and Industrial Engineering,
https://doi.org/10.1007/978-3-030-95967-8_4

coordination of organisational marketing tasks, sales, R&D, production, logistics, IT, finances and customer services within and between business units and organisations to minimise the system's input resources and loss of waste and emissions, improve its operational effectiveness and efficiency and create competitive advantages.

Simultaneously, digital technologies enable SC management in the current Industry 4.0 (I4.0) context to evolve [18]. In this context, Nascimento et al. [28] focused on sustainable SC management, additive manufacturing and smart production systems to address how I4.0 technologies can be integrated with CE practices. Daú et al. [9] simultaneously define sustainable SC and 4.0 as that which uses I4.0 tools to close materials and energy cycles and to help information and activities to flow. Manavalan and Jayakrishna [24] recommend using I4.0-related technology in the transition from a linear to a circular SC.

We herein present a summary of a more general review on SC 4.0 and on CE reviews, conceptual and analytical approaches. This review aims to act as a starting point for novel optimisation and simulation models for SC 4.0 according to CE criteria.

4.2 Review Methodology

The review methodology was based on exploring articles by employing mostly the Scopus scientific and technical database and the Web of Science to a lesser extent. No time restrictions were set. Articles were searched for and observed by combining several keywords, which were later selected in line with the selection criterion of these keywords. We now summarise the number of articles selected by searches based on these keywords: supply chain, circular economy, Industry 4.0: four results; supply chain, Industry 4.0: 13 results; supply chain, circular economy: 13 results; optimisation, supply chain, circular economy: two results; supply, chain, circular, economy: seven results; and Industry 4.0, supply chain, green, circular economy: one result with a total of 40 results.

The scientific sources of the selected articles are recognised in the research area. Table 4.1 presents the scientific sources from which 24 articles were taken.

4.3 Literature Review

We classify the reviewed works in terms of objectives and context, research methodology, benefits and limitations.

Table 4.1 Scientific source and selected references

Source	References
International Journal of Production Research	5
Sustainability	5
Journal of Cleaner Production	4
Procedia Manufacturing	3
Resources, Conservation and Recycling	3
Omega	2
Process Safety and Environmental Protection	2
Others	16

4.3.1 Objectives and the Application Context

Table 4.2 summarises the objectives and the application context of the reviewed papers.

Regarding the main objectives of the reviewed works, it is important to highlight the aim to integrate economic and environmental objectives into SC management [36]. Indeed the economic SCOR SC management model has been adopted to extend it towards circular SC practices [32]. Sustainability principles in SCs have also been addressed with optimisation models [27]. SC configurations according to sustainable [20] and circular principles have been analysed by several works [25, 30, 35], whereas the necessity for cooperation practices from consumers to suppliers has been studied [41, 42]. The government regulation to implement CE practices has been addressed [40, 42]. Several performance measurement systems or indicators have been proposed to evaluate circular SCs [16] to, for instance, compare linear and circular SC models [9, 12, 15, 24, 29].

On more ecological issues, direct, indirect and total life cycle emissions, waste recovered [17], virgin resources use, gas emissions [14] and carbon maps have been addressed by several works with circular principles [5, 13]. Finally, I4.0 technologies, such as IoT [4], optimisation algorithms, big data [34] and RFID [23], among others, have been applied to support digital SCs [2, 7, 18, 19, 33], logistics 4.0 [3, 37], sustainable SC management [1, 11, 22] and circular principles [10, 28, 38, 39].

According to the contexts to which the analysed papers apply, some works focus on automotive [5, 36], chemical and petrochemical, construction [7, 14, 29], electronic, food [13, 20], footwear and apparel [23], health care and mechanical, mobile [2], pharmaceutical [11] and steel [30] SCs. SCs, and global SCs in general, are also widely considered by several works [3, 18, 19, 32]. The interest shown in studies on I4.0 and logistics 4.0 has recently increased. Finally, it is important to highlight that the SC nationalities where case studies are conducted are basically Brazilian [9], Chinese [39–43,], European [30], Indian [7, 22, 24] with electronic and food [13, 20, 38] and Italian [17] and South African [1].

Table 4.2 Objectives and application context

References	Objectives	Context
Winkler and Kaluza [36]	To integrate economic and ecological objectives towards sustainability by creating networks of sustainable SCs and taking a circular approach	SC networks and automotive industry
Zhu et al. [41, 42]	To study how environmental SC cooperation practices influence CE and environmental and economic performance	Various Chinese sectors
Zhu et al. [43]	To examine the influence of international/national regulating politics on environmental management innovation on production operations	Chinese manufacturers
Schrödl and Simkin [32]	To integrate CE principles into the economic SCOR SC (SC operations reference) management model	SC management
Ivanov et al. [19]	To assign jobs to machines in a multistage flow shop scheduling problem	Flexible flow shop with alternative machines
Barreto et al. [3]	To deal with efficiency in organisations by using logistics 4.0	I4.0 and logistics 4.0
Butzer et al. [6]	To develop a performance measurement system for sustainable SCs	International reverse SCs
Genovese et al. [13]	To compare the performances of traditional and circular production using: direct, indirect and total life cycle emissions, waste recovered, virgin resources use and carbon maps	Chemical and food industries
Ghani et al. [14]	To focus on tracing greenhouse gas (GHG) emissions across the building SC industries and provide an optimised GHG reduction policy plans for sustainable development	SCs of building structures

(continued)

Table 4.2 (continued)

References	Objectives	Context
Majeed and Rupasinghe [23]	To improve inbound and outbound operations in an enterprise resource planning (ERP) system by using radio frequency identification (RFID) technology and business application programming interface technology in SAP®	Apparel and footwear industries
Masi et al. [25]	To identify the goals and assumptions about CE at the meso-level and assess the state of the art on CE SC configurations	SC configurations
Nasir et al. [29]	To assess and compare the environmental impacts associated with circular SCs to those related to traditionally manufactured products (linear SCs)	Construction SCs
Tjahjono et al. [33]	To analyse the impact of I4.0 on SCs on the whole, particularly on procurement, transport logistics, warehouse and order fulfilment	SC 4.0
Witkowski [37]	To present smart solutions to improve firms' competitiveness using I40 tools in logistics	Logistics and I4.0
Zeng et al. [40]	To test the mechanism and relations among institutional pressure, SC relationship management, sustainable SC design and CE capability	Chinese eco-industrial park firms
Bag et al. [1]	To identify the I4.0 enablers of SC sustainability and further attempt to propose a research framework that bridges theoretical gaps	South African manufacturing sector
Barata et al. [2]	To identify avenues for future research into mobile SC management in the advent of I4.0	Mobile SC

(continued)

Table 4.2 (continued)

References	Objectives	Context
Braun et al. [5]	To assess material efficiency within company borders in the SC by measuring the material demand and the amount of waste materials both with and without CE aspects	Manufacturer of surface-coated automotive parts
Dallasega et al. [7]	To introduce the proximity theory and deal with the impact of I4.0 concepts in SCs	Construction SCs
Ding [11]	To identify the sustainability barriers of pharmaceutical SCs and investigate how I4.0 can be applied to sustainable SC paradigms	Pharmaceutical SCs
Geissdoerfer et al. [12]	To propose a conceptual framework to circulate business models and SCs	Business models and SCs
Govindan and Hasanagic [15]	To identify the main drivers, practices and barriers to implement CE	SC management
Ivanov et al. [18]	To discuss and develop a framework for SC digital technology and disruption risk effects	SC management
Lopes de Sousa Jabbour et al. [21]	To develop a roadmap to improve the application of CE principles by I4.0 approaches	CE and I4.0
Luthraa and Mangla [22]	To identify, analyse and prioritise challenges for I4.0 initiatives to be efficient for the sustainability of SCs in emerging economies	Indian SCs
Tseng et al. [34]	To present CE and big data as a solution to optimise industrial symbiosis practices	I4.0
Yang et al. [39]	To explore the relation between the business model innovation of product-service systems and circularity in SCs	Multi-business manufacturing company
Bendaya et al. [4]	To summarise the role and the impact of the Internet of Things (IoT) on the main SC processes	SC management

(continued)

Table 4.2 (continued)

References	Objectives	Context
Daú et al. [9]	To analyse sustainable health 4.0 SCs by means of the conceptual framework of CE transition mirrored by CSR	Brazilian healthcare SC
Howard et al. [16]	To present a framework for developing CE indicators	Global companies
Isernia et al. [17]	To evaluate the extent to which waste electrical/electronic equipment management is able to meet the targets defined by the European Union (EU) by focusing on the collection centres that are the initial point of the reverse logistic cycle	Italian companies
Manavalan and Jayakrishna [24]	To analyse a case example in an SC organisation to meet I4.0 requirements and to enable CE	Indian paper manufacturer
Muñoz-Torres et al. [27]	To propose an assessment framework that extends sustainability principles to other SC members	SC management
Nascimento et al. [28]	To explore how the technologies arising from I4.0 can be integrated with CE practices	Manufacturing
Dev et al. [10]	To propose a roadmap for the excellence of operations for sustainable reverse SC/logistics by the joint implementation of I4.0 principles and CE approaches	Refrigerator company
Van Engeland et al. [35]	To provide an overview of strategic network design in waste reverse SCs by means of combinatorial optimisation models	Waste management and reverse logistics
Krishnan et al. [20]	To identify any operational and resource inefficiencies present in the food SC by environmental impact assessments and propose a framework for its redesign	Indian food SC

(continued)

Table 4.2 (continued)

References	Objectives	Context
Pinto and Diemer [30]	To determine the way in which applying different SC integration strategies based on raw material self-sufficiency and retaining the property of resources can affect CE	European steel industry
Yadav et al. [38]	To develop a framework to overcome sustainable SC challenges through I4.0 and CE-based solution measures	Indian automotive SC

4.3.2 Research Methodology and Methods

According to Dangayach and Desmukh [8], the research methodologies classification of the reviewed works is as follows: (1) conceptual, basic or fundamental concepts; (2) descriptive, an explanation or description of the process or content and performance measurement issues; (3) empirical, study data from existing databases, literature reviews, case studies and taxonomy or typology approaches; (4) exploratory cross-sectional by surveying at one time point; (5) exploratory longitudinal, where data collection was done at two time points or more. Table 4.3 summarises the research methodology followed in the reviewed works.

Here conceptual approaches are based mainly on systematic literature reviews. Descriptive methodologies generally propose frameworks validated through performance measure systems or indicators. Empirical approaches address case studies and practical examples. In this empirical methodology, the LCA methodology has been applied by several works. Finally, it is worth highlighting that analytical approaches and real-world applications scarcely appear in the reviewed papers.

4.3.3 Benefits

Table 4.4 presents the main benefits identified throughout the reviewed papers.

4.3.4 Limitations and Critical Points

Table 4.5 summarises the main limitations and critical points identified throughout this review.

Table 4.3 Research methodology (RM) and methods

References	RM	Methods
Winkler and Kaluza [36]	Conceptual	Generic structure, principles and economic and ecological measures
Zhu et al. [41, 42]	Exploratory cross-sectional	Questionnaires (396) and statistical techniques for data analysis
Zhu et al. [43]	Exploratory cross-sectional	Questionnaires (374) and statistical techniques for data analysis
Schrödl and Simkin [32]	Descriptive	Design science: problem recognition, suggestion, artefact development, demonstration and evaluation, conclusion
Ivanov et al. [19]	Empirical	Dynamic decomposition of a scheduling problem with the optimal program control theory and continuous variables and performance indicators
Barreto et al. [3]	Conceptual	Analysis and discussion
Butzer et al. [6]	Descriptive	Literature review and the balanced scorecard approach
Genovese et al. [13]	Empirical	The combination of a life cycle assessment (LCA) methodology and the environmentally extended multiregional input–output hybrid model
Ghani et al. [14]	Empirical	A two-step hierarchical approach: (i) economic input–output-based LCA (EIO-LCA) to quantify GHG emissions; (ii), a mixed integer linear programming (MILP) to identify optimal GHG emissions' reduction
Majeed and Rupasinghe [23]	Exploratory cross-sectional	A survey with 30 SAP consultants who are experienced in both technical and functional areas and 10 SAP users
Masi et al. [25]	Conceptual	Systematic literature review
Nasir et al. [29]	Empirical	LCA methodology utilising a combination of data provided by the industry and a reliable database

(continued)

Table 4.3 (continued)

References	RM	Methods
Tjahjono et al. [33]	Descriptive	A desk-based study utilising various types of literature, key performance identifiers and hypothetical example
Witkowski [37]	Conceptual	Presentation of concepts and examples
Zeng et al. [40]	Exploratory cross-sectional	Questionnaires (363) and structural equation modelling
Bag et al. [1]	Conceptual	Systematic literature review
Barata et al. [2]	Conceptual	Systematic literature review
Braun et al. [5]	Empirical	Deterministic calculations and simulations performed by the SC of the studied company
Dallasega et al. [7]	Empirical	Systematic literature review and case studies
Ding [11]	Conceptual	Systematic literature review
Geissdoerfer et al. [12]	Empirical	Four case studies
Govindan and Hasanagic [15]	Conceptual	Systematic literature review
Ivanov et al. [18]	Empirical	Literature review and analysis of practical examples
Lopes de Sousa Jabbour et al. [21]	Descriptive	Extension of state of the art by an agenda and roadmap proposal
Luthraa and Mangla [22]	Exploratory cross-sectional	Literature review, survey with 96 Indian manufacturers, research techniques based on analyses of explanatory factors and the analytical hierarchy process
Tseng et al. [34]	Conceptual	Literature review
Yang et al. [39]	Empirical	Case study
Bendaya et al. [4]	Conceptual	Systematic literature review
Daú et al. [9]	Empirical	Case study
Howard et al. [16]	Empirical	Literature review, company reports and case studies
Isernia et al. [17]	Empirical	Probability transition matrix methodology
Manavalan and Jayakrishna [24]	Empirical	Case study
Muñoz-Torres et al. [27]	Descriptive	Systematic literature review and framework proposal

(continued)

Table 4.3 (continued)

References	RM	Methods
Nascimento et al. [28]	Descriptive	Literature review, conceptual framework and focus group validation
Dev et al. [10]	Empirical	Taguchi experimental design
Van Engeland et al. [35]	Conceptual	Literature review
Krishnan et al. [20]	Empirical	Literature review and LCA methodology
Pinto and Diemer [30]	Empirical	LCA and system dynamics
Yadav et al. [38]	Descriptive	Literature review and seeking expert opinion

4.4 Discussion and Conclusions

The CE concept promotes continuous economic development without implying significant environmental and resources challenges. This is supported so that economic systems can, and must, operate according to the principles of the materials and energy cycles that sustain natural systems by emphasising the capacity of an organisation's waste being used as another organisation's resource via self-organisation capacity [41, 42]. Thus, one of the main challenges of implementing CE initiatives is to pay attention to economic implications [13]. So finding ways to align sustainable SC strategies with CE principles and understanding all the economic–environmental implications to do so are most important if we wish to broaden the limits of environmental sustainability, especially in intensive energy and materials industries [29]. In other domains for example, the main problems that prevent sustainability involve high costs and long times, little experience and training, the application of regulations, few commercial incentives, inefficient collaboration and coordination, lack of objective target reference points and barely any knowledge about end customers [11]. Govindan and Hasanagic [15] are in favour of promoting CE by a governmental perspective, i.e. laws, policies, risk reduction (through tax levies) and strict governance.

This CE paradigm will, in turn, be applied to an I4.0 context, specifically a logistics 4.0 context, which is understood as the combination of logistics with the innovations and applications of cyberphysical systems [3]. Industrial production systems must be balanced from the environmental, social, economic and technological points of view. Here it is important to highlight the work by Lopes de Sousa Jabbour et al. [21], who propose a first roadmap towards I4.0 and CE based on the ReSOLVE business model; or Dallasega et al. [7] distinguish different proximity dimensions: technological, organisational, geographical and cognitive. These findings demonstrate that not all I4.0 concepts affect all proximities to the same extent.

Table 4.4 Benefits

References	Benefits
Winkler and Kaluza [36]	A sustainable supply chain network (SSCN) approach makes possible to move from a flow economy to a closed circuit one what would benefit economic–ecological development by leading to improved competitive positions
Zhu et al. [41, 42]	Environmental-oriented supply chain cooperation is set up to reduce the use of material, water and energy throughout the SC by cooperating with suppliers and customers
Zhu et al. [43]	To opt for coercive equilibrium (command and control) with voluntary incentives and mechanisms as a favourable hybrid approach to apply effective ecological modernisation theory policies. International policies can influence environmental management practices to be set up by developing countries
Schrödl and Simkin [32]	The proposed model acts as a reference for up-to-date sustainable SC management issues and as a basis for the future development of an inter-organisational generic approach that depends on production
Ivanov et al. [19]	The results could be included in general iterative search procedures for optimal schedule programmes. It also extends the scope of the SC analysis by obtaining analytical solutions or robustness properties and investigating different adaptation policies in a scheduling problem
Barreto et al. [3]	The main challenges to apply the logistics 4.0 concept are identified
Butzer et al. [6]	Six perspectives are defined to assess reverse SCs: citizen and legislation, financial, interested parties, process, innovation and growth and flexibility
Genovese et al. [13]	The results suggest a positive influence from improving business sustainability by reinserting waste into the SC to manufacture products on demand
Ghani et al. [14]	This paper integrates EIO-LCA and MILP frameworks to identify the most pollutant industries in SCs of building structures
Majeed and Rupasinghe [23]	Proposal of a framework to integrate RFID with SAP ERP into an I4.0 context
Majeed and Rupasinghe [25]	To identify CE definitions and understandings, and to study its three SC configurations: eco-industrial parks, environmental and closed loop SCs
Nasir et al. [29]	The results show that transport elements dominate a larger proportion of the total emissions of a circular SC compared to a linear one
Tjahjono et al. [33]	To provide initial thought about SC 4.0

(continued)

Table 4.4 (continued)

References	Benefits
Witkowski [37]	To identify the benefits IoT, big data and I4.0 to cover customer demands and the development of SCs' logistics and management
Zeng et al. [40]	Theoretical and practical contributions for eco-industrial park firms to achieve sustainable SCs and to study CE capability
Bag et al. [1]	The identification of 13 key enablers of I4.0 that play a key role in driving SC sustainability: government support; support of research institutes and universities; law and policy on employment; among others
Barata et al. [2]	The identification of six avenues for future work: update existing studies to account for new technologies; to apply mobile technologies; produce samples of mobile SC cases; study the implications of regulatory compliance; develop maturity models for diagnostic and guidance; explore social aspects
Braun et al. [5]	It shows the routine for material efficiency improvements of SC elements. Moreover, investigating possible optimisation potentials in the SC of manufacturing enterprise is performed. The impact of CE activities is assessed
Dallasega et al. [7]	It identifies the proximity thresholds on its various dimensions that lead to loss of SC efficiency
Ding [11]	It proposes and analyses the Pharma I4.0 concept to adopt new technologies based on I4.0 to obtain a higher level of control and management throughout the product's life cycle using predictive analysis and proactive actions
Geissdoerfer et al. [12]	It identifies a change that requires changing conduct in consumers/suppliers where economic, environmental and social objectives must be promoted
Govindan and Hasanagic [15]	To implement CE into an SC, 13 drivers, 34 practices and 39 barriers are identified
Ivanov et al. [18]	The proposal of several conceptual frameworks regarding digital SC by providing the main technologies (big data analytics, I4.0, additive manufacturing, advanced tracking and tracing), applications and challenges
Lopes de Sousa Jabbour et al. [21]	The proposed roadmap combines novel concepts from the ReSOLVE business mode that act as a guide to implement CE principles and I4.0 technologies
Luthraa and Mangla [22]	18 challenges are identified and evaluated to contribute to I4.0 diffusion to lead towards smarter and sustainable manufacturing SCs
Tseng et al. [34]	The data-based analysis can be potentially used to optimise sustainable solutions that intend to reduce the intensities of resources and emissions of industrial systems

(continued)

Table 4.4 (continued)

References	Benefits
Yang et al. [39]	The dimensions of former research works about PSS are extended by incorporating the circularity of SCs
Bendaya et al. [4]	Avoidable waste production, security and efficiency throughout the food SC are identified as a potential convincing line to apply the IoT
Daú et al. [9]	The link among the triple bottom line, I4.0 and CSR allows the transition of linear to circular models and can improve the sustainable health SC 4.0
Howard et al. [16]	This paper reviews and analyses nine global companies and four practical cases that form a significant group of companies to investigate current and future practices in the development of CE indicators
Isernia et al. [17]	This article heads up the research community for defining customised policies to support the achievement of balanced e-waste collection performance
Manavalan and Jayakrishna [24]	This paper uses case-driven suggestions based on the 6Rs (recover, reuse, remanufacture, recycle, redesign, reduce) to leverage technology across SC so that organisations become more sustainable
Muñoz-Torres et al. [27]	The corporate sustainability assessment tool is implemented by a company into its SC management framework
Nascimento et al. [28]	The recommendation of a circular model to reuse scrap electronic devices by integrating web technologies, reverse logistics and additive manufacturing
Dev et al. [10]	The proposed model that integrates I4.0 and CE represents a real-time decision model for the sustainable reverse logistics system
Van Engeland et al. [35]	This paper provides a more updated vision of optimisation models in waste reverse SCs
Krishnan et al. [20]	Applying the conceptual framework to other industries can be done by this approach, wherein opportunities for improving operational efficiency and resource recovery for environmental sustainability can be similarly identified
Pinto and Diemer [30]	To provide new branches of strategic decision- and policy-making discussion in the European steel industry
Yadav et al. [38]	Proposal of a framework to overcome sustainable SC challenges

In short, this paper presents a preliminary literature review on research work oriented to SC 4.0 initiatives with CE principles. Here the objectives and application context and the research methodology of the reviewed works are analysed. An overview of the main benefits and limitations is addressed. This work identifies the

Table 4.5 Limitations and critical points

References	Limitations and critical points
Winkler and Kaluza [36]	Empirical projects to validate the SSCN proposal are lacking
Zhu et al. [41, 42]	The results cannot be generalised to other countries or specific industry sectors
Zhu et al. [43]	This study is limited to only one region of China
Schrödl and Simkin [32]	Green SC management, based on the SCOR model, lacks generality and covers only some parts of relevant aspects
Ivanov et al. [19]	Strong centralised control and lack of software tools to perform a comparative analysis with existing benchmark solutions. A higher level of heuristics is also required to improve computational efficiency
Barreto et al. [3]	It deals with the online transactions, the integration of new technologies and their better access for third parties and lack of computer security
Butzer et al. [6]	Lack of knowledge about evaluating international reverse SCs
Genovese et al. [13]	More relevant environmental indicators can be considered to make a comparison between linear and circular systems
Ghani et al. [14]	The application area of the proposed integrated approach is context specific
Majeed and Rupasinghe [23]	The results focus on a case study based on SAP ERP and the footwear and apparel industry
Masi et al. [25]	At the microlevel, it was impossible to differentiate among SC, product design, commercial strategy and the broader business model perspective
Nasir et al. [29]	Reliance on secondary data
Tjahjono et al. [33]	This is a preliminary work that requires further conceptual and empirical research
Witkowski [37]	Empirical research works that focus on logistics 4.0 strategies are needed
Zeng et al. [40]	Data should be collected from other countries to further validate the research hypotheses and conclusions of this study
Bag et al. [1]	The identified enablers need empirical testing in real-world SCs
Barata et al. [2]	Its limitations are the normal ones when performing a literature review as the analysis is delimited by the language, the keywords and the databases
Braun et al. [5]	CE scenarios for the example manufacturing enterprise and its SC are not examined in this analysis. Only the effect of flat 50% waste reintroduction into the economic circle for material supply is explored

(continued)

Table 4.5 (continued)

References	Limitations and critical points
Dallasega et al. [7]	It deals with the seasonality of projects and the large number of tasks and processes, which prevent productivity from improving
Ding [11]	It is based on only 33 scientific articles, of which only 1/3 are from high-impact journals. It mainly targets the downstream pharmaceutical SC
Geissdoerfer et al. [12]	A limited number of case studies and data collection are based only on a single interview per case
Govindan and Hasanagic [15]	The inclusion of multiple stakeholders and other theoretical frameworks, and considering the specific needs of a country or an industrial sector
Ivanov et al. [18]	Empirical results of the proposed conceptual frameworks are needed
Lopes de Sousa Jabbour et al. [21]	Organisations can face other difficulties when following the proposed roadmap due to distrust when integrating technological systems among SC partners and lack of technical knowledge about CE and I4.0
Luthraa and Mangla [22]	Some of the other challenges in different country contexts can be included in future studies. The Indian findings can be extended to other developing nations with marginal modifications
Tseng et al. [34]	The literature on "I4.0" and "CE" is scarce
Yang et al. [39]	Research in the field of circular SCs and business models is not yet mature, as indicated by it lacking agreed concepts and practices
Bendaya et al. [4]	Most studies have centred on conceptualising the IoT impact with analytical models and very few empirical studies. Moreover, most studies have focused on the delivery process and food SCs
Daú et al. [9]	More empirical studies must be done to analyse the replicability of the conceptual model proposed in other environments
Howard et al. [16]	The proposed framework is very concise and targets global companies
Isernia et al. [17]	The first limit lies in the analysed country-specific context, while the second one lies in the fact that it focuses on the collection part of the WEEE management system without considering other SC processes
Manavalan and Jayakrishna [24]	The paper focuses on a particular case study, and the suggestions to transform the linear SC into a circular one are specific and cannot be applied universally to any SC/organisation
Muñoz-Torres et al. [27]	The framework is not validated in a real-world case study

(continued)

Table 4.5 (continued)

References	Limitations and critical points
Nascimento et al. [28]	The perceptions of focus groups introduce subjectivity. In addition, the sample of experts is small and does not allow the results to be generalised
Dev et al. [10]	The results are context specific and deserve a more detailed analysis in terms of parameterisation and SC network structures for generalisation purposes. The study can be extended to multiple suppliers
Van Engeland et al. [35]	The framework is not validated in a real-world case study
Krishnan et al. [20]	In order to generalise the findings, further data collection is necessary. In addition, the specific focus on the cultivation, processing, packaging and transportation of a food SC is also a limiting factor
Pinto and Diemer [30]	Empirically addresses capital requirements for different investments in SC integration, the implications of investing in logistics and the specific environmental impacts that derive from these strategic changes
Yadav et al. [38]	The challenges and solution measures included in this study especially lie in the context of developing economies

need for more conceptual and empirical research to provide a framework and analytical models towards SC 4.0 with CE principles. Hence, optimisation and simulation models are required to support decision-making to optimise CE and I4.0 practices in SCs. Finally, the validation of these new optimisation and simulation circular SC 4.0 models should be validated with real-world applications.

Acknowledgements This work was supported by the Spanish Ministry of Science, Innovation and Universities project entitled "Optimisation of zero-defects production technologies enabling supply chains 4.0 (CADS4.0)" (RTI2018-101344-B-I00) and the Operational Program of the European Regional Development Fund (ERDF) of the Valencian Community 2014-2020 IDIFEDER/2018/025.

References

1. Bag S, Telukdarie A, Pretorius JHC, Gupta S (2018) Industry 4.0 and supply chain sustainability: framework and future research directions. Benchmark Int J 28(5): 1410–1450. https://doi.org/10.1108/BIJ-03-2018-0056
2. Barata J, Da Cunha PR, Stal J (2018) Mobile supply chain management in the Industry 4.0 era. J Enterp Inf Manage 31(1):173–192. https://doi.org/10.1108/JEIM-09-2016-0156
3. Barreto L, Amaral A, Pereira T (2017) Industry 4.0 implications in logistics: an overview. Procedia Manuf 13:1245–1252
4. Bendaya M, Hassini E, Bahroun Z (2019) Internet of things and supply chain management: a literature review. Int J Prod Res 57(15–16)

5. Braun AT, Kleine-Moellhoff P, Reichenberger V, Seiter S (2018) Case study analysing potentials to improve material efficiency in manufacturing supply chains, considering circular economy aspects. Sustainability 10(3):880
6. Butzer S, Schötz S, Petroschke M, Steinhilper R (2017) Development of a performance measurement system for international reverse supply chains. Procedia CIRP 61:251–256
7. Dallasega P, Rauch E, Linder C (2018) Industry 4.0 as an enabler of proximity for construction supply chains: a systematic literature review. Comput Ind 99:205–225
8. Dangayach GS, Deshmukh SG (2001) Manufacturing strategy literature review and some issues. Int J Oper Prod 21(7):884–932
9. Daú G, Scavarda A, Scavarda LF Julianelli V (2019) The healthcare sustainable supply chain 4.0: the circular economy transition conceptual framework with the corporate social responsibility mirror. Sustainability 11(12):3259
10. Dev NK, Shankar R, Qaiser FH (2020) Industry 4.0 and circular economy: operational excellence for sustainable reverse supply chain performance. Resour Conserv Recycl 153:104583
11. Ding B (2018) Pharma Industry 4.0: literature review and research opportunities in sustainable pharmaceutical supply chains. Process Saf Environ 119:115–130
12. Geissdoerfer M, Morioka SN, de Carvalho MM, Evans S (2018) Business models and supply chains for the circular economy. J Clean Prod 190:712–721
13. Genovese A, Acquaye AA, Figueroa A, Koh SCL (2017) Sustainable supply chain management and the transition towards a circular economy: evidence and some application. Omega 66:344–357
14. Ghani NMAMA, Egilmez G, Kucukvar M, Bhutta MKS (2017) From green buildings to green supply chains: an integrated input-output life cycle assessment and optimization framework for carbon footprint reduction policy making. Manage Environ Qual Int J 28(4):532–548
15. Govindan K, Hasanagic M (2018) A systematic review on drivers, barriers, and practices towards circular economy: a supply chain perspective. Int J Prod Res 56:278–311
16. Howard M, Hopkinson P, Miemczyk J (2019) The regenerative supply chain: a framework for developing circular economy indicators. Int J Prod Res 57(23):7300–7318
17. Isernia R, Passaro R, Quinto I, Thomas A (2019) The reverse supply chain of the e-waste management processes in a circular economy framework: evidence from Italy. Sustainability 11(8):2430
18. Ivanov D, Dolgui A, Sokolov B (2018) The impact of digital technology and Industry 4.0 on the ripple effect and supply chain risk analytics. Int J Prod Res 57(3):829–846
19. Ivanov D, Dolgui A, Werner W (2016) A dynamic model and an algorithm for short-term supply chain scheduling in the smart factory industry 4.0. Int J Prod Res 54(2):386–402
20. Krishnan R, Agarwal R, Bajada C, Arshinder K (2020) Redesigning a food supply chain for environmental sustainability. An analysis of resource use and recovery. J Clean Prod 242:118374
21. Lopes De Sousa-Jabbour AB, Chiappetta-Jabbour CJ, Godinho-Filho M, Roubaud D (2018) Industry 4.0 and the circular economy: a proposed research agenda and original roadmap for sustainable operations. Ann Oper Res 270:273–286
22. Luthraa S, Mangla SC (2018) Evaluating challenges to Industry 4.0 initiatives for supply chain sustainability in emerging economies. Process Saf Environ 117:168–179
23. Majeed AMA, Rupasinghe TD (2017) Internet of Things (IoT) embedded future supply chains for industry 4.0: An assessment from an ERP-based fashion apparel and footwear industry. Int J Supply Chain Manage 6(1):25–42
24. Manavalan E, Jayakrishna K (2019) An analysis on sustainable supply chain for circular economy. Procedia Manuf 33:477–484
25. Masi D, Day S, Godsell J (2017) Supply chain configurations in the circular economy: a systematic literature review. Sustainability 9(9):1602
26. McDonough W, Braungart M (2000) A world of abundance. Interfaces 30(3):55–65
27. Muñoz-Torres MJ, Fernández-Izquierdo MA, Rivera-Lirio JM, Ferrero-Ferrero I, Escrig-Olmedo E, Gisbert-Navarro JV, Marullo MC (2019) An assessment tool to integrate sustainability principles into the global supply chain. Sustainability 10(2):535

28. Nascimento D, Alencastro V, Quelhas O, Caiado R, Garza-Reyes J, Rocha-Lona L, Tortorella G (2019) Exploring Industry 4.0 technologies to enable circular economy practices in a manufacturing context: a business model proposal. J Manuf Technol Manage 30:607–627
29. Nasir MHA, Genovese A, Acquaye AA, Koh SCL, Yamoah F (2017) Comparing linear and circular SCs: a case study from construction industry. Int J Prod Econ 183:443–457
30. Pinto JT, Diemer A (2020) Supply chain integration strategies and circularity in the European steel industry. Resour Conserv Recy 153:104517
31. Sarkis J, Zhu Q, Lai KH (2011) An organizational theoretic review of green supply chain management literature. Int J Prod Econ 130(1):1–15
32. Schrödl H, Simkin P (2014) Bridging economy and ecology: a circular economy approach to sustainable supply chain management. In: Thirty fifth international conference on information systems, Auckland
33. Tjahjono B, Esplugues C, Ares G, Pelaez E (2017) What does industry 4.0 mean to supply chain? Procedia Manuf 13:1175–1182
34. Tseng ML, Tan RR, Chiu ASF, Chien CF, Ku TC (2018) Circular economy meets industry 4.0: can big data drive industrial symbiosis? Resour Conserv Recy 131:146–147
35. Van Engeland J, Beliën J, De Boeck L, De Jaeger S (2020) Literature review: strategic network optimization models in waste reverse supply chains. Omega 91:102012
36. Winkler H, Kaluza B (2006) Sustainable supply chain networks—A new approach for effective waste management. WIT Trans Ecol Environ 92:10
37. Witkowski K (2017) Internet of things, big data, industry 4.0. Innovative solutions in logistics and supply chains management. Procedia Eng 182:763–769
38. Yadav G, Luthra S, Jakhar SK, Mangla SK, Rai DP (2020) A framework to overcome sustainable supply chain challenges through solution measures of industry 4.0 and circular economy: an automotive case. J Clean Prod 254(1):120112
39. Yang M, Smart P, Kumar M, Jolly M, Evans S (2018) Product-service systems business models for circular supply chains. Prod Plann Control 29(6):498–508. https://doi.org/10.1080/09537287.2018.1449247
40. Zeng H, Chen X, Xiaoa X, Zhou Z (2017) Institutional pressures, sustainable supply chain management, and circular economy capability: empirical evidence from Chinese eco-industrial park firms. J Clean Prod 155(2):56–65
41. Zhu Q, Geng Y, Lai KH (2010) Circular economy practices among Chinese manufacturers varying in environmental-oriented supply chain cooperation and the performance implications. J Environ Manage 91(6):1324–1331
42. Zhu Q, Geng Y, Lai KH (2011) Environmental supply chain cooperation and its effect on the circular economy practice-performance relationship among Chinese manufacturers. J Ind Ecol 15:405–419
43. Zhu Q, Geng Y, Sarkis J (2011) Evaluating green supply chain management among Chinese manufacturers from the ecological modernization perspective. Transport Res E-Log 47(6):808–821

Part II
Supply Chain Management

Chapter 5
Conceptual Framework of Supply Chain Competition Based on a System of Systems Approach

M. Gutierrez and L. Urciuoli

Abstract System of Systems (SoS) paradigm has been extensively applied to a wide variety of fields. In recent years, some works have shown that the supply chain can be conceptualized as a SoS, yet they do not consider market competition among supply chains. We develop a competitive supply chain SoS framework that extends existing approaches to incorporate multi-chain market competition, yielding an illustrative case of an uncommon SoS with competitive constituents. While satisfaction of customer needs in a certain market is a key objective for supply chain management, it is only achieved by the set of competitive supply chains.

Keywords System of Systems · Supply chain management · Supply chain competition

5.1 Introduction

Since the beginning of the century, there has been a growing interest in the development and application of the System of Systems (SoS) paradigm to different fields [12, 29]. The development of the systems approach traces back almost a century when in 1926 Smuts introduced "holons" from a nature point of view and broadened the idea of a "whole that is more than the sum of its parts" [3, 43]. Von Bertalanffy [47] extended this concept and developed the grounds of the general system theory. Regarding the current System of Systems approach, we can underline the work of Ackoff [1] as a milestone. Ackoff [1] organizes and gives coherence to previous knowledge on the conceptualization of a system; at the same time, the author provides

M. Gutierrez (✉)
Universidad Politécnica de Madrid, c/José Gutiérrez Abascal 2, 28006 Madrid, Spain
e-mail: miguel.gutierrez@upm.es

L. Urciuoli
KTH Royal Institute of Technology, Lindstedtsvägen 30, SE-114 28 Stockholm, Sweden
e-mail: luca.urciuoli@indek.kth.se

MIT-Zaragoza International Logistics Program, Zaragoza Logistics Center, Zaragoza, Spain

C. Avilés-Palacios and M. Gutierrez (eds.), *Ensuring Sustainability*,
Lecture Notes in Management and Industrial Engineering,
https://doi.org/10.1007/978-3-030-95967-8_5

a reference framework through a set of basic definitions, starting with "a system is a set of interrelated elements." A system composed of other systems arises naturally in this framework, although, for a set of systems to become a SoS, additional features and behaviors are required.

Jaradat et al. [29] provide a historical approach to SoS. Through the study of hundreds of sources, the authors define the main stages in the development of the SoS concept and review its main characteristics. The general concept of a SoS is approached as follows: "At a most basic level, SoS is concerned with the integration and coordination of multiple autonomous systems, considered as a unity, that functions to achieve performance, purpose or behavior that none of the individual constituent systems is capable of independently" [29].

Several definitions of SoS have been proposed, as well as several attempts to identify the distinguishing features of a SoS [4, 19, 24, 29, 30]. Of particular importance is the so-called ABCDE characterization proposed by Boardman and Sauser [8]. It is the result of a comprehensive review of the literature and the analysis of more than 40 SoS definitions, producing a set of five main features of the constituent systems [8, 24]: *Autonomy* refers to the freedom of the constituent systems to set and follow their goals under some restrictions but without being subject to external control; *belonging* (B) refers to the voluntary decision of the constituent systems to take part in the SoS; *connectivity* (C) refers to the ability of the constituent systems to dynamically establish links among them, typically forming a network-based architecture; *diversity* (D) refers to the variety of the SoS capability compared to the requirements-driven functionality of the systems; *emergence* (E) refers to the capability of a SoS to show unforeseeable behavior, evolve and adapt dynamically to new conditions in ways that are not the result of a previous design. Boardman and Sauser [8] emphasize the differentiation between a system of subsystems and a SoS. Characterization is further developed in Gorod et al. [24], with a detailed historical path to the development of the concept of SoS from an engineering point of view, as well as a confrontation with the systems engineering paradigm. Different types of SoS can be identified depending on how much the constituent systems fit into the profile of each SoS defining characteristic.

The standard conceptualization of a supply chain is a network of "all the parties involved, directly or indirectly, in fulfilling a customer request" [15]. These parties include suppliers, manufacturers, distributors, warehousers and retailers. Over the last few years, some authors have shown the applicability of System of Systems to supply chain management (SCM) [11, 14, 28, 32]. The supply chain is formed by a set of independent systems (*autonomy*), each one with its own purpose that aims to jointly achieve a supra-purpose: the satisfaction of customer needs (*belonging*). The systems dynamically create a network that evolves according to the global objectives and through the expanding possibilities of technology (*connectivity*) to offer an increasing variety of products and services (*diversity*) while adapting to a changing environment and varying customer needs (*emergence*). From the viewpoint of some of the ABCDE characteristics, clearly a generic supply chain falls into the SoS profile (*autonomy, belonging, emergence*), whereas some of the characteristics also show behavior that is partly characteristic of a system composed of subsystems (*connectivity, diversity*).

Although there are several works that justify the conceptualization of a supply chain as a SoS, we identify an existing gap in the literature: They do not consider the implications of multi-chain market competition. In this paper, we incorporate the results of the referred works within the framework of a SoS approach to SCM and propose to consider a new level of abstraction in the conceptualized system to include market competition among supply chains. The remainder of the chapter is organized as follows. In the second section, we summarize the existing SoS approaches to supply chain management. Then, in the third section the proposed conceptual framework of SoS approach to supply chain competition is developed. In the fourth section, the resulting system is characterized as a SoS, showing how it can describe relevant aspects of supply chain competition through the analysis of the ABCDE behavior. Finally, the main conclusions are summarized in Sect. 5.5.

5.2 Literature Review

Some authors have shown the applicability of the SoS paradigm to supply chain management (SCM) based on different SoS characterizations. A preliminary work of Hassan [26] delves into the conceptualization of the supply chain as a system composed of subsystems, posing some relevant considerations regarding the systems approach to the supply chain but without considering the specificities of the SoS behavior. Mastrocinque et al. [32] underline the intrinsic fit of some features of a supply chain to the concept of SoS based on the work of Bjelkemyr et al. [7] and show the interest in applying this paradigm to the design of a supply chain. Jaradat et al. [28] analyze the convergence of SoS attributes and the principles and concepts of SCM. The authors show how the SoS paradigm can complement SCM practices, placing particular emphasis on the satisfaction of customer needs through the integration and collaboration among supply chain participants. Choi et al. [14] demonstrate that the sustainable fashion supply chain (extendable to a supply chain in general) is a SoS based on ABCDE criteria. Bondar et al. [9] analyze the emergence behavior that is characteristic of SoS from the perspective of information systems architecture and present the collaborative concurrent engineering process in the automotive supply chain as an example of an agile SoS. Darabi et al. [18] propose a new approach to governance of a system specifically aimed at SoS and apply their proposal to a supply chain to illustrate the characteristics of the framework. Since the publication of the fundamental works of Christopher and Peck [16] and Sheffi [41], the topic of supply chain resilience has gained increasing attention [27, 36]. It is an issue that has also been addressed with an SoS approach. Bukowski [11] analyzes the dependability feature in SoS and presents a case of disruption in a supply chain as an illustrative example to demonstrate the influence of dependability on the resilience of the system.

Table 5.1 summarizes the applications of the SoS paradigm to the supply chain, including the criteria to characterize a supply chain as a SoS and the focus of the application. As aforementioned, none of the works consider the set of competitive supply chains when defining the System of Systems. Furthermore, there are few examples of

Table 5.1 Supply chain as a System of Systems

Author	SoS characterization	Supply chain focus
Mastrocinque et al. [32]	Evolutionary behavior, self-organization, heterogeneity, emergent behavior, small-world/scale-free networks [7]	SC design and optimization
Jaradat et al. [28]	Integration, interconnectivity, emergence, complexity, evolutionary development, ambiguity	SC vertical integration and collaboration
Choi et al. [14]	Autonomy, belonging, emergence, connectivity, diversity (ABCDE) [8]	Sustainable fashion SCM
Bondar et al. [9]	Information systems architecture with emergence behavior	SC emergence in concurrent engineering
Darabi et al. [18]	Purpose integration, belonging regulation, incentivizing device, interactions protocol, and principles dissemination and perception distortion	SC governance
Bukowski [11]	Multidimensional complexity, independence, emergence behavior, evolutionary development	SC resilience

competitive SoS. In the standard approach of a generic SoS, the constituent systems join the SoS and establish connections among them to collaborate in the pursuit of a supra-purpose. Thus, most of the previous works are focused on collaborative SoSs [4, 12]. The work of Darabi and Mansouri [19] is an exception and provides valuable insights into the roles of collaboration and competition in SoSs, studying its influence on the autonomy and belonging characteristics of the constituent systems. The study is supported by an experiment with an agent-based simulated system which shows the relevance of competition to autonomy and behavior when resources are scarce. Collaboration is established not only as a voluntary decision to achieve a mutual benefit but also as a necessity to continue belonging to the SoS.

We attempt to fulfill the identified gap in the literature with the proposal of a conceptual framework of competitive supply chain SoS, that, not only contributes from the supply chain management point of view, but also constitutes an illustrative example of a competitive SoS.

5.3 Development of the Proposed Framework

To develop the proposed framework, we use a bottom-up nested conceptual modeling approach, often referred to as a "Russian doll" approach [45], in which each model constitutes a part of a broader scope model. It has been applied to a wide variety of domains [20, 38].

As the starting point for the modeling process, we take the standard supply chain conceptualization as a SoS (Table 5.1) and complement it with the consideration of the trend toward SC 4.0 and the explicit characterization of the customers SoS (Sect. 3.1). Subsequently, the model is enhanced by widening the scope, under a SoS approach, to consider the whole competitive market in which different supply chains concur (Sect. 3.2).

5.3.1 *Supply Chain 4.0 SoS*

As mentioned above, the development of the framework takes the Supply Chain SoS (SC SoS) as the starting point of the conceptual modeling process and complements it taking a SoS approach to the Supply Chain 4.0 or digital supply chain. We identify three main SoSs as depicted in Fig. 5.1:

- **Supply Chain SoS**. It corresponds to the standard supply chain conceptualization as a network of agents involved in satisfying customer orders [15]. We represent the supply chain, excluding customers, as a simplified network of industrial nodes. As detailed in the introduction (Table 5.1), different authors show that the supply chain is a SoS based on a variety of criteria. SC SoS includes a constituent control SoS as described next.
- **Control SoS**. For a supply chain to be competitive nowadays, it is necessary to achieve an important level of coordination [15]. Panetto et al. [34] provide a framework for the application of the new technologies that are behind the so-called SCM 4.0 and propose the suitability of SoS as a reference to build a cyber network

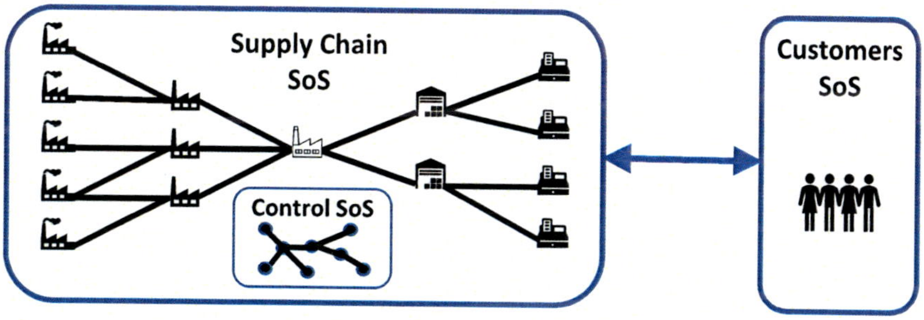

Fig. 5.1 Supply chain 4.0: System of Systems approach

of sensors and control systems that will enable SCM. Referring to the ABCDE SoS characterization, new technologies advocate for a decentralized network (C) of autonomous intelligent agents (A) that are loosely integrated so they can join or leave the system (B), thus providing the necessary resilience (D) while being able to adapt dynamically to changing conditions (E) [34]. Choi [13] outlines the applicability of the SoS approach to SCM that incorporates big data and related technologies. Therefore, we can include in each supply chain SoS a constituent controlling SoS that will evolve alongside digital technologies. The control SoS is schematically depicted as a network of lines connecting dots in Fig. 5.1.

- **Customers SoS.** Customers will form a social SoS as defined by Bar-Yam et al. [5]. Regarding the ABCDE characterization, the profile of the constituent systems (i.e., the customers) shows some characteristics that are directly aligned with a pure SoS, whereas others differ to some extent. It becomes evident that the customers are autonomous (A). Even though they do not group explicitly, their belonging is the result of individual interest, which in general aligns with the global purpose (satisfaction of their needs) (B). Since we are referring to customers of a specific supply chain, connectivity is only relevant in some cases in which customers use network communications to create a community (C). The constituent customers will share common facets and present a certain homogeneity, although they will be diverse in the sense that each individual is inherently unique (D). The group of customers will show some degree of emergence since their behavior can only be approximately predicted in the short term (E).

5.3.2 *Supply Chain Competition: Competitive Supply Chain SoS*

Stock and Boyer [44] analyze 173 definitions of supply chain management and propose the following definition that encompasses the main aspects and elements identified: "The management of a network of relationships within a firm and between interdependent organizations and business units consists of material suppliers, purchasing, production facilities, logistics, marketing and related systems that facilitate the forward and reverse flow of materials, services, finances and information from the original producer to final customer with the benefits of adding value, maximizing profitability through efficiencies and achieving customer satisfaction." [44] Satisfying customers' needs is essential in the usual current conceptualization of SCM. However, it should be noted that in general, customers' needs are not satisfied by a single supply chain, but through the existence of a set of competitive supply chains that offers a variety of substitute products. This fact is stressed under the proposed conceptual model. In addition, supply chains compete for market share, and competition encourages the development of products that meet the needs and preferences of customers.

The core proposal of this work is to conceptualize supply chain competition based on a SoS approach. According to the "Russian doll" conceptual modeling approach

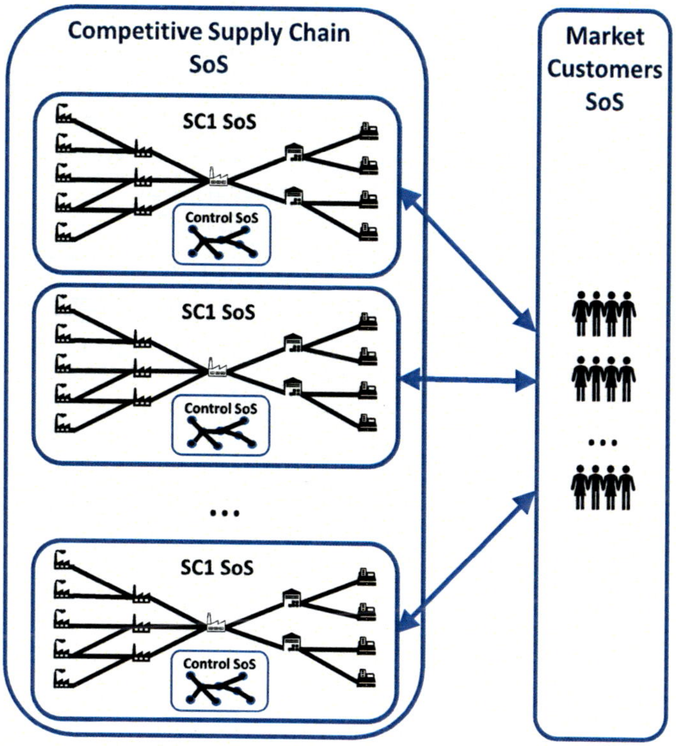

Fig. 5.2 Supply chain competition: System of Systems approach

adopted [45], we develop it as an evolution of the framework of Fig. 5.1, as depicted in Fig. 5.2, in which we can define two new SoSs:

- **Market Customers SoS**. Formed by all actual and potential customers of the market. Significantly, it fits better as a SoS according to the ABCDE characterization than the customer SoS of Fig. 5.1. Customers benefit from the fact of participating in a system closer to a SoS in terms of satisfaction of their needs. Connectivity in the market customers SoS is much more relevant than in the customer SoS (Fig. 5.1), with a direct effect in the *emergence* as some customers influence the market. Communication through social networks and opinions shared in Internet forums have grown enormously in recent years, becoming crucial to customer buying decisions in some sectors [37, 39, 46].
- **Competitive Supply Chain SoS**. Formed by the set of all supply chains that operate in a certain market. Each supply chain constitutes a SoS itself, as previously defined. Satisfaction of customers' needs acquires full sense as the supra-purpose of the competitive SC SoS. The competitive SC SoS illustrates the conceptual modeling process that starts from the Supply Chain 4.0 SoS and provides an explicit representation of supply chain competence through the interaction with the customers. However, the actual structure of the set of competitive supply chains

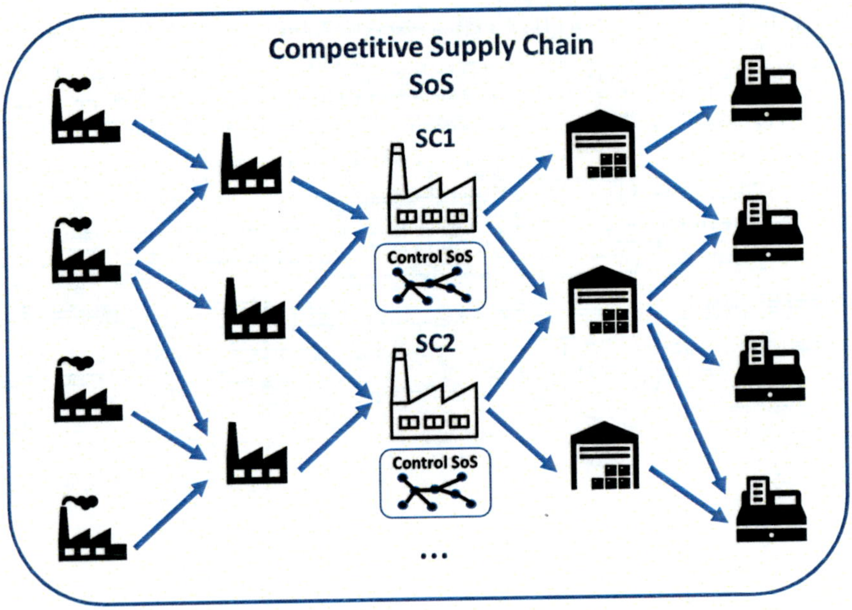

Fig. 5.3 Competitive Supply Chain SoS

is generally a very complex single network, with some of the nodes belonging to different supply chains. Figure 5.3 represents this single competitive SC SoS network. The characterization of this network as a SoS is analyzed in the following subsection.

5.4 Characterization of the Competitive Supply Chain as a SoS

We complete the proposal of a competitive supply chain framework with its characterization as a SoS. It is carried out on the basis of the ABCDE characterization and by paying special attention to the comparison with the single supply chain SoS (summarized in Table 5.2). In fact, it is shown that it fits in the SoS paradigm even better than the single supply chain.

- **Autonomy**. Clearly, in general, the constituent agents of the network of all competitive supply chains are autonomous and have managerial and operational independence [40]. The same applies to each supply chain [14], but it will be less applicable as the supply chain shows a high degree of vertical integration [25]. When a company encompasses many of the steps of the supply chain, the autonomy of the constituent systems is reduced. Even for a specific industry, we can find different autonomy behaviors among different supply chains, such is the case of the fashion

Table 5.2 ABCDE characteristics of the Competitive Supply Chain SoS versus single Supply Chain SoS

Criteria	Competitive supply chain SoS versus single supply chain SoS
Autonomy (A)	More proportion of independent constituent systems
Belonging (B)	More stable and stronger linkage of the constituent systems to the SoS
Connectivity (C)	Denser network structure and consideration of horizontal collaborations
Diversity (D)	Variety of products and drastic increment of resilience
Emergence (E)	Improvement of customer satisfaction and adaptation to changing markets

industry studied by Choi et al. [14] in which the case of H&M analyzed by the authors differs from its competitor Inditex-Zara [2, 21].

- **Belonging**. Constituent agents freely join other agents, establishing collaboration relationships, accepting the necessary rules and aligning their own purpose as autonomous systems to achieve the SoS supra-purpose of satisfying customer needs [40]. Since a company (supplier, manufacturer, distributor) can participate in multiple supply chains of a certain market, the belonging feature gains more meaning under the proposed competitive supply chain SoS framework than the single supply chain approach taken in the literature (Table 5.1). Companies can exercise their autonomy and abandon the SoS, but in general their linkage to the Competitive SC SoS is more stable than the commitment to a single supply chain. A supplier might break the relationship with some manufacturer—with a single supply chain—while maintaining its belonging to the Competitive SC SoS by delivering materials to other manufacturers (the same applies to a distributor).
- **Connectivity**. However connectivity is a feature of each supply chain considered as a SoS, it also gains relevance and stability in Competitive SC SoS, since, in general, there are more connections—that is, more suppliers and/or customers—per node, so the resulting network is much closer to the characteristic structure and behavior of the SoS [40]. Frequently, some suppliers deliver materials to competitor manufacturers, whereas some distributors will consolidate deliveries from competitor manufacturers. Since the practical expansion of computer network protocols and multi-tier architectures, information technologies have played an essential role in enabling efficient inter-company network consortiums [6]. The increasing role of technology and the trend toward Supply Chain 4.0 favor the efficacy and efficiency of multi-chain material flow [33]. On the other hand, and of particular relevance, the competitive SC SoS framework elicits another type of links that are receiving increasing importance: those among competitors at the same tier of the supply chain that lead to horizontal collaboration or cooperation [17, 23, 31]. Polenske [35] provides a formal distinction between collaboration and cooperation, followed by an in-depth analysis of the interrelation of the so-called 3C—collaboration–cooperation–competition. Although there is a variety of particular cases, and no generalization is possible, vertical relationships would be typical examples of collaboration, whereas horizontal relationships would be typical examples of cooperation. Another term is being increasingly used

to refer to relationships between firms that compete in their core processes, while cooperating in non-core processes: coopetition [48, 50].

- **Diversity**. Diversity "is a necessary condition for making the overall SoS resilient" [10]. As mentioned in the introduction, the SoS approach to the supply chain is particularly well suited to analyze the dynamics of supply chain disruptions and, consequently, the design for resilience [11]. It is noteworthy that the consideration of the set of competitive supply chains leads to a highly relevant shift in terms of the diversity provided by the SC SoS. The evolving variety of competitive products offered in the market not only reflects better customer needs satisfaction, but it is also the key to overcome supply chain disruptions from the customer's perspective. If a disruption affects a specific supply chain, competitors can fulfill market demand until the affected supply chain recovers from the disruption. If they cannot wait until the recovery, some of the customers will find the alternative products as a temporary solution, whereas others might even change their preferences in the absence of the product usually acquired. When the disruption has a global effect, such as in the case of natural disasters, transportation strikes, political disorders or a pandemic such as COVID-19, and all competitive supply chains in a certain market are affected, the impact will generally be drastically mitigated by the set of competitive supply chains compared to a single supply chain. The time during which there is no product of a certain market available is drastically reduced.
- **Emergence**. This feature is significantly more present in the competitive SC SoS than in the single SC SoS. On many occasions, the changing market and/or irruption of new technologies force some competitors to abandon the market leading pace to others. It is the adapting nature of the competitive SC SoS that makes it possible to evolve alongside with the market and provide the customers with the products they demand. On the other hand, when faced with global disruptions, the offer of new substitutive products and the alternatives identified by the set of competitive supply chains will mitigate the effect for customers. Sheffi describes numerous examples of innovative successful alternatives that companies developed in response to the COVID-19 pandemic [42].

5.5 Conclusions

The study of literature shows that a single supply chain constitutes a System of Systems (SoS), although there has been no attempt to conceptualize the system formed by the set of market competitor supply chains. We show how the System of Systems (SoS) approach can be applied to conceptualize this set of competitor supply chains and that the resulting competitive supply chain SoS fits as an SoS better than a single supply chain according to the autonomy, belonging, connectivity, diversity and emergence (ABCDE) characterization. On the other hand, the SoS thus defined constitutes an illustrative example of competitive SoS.

The analysis of the different SoS aspects shows that it can be applied to describe supply chain market competition dynamics. Specifically, it elicits the fact that customer needs are satisfied not by a single supply chain but by the set of competitive supply chains. The analysis of the SoS allows to characterize supply chain resilience issues as well as to identify ways of improving its performance. With this regard, the role of collaboration among market competitors appears as a promising aspect to be studied in order to systematically find ways of global performance improvement.

The proposed framework can incorporate open-loop supply chains when the material flows are handled by companies of one sector. It can be expanded with the interaction of external systems to model the generic circular supply chain [22] and complement existing conceptualizations [49].

References

1. Ackoff RL (1971) Towards a System of Systems concepts. Manag Sci. https://doi.org/10.1287/mnsc.17.11.661
2. Aftab MA, Yuanjian Q, Kabir N, Barua Z (2018) Super responsive supply chain: the case of Spanish fast fashion retailer Inditex-Zara. Int J Bus Manage 13:212. https://doi.org/10.5539/ijbm.v13n5p212
3. Ansbacher HL (1994) On the origin of holism. Individ Psychol: J Adlerian Theory Res Pract 50:486–492
4. Baldwin WC, Sauser BJ, Boardman J (2017) Revisiting "The Meaning of Of" as a theory for collaborative System of Systems. IEEE Syst J 11:2215–2226. https://doi.org/10.1109/JSYST.2015.2430755
5. Bar-Yam Y, Allison MA, Batdorf R, et al (2004) The characteristics and emerging behaviors of System of Systems. In: NECSI: complex physical, biological and social systems project. http://necsi.edu/education/oneweek/winter05/NECSISoS.pdf. Accessed 30 Aug 2020
6. Batista de Chambers D, Sastrón F, Gutiérrez M (2001) Production planning and control information system for the engineering and make to order environment. A virtual enterprise approach. IFIP Adv Inf Commun Technol 56:333–340. https://doi.org/10.1007/978-0-387-35399-9_32
7. Bjelkemyr M, Semere DT, Lindberg B (2008) Definition, classification, and methodological issues of System of Systems. In: Jamshidi M (ed) Systems of systems engineering. CRC Press, Boca Raton, pp 191–206
8. Boardman J, Sauser B (2006) System of Systems—the meaning of of. In: 2006 IEEE/SMC international conference on System of Systems engineering, pp 118–123
9. Bondar S, Hsu JC, Pfouga A, Stjepandić J (2017) Agile digital transformation of System-of-Systems architecture models using Zachman framework. J Ind Inf Integr 7:33–43. https://doi.org/10.1016/j.jii.2017.03.001
10. Bourne M, Franco-Santos M, Micheli P, Pavlov A (2018) Performance measurement and management: a System of Systems perspective. Int J Prod Res 56:2788–2799. https://doi.org/10.1080/00207543.2017.1404159
11. Bukowski L (2016) System of Systems dependability—Theoretical models and applications examples. Reliab Eng Syst Saf 151:76–92. https://doi.org/10.1016/J.RESS.2015.10.014
12. Cadavid H, Andrikopoulos V, Avgeriou P (2020) Architecting systems of systems: a tertiary study. Inf Softw Technol 118:106202. https://doi.org/10.1016/J.INFSOF.2019.106202
13. Choi T (2018) A System of Systems approach for global supply chain management in the big data era. IEEE Eng Manage Rev 46:91–97. https://doi.org/10.1109/EMR.2018.2810069
14. Choi T, Cai Y, Shen B (2018) Sustainable fashion supply chain management: a System of Systems analysis. IEEE Trans Eng Manage:1–16. https://doi.org/10.1109/TEM.2018.2857831

15. Chopra S (2019) Supply chain management: strategy, planning, and operation, 7th edn. Pearson, Boston
16. Christopher M, Peck H (2004) Building the resilient supply chain. Int J Logist Manage 15:1–14. https://doi.org/10.1108/09574090410700275
17. Cruijssen F, Dullaert W, Fleuren H (2007) Horizontal cooperation in transport and logistics: a literature review. Transp J 46:22–39. https://doi.org/10.2307/20713677
18. Darabi HR, Gorod A, Mansouri M (2012) Governance mechanism pillars for systems of systems. In: Proceedings 7th international conference on System of Systems engineering, SoSE 2012, pp 374–379
19. Darabi HR, Mansouri M (2013) The role of competition and collaboration in influencing the level of autonomy and belonging in System of Systems. IEEE Syst J 7:520–527. https://doi.org/10.1109/JSYST.2013.2256972
20. de Waal FBM (2007) The "Russian doll" model of empathy and imitation. In: Braten S (ed) On being moved: from mirror neurons to empathy. John Benjamins, Amsterdam, pp 49–69
21. Dumbrill W, Wang Y (2021) Implementing vertical integration in the fashion industry. In: Lecture notes in electrical engineering. Springer, pp 401–407. https://doi.org/10.1007/978-981-33-6318-2_50
22. Farooque M, Zhang A, Thürer M et al (2019) Circular supply chain management: a definition and structured literature review. J Clean Prod 228:882–900. https://doi.org/10.1016/j.jclepro.2019.04.303
23. Ferrell W, Ellis K, Kaminsky P, Rainwater C (2020) Horizontal collaboration: opportunities for improved logistics planning. Int J Prod Res 58:4267–4284. https://doi.org/10.1080/00207543.2019.1651457
24. Gorod A, Sauser B, Boardman J (2008) System-of-Systems engineering management: a review of modern history and a path forward. IEEE Syst J. https://doi.org/10.1109/JSYST.2008.2007163
25. Guan W, Rehme J (2012) Vertical integration in supply chains: driving forces and consequences for a manufacturer's downstream integration. Supply Chain Manage 17:187–201. https://doi.org/10.1108/13598541211212915
26. Hassan MMD (2006) Engineering supply chains as systems. Syst Eng 9:73–89. https://doi.org/10.1002/sys.20042
27. Hohenstein N-O, Feisel E, Hartmann E, Giunipero L (2015) Research on the phenomenon of supply chain resilience. Int J Phys Distrib Logist Manage 45:90–117. https://doi.org/10.1108/IJPDLM-05-2013-0128
28. Jaradat R, Adams F, Abutabenjeh S, Keating C (2017) The Complementary perspective of System of Systems in collaboration, integration, and logistics: a value-chain based paradigm of supply chain management. Systems 5:50. https://doi.org/10.3390/systems5040050
29. Jaradat R, Keating C, Bradley J (2014) A histogram analysis for System of Systems (SoS). Int J Syst Syst Eng 5:193–227. https://doi.org/10.1504/IJSSE.2014.065750
30. Maier MW (1998) Architecting principles for systems-of-systems. Syst Eng 1:267–284. https://doi.org/10.1002/(SICI)1520-6858(1998)1:4%3c267::AID-SYS3%3e3.0.CO;2-D
31. Martin N, Verdonck L, Caris A, Depaire B (2018) Horizontal collaboration in logistics: decision framework and typology. Oper Manage Res 11:32–50. https://doi.org/10.1007/s12063-018-0131-1
32. Mastrocinque E, Yuce B, Lambiase A, Packianather MS (2014) A System of Systems approach to supply chain design. In: Frontiers of manufacturing and design science IV. Trans Tech Publications Ltd., pp 2807–2814
33. Melander L, Pazirandeh A (2019) Collaboration beyond the supply network for green innovation: insight from 11 cases. Supply Chain Manag 24:509–523. https://doi.org/10.1108/SCM-08-2018-0285
34. Panetto H, Iung B, Ivanov D et al (2019) Challenges for the cyber-physical manufacturing enterprises of the future. Annu Rev Control. https://doi.org/10.1016/J.ARCONTROL.2019.02.002

35. Polenske KR (2004) Competition, collaboration and cooperation: an uneasy triangle in networks of firms and regions. Reg Stud 38:1029–1043. https://doi.org/10.1080/0034340042000292629
36. Ponomarov SY, Holcomb MC (2009) Understanding the concept of supply chain resilience. Int J Logist Manag 20:124–143. https://doi.org/10.1108/09574090910954873
37. Reyes-Menendez A, Correia MB, Matos N (2020) Understanding online consumer behavior and eWOM strategies for sustainable business management in the tourism industry. Sustainability 12:1–14. https://doi.org/10.3390/su12218972
38. Sadon SK, Din NM, Al-Mansoori MH et al (2012) Dynamic hierarchical bandwidth allocation using Russian Doll Model in EPON. Comput Electr Eng 38:1480–1489. https://doi.org/10.1016/j.compeleceng.2012.05.002
39. Safi R, Yu Y (2017) Online product review as an indicator of users' degree of innovativeness and product adoption time: a longitudinal analysis of text reviews. Eur J Inf Syst 26:414–431. https://doi.org/10.1057/s41303-017-0045-2
40. Sauser B, Boardman J, Gorod A (2009) System of Systems management. In: Jamshidi M (ed) System of Systems engineering: innovations for the 21st century. Wiley, pp 191–217
41. Sheffi Y (2005) The resilient enterprise: overcoming vulnerability for competitive advantage. The MIT Press, Cambridge, MA
42. Sheffi Y (2020) The new (ab)normal: reshaping business and supply chain strategy beyond covid-19. MIT CTL Media, Cambridge
43. Smuts JC (1927) Holism and evolution, 2nd edn. Macmillan and Co., London
44. Stock JR, Boyer SL (2009) Developing a consensus definition of supply chain management: a qualitative study. Int J Phys Distrib Logist Manage 39:690–711. https://doi.org/10.1108/09600030910996323
45. Thalheim B (2010) Towards a theory of conceptual modelling. J Univ Comput Sci 16:3102–3137. https://doi.org/10.1007/978-3-642-04947-7_7
46. Tien DH, Amaya Rivas AA, Liao YK (2019) Examining the influence of customer-to-customer electronic word-of-mouth on purchase intention in social networking sites. Asia Pac Manage Rev 24:238–249. https://doi.org/10.1016/j.apmrv.2018.06.003
47. Von Bertalanffy L (1950) An outline of general system theory. Br J Philos Sci 1:134–165. https://doi.org/10.1093/bjps/I.2.134
48. Walley K (2007) Coopetition: an introduction to the subject and an agenda for research. Int Stud Manage Organ 37:11–31. https://doi.org/10.2753/imo0020-8825370201
49. Weetman C (2017) A circular economy handbook for business and supply chains: repair, remake, redesign, rethink. Kogan Page, London
50. Wilhelm MM (2011) Managing coopetition through horizontal supply chain relations: linking dyadic and network levels of analysis. J Oper Manage 29:663–676. https://doi.org/10.1016/J.JOM.2011.03.003

Chapter 6
Supply Chain Response: Proposal for a General Definition

R. A. Díaz and E. Benedito

Abstract This research on supply chain response (SCR) is a topic of interest to academics and professionals that helps to meet customer expectations in a timely manner and contributes to the achievement of supply chain (SC) objectives. Various definitions of SCR are found in the scientific literature, each valid for the purpose of the research that proposes it. The wide range of definitions prevents from establishing a framework for analysis and improvement of the SCR that can be applied to any type of SC. For the study of the definitions of SCR, a qualitative content analysis methodology was applied. The level of analysis of the SC and the research method applied by the author of the definition were also taken into account in the analysis. The common characteristics of the various definitions, the differences between them, and the shortcomings of each have been determined. The main contribution of this research is the proposal of a general definition of SCR that has the common elements of existing ones and is useful for any type of SC. A general definition will allow addressing the supply chain response with a common framework for any type of supply chain and not only those mentioned in the academic literature.

Keywords Supply chain response · Supply chain responsiveness · Definition responsiveness

6.1 Introduction

Two major challenges facing supply chains (SC) today are responding to customer changes in the shortest possible time [5] and gaining competitive advantage [18]. In

R. A. Díaz (✉)
Department of Social Sciences, Universidad Nacional de Colombia, Cra. 32 # 12–00, Palmira, Valle del Cauca, Colombia
e-mail: radiazpa@unal.edu.co

E. Benedito
Department of Management and Institute of Industrial and Control Engineering, Universidad Politécnica de Catalunya. Avinguda Diagonal, 647 Barcelona, España
e-mail: ernest.benedito@upc.edu

C. Avilés-Palacios and M. Gutierrez (eds.), *Ensuring Sustainability*, Lecture Notes in Management and Industrial Engineering,
https://doi.org/10.1007/978-3-030-95967-8_6

this sense, [27] defines the supply chain response (SCR) as the SC action capacity to respond to customer demand and market variability in a reasonable time, gaining and maintaining competitive advantage. In this definition, three aspects are combined: firstly, a change in some external element to the SC (change in customer demand), secondly, the modification of the behavior of the SC to adapt to the external change (action to give response), and thirdly, the objective pursued when modifying the behavior of the SC (gaining and maintaining competitive advantage). The current SC also has to adapt to various types of changes both external and internal of the SC itself and with different objectives. Thus, for example, humanitarian SC, often has to respond to natural disasters in order to meet the urgent needs of the affected people [13], or perishable food SC that must respond with delivery of product on time and fresh, and late arrival of an ingredient may affect other perishable ingredients by delaying product preparation and delivery to the customer [3]. SC of industrial and service products must respond to internal changes such as the unavailability of workers or changes in production scheduling [19]. According to [26], the internal changes of the supplier directly affect the response of the buyer to meet customer requirements.

The research carried out on the SCR aims at proposing improvements in SC management [24] and performance [12]. However, each research proposes a definition according to the improvement under investigation; for example, [30] propose a coordinated SCR definition with the aim of improving the coordination of different response strategies, and [7] define the SCR in relation to the demand that will be met in an area with the aim of improving the design of supplying networks. Therefore, in SCR research, there is no definition that can be applied to all SC or generalized for all processes and functions of the chain. A general definition should allow developing the theory, investigation, and practice of the SCR, thus facilitating to define and to establish relations between the managers of the answer.

The aim of this research is to provide a definition of supply chain response that is applicable to any type of SC that has to adapt to any type of change. The methodology applied to development this research was qualitative content analysis. According to Arbeláez and Onrubia [1] through qualitative content analysis, it is possible to "verify the presence of topics, words, or concepts in a content and their meaning within a text". Krippendorff [17] claims that content analysis allows determining patterns through secondary data. The content analysis methodology has been used in various areas of research and particularly that of SC. Three reference applications of the content analysis methodology in SC were carried out by [28, 23, 14]. The first applies content analysis in the measurement of the SC construct. The second is applied to document the use of content analysis in SC literature. The third one applies content analysis to innovation in SC.

This article is composed of the following sections: Sect. 6.2 details the methodology applied for the development of this research, Sect. 6.3 deals with the definitions of SCR and the description of its characteristics, in Section number 4, the general definition of the SCR is written, in Sect. 6.5, the general definition of the SCR is validated, and it finishes with Sect. 6.6 conclusion and future research.

6.2 Methodology

In order to carry out this research, the qualitative content analysis method proposed by Marying (2014) [21] was applied. The method was performed in several steps. First, identification of the mínimum unit of analysis. Second, selection of the documents that would be part of this investigation. Third, identification of the common characteristics, differences, and deficiencies of the analyzed definitions.

The minimum unit of analysis was identified in the academic literature on SCR corresponding to the period 1996–2019. The databases reviewed were Scopus and Web of Science.

Searches were performed combining the keywords *supply chain, supply chain responsiveness, definition responsiveness, and* 858 documents were obtained. The documents were filtered verifying that they addressed the SCR concept, and a list of 191 articles was generated. For each of them, the title, summary, introduction, and conclusions were reviewed to determine if they proposed a definition of supply chain response. Finally, 20 articles with SCR definitions that form the basis of this research were selected.

The common characteristics, differences, and dificiencies between the selected definitions were analyzed.

Also, the context in which the research was developed was analyzed. The context of the analyzed definitions of SCR is made up of the purpose of the research, the level of analysis of the SC in which it was defined and the research method that was applied. With the results of the analysis of the definitions mentioned above, the general definition of SCR was drawn up.

Finally, the proposed definition was validated. The validation was carried out following two steps. First, the general definition of SCR proposed here overlapped on the definition that was initially identified in the document. Second, the components of the general definition of SCR proposed here in the analyzed documents were identified..

6.3 Selected SCR Definitions

This section presents 15 definitions of SCR found in the 20 articles finally selected from the literature review. Taking into account that the definitions of the responsiveness, responsive supply chain, and supply chain responsiveness encompass the response to the supply chain, these terms have been consolidated as supply chain response. These definitions provide differentiating elements among them.

- Catalan and Kotzab [5] defined the SCR as "the ability to respond and adapt in a timely manner to current market signals based on the ability to read and understand in real time and back from the SC the changes in demand generated by the final user".

- According to [15], the SCR is "the ability of a company together with the participants of the SC to respond to changes in the needs of the market and those of the competitive environment".
- For [25], the SCR is "the speed with which a system (manufacturing system or a SC) can adjust its outputs to achieve any of the four types of flexibility: product, volume, product combination, and delivery in response to external stimuli such as the order of a customer".
- Gunasekaran et al. [9] extend the previous definitions of SCR to "a network of companies capable of creating profit for shareholders in a competitive environment through rapid reaction and the effective cost to face changes in market requirements".
- Squire et al. [26] defined responsiveness as "the speed with which the supplier reacts to information from the buyer firm and more generally from the overall market".
- Klibi et al. [16] defined the SCR of an SC network as the ability to respond positively to changes in business conditions.
- Williams et al. [29] defined the response of the SC in terms of four external flexibilities: flexibility of new products, volume of production, variety in production and delivery, and flexibility to modify the product or service (customization). The four types of flexibility include changes in demand and the supply.
- Hum and Parlar [11] in their research related to SC measurement through optimization define responsiveness as the "probability that a placed order can now be fulfilled within t time units".
- Dreyer et al. [6] defined the SCR as the change of appropriate behavior of the system to respond to external stimuli.
- L'Hermitte et al. [20] in the context of agile humanitarian SC define the SCR as the ability to respond in a timely manner to the identification of operational risks and opportunities.
- Rajagopal et al. [24] defined responsiveness in the context of the supply chain as "the ability to react persistently and within an apt period to client's demand or changes in the marketplace, besides to generate or sustain a competitive advantage as the way forward".
- Fattahi et al. [7] researching the design of resilient and responsive supply chain networks define the SCR as a percentage of the potential demand of customers in an area that can attend a SC.
- Kristianto et al. [18] defined responsiveness as "the ability to dampen the effects of demand changes thorough purposeful reaction within a specified response time window".
- Cannella [4] defined the supply chain responsiveness as "the ability of a system to deliver the same product within a shorter lead time".
- Hum et al. [12] defined supply chain responsiveness "as the probability of fulfilling a customer order within a quoted lead time".

The SCR definitions of the selected articles have four characteristic elements: first, the occurrence of a change in some external element to the SC. External change affects the operation of the SC. Second, the adaptation of the SC behavior to that change. The adaptation of the SC activities encompasses the change in the SC´s strategic, tactical, or operational planning. Examples of the activities that can be adapted are product design, the material acquisition, production, or delivery. The adaptation helps to reduce the impact of change in SC. Third, the objective pursued by the SC in modifying its behavior and respond to the change. Fourth, the criterion used to assess the response. The criterion applied to evaluate the SCR is the response time variable. Table 6.1 shows these elements to all definitions analyzed in this research.

Regarding the first element, external changes are beyond the control of the SC and direct the adaptation of SC activities. To refer to these changes, generic terms are applied for instance: changes in demand, changing business, and market conditions

Table 6.1 SCR definition features

Article	External change	AT	Obj	RA	Al	Rm
Catalan and Kotzab [5]	Current market signals	E		X	Ch	I
Kim et al. [15]	Needs of the market	E			Fm	Su
Reichart and Holweg [25]	Stimuli	E/I		X	NA	Sd
Gunasekaran et al. [9]	Market requirements	E	X	X	Fm	Sd
Squire et al. [26]	Overall market	E		X	Dy	Su
Klibi et al. [16]	Business conditions	E			Net	Sd
Williams et al. [29]	Demand and the supplier	E		X	Fm	Su
Hum and Parlar [11]	Placed order	E		X	Ch	Om
Dreyer et al. [6]	External stimuli	E		X	Ch	Cs
L'Hermitte et al. [20]	Operational risks and opportunities	E/I		X	Fu	Sd
Rajagopal et al. [24]	Demand or marketplace	E	X	X	Fu	Su
Fattahi et al. [7]	Potential demand	E	X		Net	Om
Kristianto et al. [18]	Demand	E		X	Fu	Sd
Cannella et al. [4]	Deliver the same product	E		X	Ch	Om
Hum et al. [12]	Customer order	E		X	Net	Om

Caption
External Change: External changes that occur outside the SC.
Adaptation type: adaptation type (AT) of the SC external (E) or internal (I).
Objective: objective of the supply chain (Obj).
Assess: response assessment (RA).
For each definition, the external change that takes into account is shown in the column external change, if the adaption of the SC is internal (I) or external (E) in the AT column, if it includes any objective of the SC (marked with X) in the Obj column, and if it includes an evaluation factor of the answer (marked with X) in the RA column.
Analysis level (Al): Function (Fu), Firm (Fm), Dyad (Dy), Chain (Ch), Network (Net), Does not apply (NA)
Research method (Rm) Case study (Cs), Interview (I), Optimization models (Om), Secondary data (Sd), Survey (Su)

[16, 24]. Respecting the second element, the adaptation of SC activities to external changes, it was identified in the definitions of SCR analyzed that the adaptation of the SC behavior is associated with the flexibility of the SC either internal [25] or external [29]. Concerning the third element, the objectives pursued by the SC in adapting to external changes are varied in the different definitions. Thus, we find objectives such as satisfying the client [7] or gaining and maintaining competitive advantage [24, 27].

Regarding the fourth element, the assessment of the response, it is observed that it is always done by means of the time variable. Time is described with the following terms: in a timely manner [5], quick reaction (Gunasekaran et al. [9]), speed [25, 26], adequate period or reasonable period [24, 18, 27], t time units [11], and agreed delivery time [12].

The context in which the characteristics of the previously analyzed SCR definitions were identified is described below. The description of the context allowed us to recognize for which specific level of SC, the analyzed definitions were proposed. According to [10], the levels of analysis of the SC are function, firm, dyad, chain, network. The function level covers a specific area of the organization, for example, manufacturing. Dyad includes two or more companies, e.g., supplier–buyer. Chain groups are different component of the chain such as supplier, buyer and consumer. Net connects several links in the chain, that is, to say several suppliers connected to a buyer [8]. The analysis level column in Table 6.1 shows at which level of analysis the definition was proposed. The proposed definitions of SCR have been generated for particular cases and referring to a specific level of SC.

The research method applied in the documents containing the definitions of SCR that were analyzed is also part of the context. The research method allows us to understand how knowledge has been generated in SCR and how the SCR was evaluated in each investigation. Research in SCR has been characterized by applying different research methods. The research methods that have been used the most are the survey and secondary data analysis. The surveys have been carried out via email and telephone. The respondents have been supervisors and managers from various areas of the SC. Secondary data is data that has not been generated by the author of the research, such as academic documentation, books, organization manuals, among others. The secondary data analysis research method has made it possible to propose conceptual models for SCR management. Followed by optimization models and case studies, the optimization models have addressed topics such as the optimization of inventory levels, queuing theory applied to the design of SC networks, multi-objective programming, stochastic, linear, and simulation. The variable that is optimized is time in the different terms discussed above. The case studies developed in the SCR research have addressed the integration and coordination issues of the SC. Information and demand flows are addressed mainly with the interview method in SCR research. The research method column in Table 6.1 shows the research methods that were identified in the analysis of the SCR definitions performed. Table 6.2 shows the relationship between the research methods at the level of SC analysis in which

Table 6.2 SCR research context

		Research method				
		Survey	Optimization model	Secondary data	Case study	Interview
SC analysis level	Function	1		3	1	
	Firm	2			1	
	Dyad	1				
	Chain	1	2		2	2
	Network		2	1		
	N/A			1		

the SCR has been investigated. It highlights that most of the definitions of SCR have been proposed by researching at the SC level and applying the survey application research method.

6.4 A General Purpose Supply Chain Response Definition

The definitions of SCR studied are applied to SC that have to adapt to the external changes mentioned in each of them. However, on the one hand, there are other types of external changes to the SC that require a response from the SC, which do not appear in the analyzed definitions. For example, natural disasters and events that cause humanitarian emergencies require an immediate response. On the other hand, the definitions set out in the literature do not take into account that SC often has to respond to internal changes such as an unplanned machine shutdown, a change of inventory, or a change of information system. In addition, to respond to external changes, SC generates internal changes, to which they must respond as well. Hence, SC presents external and internal changes that must be answered. Kritchanchai and MacCarthy [19] propose the term stimulus to refer to them. Regarding the adaptation of SC activities to the stimuli it receives, whether external or internal, it was identified that not all definitions include this component. Not recognizing the adaptation component as part of the definition of SCR contributes to increasing the negative impact of stimuli on SC. Highlighting the importance of this component, [2] state that SCR is associated with the timely change of activities in the presence of stimuli.

The objectives of the humanitarian SC are related to the satisfaction of urgent needs caused by natural disasters [13] and to save lives [22]. These objectives are not taken into account in the proposed definitions of SCR. With the exception of the definition of Gunasekaran et al. [9], which includes aspects of cost and profit, there is no evidence in the definitions proposed regarding the objective of the response and the fulfillment of the objectives of the SC.

The results described above are summarized by stating that there is no general definition of SCR in which all the SC or results that can be achieved with the identification of the response. Additionally, they do not take into account the fulfillment of the objectives of the SC. Taking into account the characteristics of the SCR definitions analyzed and their deficiencies, a general definition of supply chain response is proposed which generalizes the previous ones, as follows:

The supply chain response is the adaptation of the supply chain activity to the stimuli it receives, to simultaneously meet certain objectives and those of the supply chain, assessing the adaptation for the time elapsed between the occurrence of the stimulus and the fulfillment of the determined objective.

6.5 Validation of General Definition of Supply Chain Response

This section presents the result of the validation by superimposing the general definition of SCR proposed in Section 4 of the definition initially identified in the articles selected for this research. The validation of the definition was carried out by recognizing the inclusion of the external and internal stimuli, the adaptation of SC activities, the objectives of both the SC and the stimulus, and the evaluation of the response in the articles analyzed. All four components are contained in the proposed general definition of SCR.

Regarding the inclusion of external and internal stimuli in the researches carried out on SCR, it was observed that all the investigations explicitly include external stimulus. The second column on Table 6.3 shows the external stimuli that were identified in the validation. The response to internal stimuli of SC is not explicitly mentioned in the researches analyzed. However, [12] address the response to internal stimuli by investigating the impact of placing an order throughout the SC.

Regarding the adaptation that the SC must make of its activities to respond to the stimuli, it was identified that the most frequent adaptation is made on the strategies and manufacturing processes. The supplier´s processes, transportation, distribution, and information processing systems are also adapted. The third column on Table 6.3 shows the adaptations mentioned above.

Regarding the objectives of the SC, it was identified that the researches have included goals of the SC and the SC processes. However, SCR investigations must ensure compliance with other SC objectives that are also important, such as social and environmental responsibility objectives and civil liability objectives. Also, they ensure the fulfillment of the objectives that depend on the stimulus to which it is responding, such as meeting urgent needs, saving lives, adjusting production to a change of raw material, and modifying the delivery route due to a road closure. In any case, the response to one stimulus must guarantee the fulfillment of strategic, tactical, and operational objectives of the entire SC.

Table 6.3 Stimuli and adaptation of activities of the supply chain response

Article	Stimuli	Adaptation of activities
Catalan and Kotzab [5]	Demand in mobile phone	Strategic suppliers Vendor management inventory Postponement
Kim et al. [15]	Needs of the market	Information systems processes
Reichart and Holweg [25]	External flexibility and demand changes	Manufacturing process Manufacturing flexibility Adjust upstream supply chains
Gunasekaran et al. [9]	Market requirements	Product models Between productlines
Squire et al. [26]	Customer specifications	Supplier process
Klibi et al. [16]	Variations in business conditions	Based on capabilities
Williams et al. [29]	Changes in business environment Changes in demand and supply	Production/delivery quantities and qualities
Hum and Parlar [11]	Meet a given demand	Adjust the speed of machines
Dreyer et al. [6]	Market dynamics	Postponement
L'Hermitte et al. [20]	Operational risks and opportunities	Operating routines Delivery location The mode of transport used The transport routes
Rajagopal et al. [24]	Demand or marketplace	Postponement
Fattahi et al. [7]	Demands of customer	Manufacturing and warehouse operations
Kristianto et al. [18]	Demand changes	Shift manufacturing operations
Cannella et al. [4]	Customers' changes in the demand volume	Smoothing replenishment rules
Hum et al. [12]	Customer order	Make to order
Sundram et al. [27]	Customer demand and market	Lean process strategies

Finally, the goal that is implicit in all investigations is to achieve a response in time for the stimulus. The SC objectives column of Table 6.4 shows the SCR research objectives discussed on this paragraph.

The researches analyzed have evaluated SCR mainly through lead time. Other evaluations have been carried out in the new product development, supplier, and manufacturing processes. In addition, the key perfomance indicators have been used in the evaluation of the response. Column 3 of Table 6.4 shows how the SCR has been assessment in the researches analyzed. Also, in column 3 of Table 6.4 is observed

Table 6.4 Supply chain objectives and response assessment in the supply chain response

Article	SC objectives	Response assessment
Catalan and Kotzab [5]	Sustainable competitive advantage	Lead time, postponement, bullwhip effect, and information exchange
Kim et al. [15]	Sustainable competitive advantage, new products, and market development	Demand changes, development of new products, and relationships with partners in the chain
Reichart and Holweg [25]	Competitive advantage Cost-efficient supply chains	
Gunasekaran et al. [9]	Competitive advantages Minimum total cost	
Klibi et al. [16]	Maximization of the present value of the cash inflows Minimization of total network costs	
Squire et al. [26]	Reducing costs	Supplier abilities
Williams et al. [29]		
Hum and Parlar [11]	Meet placed order	
Dreyer et al. [6]	Competitive advantage	Base on external and internal factors proposed by [25]
L'Hermitte et al. [20]	Competitive advantage	
Rajagopal et al. [24]	Competitive advantage, lessen costs, increase quality	
Fattahi et al. [7]	Minimizing lateness of products' delivery, minimizing customers' service time, and maximizing fill rate of customers' demands in addition to economic objectives	Delivery lead time
Kristianto et al. [18]	Competitive advantage Maximize responsiveness	
Cannella et al. [4]	Customer satisfaction	Volume responsiveness
Hum et al. [12]	Satisfy customer demand	Expected lead time
Sundram et al. [27]	Competitive advantage	KPI-based measurement

that the SCR evaluation component has not been taken into account in all the research analyzed.

6.6 Conclusions and Future Research

In this work, a review of the literature on definitions supply chain response has been carried out. Twenty SCR definitions have been analyzed to define the common elements, the differences between them, and their deficiencies. Four elements have been identified that are part of most of the existing SCR definitions, and one has been proposed that contains them all. The supply chain response general definition proposal encompasses the partial definitions of this concept focused on external stimuli from the client or the market and allows to include other external stimuli (responses to various types of disasters) and internal (changes in technology or personnel). Also, the proposed definition includes the adaptation of SC activities in response to the stimulus that impacts the SC. Additionally, this definition links the response with the fulfillment of the objectives of the SC, guaranteeing the benefit for the parties involved in the response. Finally, time is included as a valid response evaluation parameter in all cases.

Regarding the elements that make up the general definition of SCR, the validation allowed identifying that both the previously proposed definitions and the research carried out cover partial aspects of SCR. Also, the applicability of the general definition of SCR was validated at various levels of SC analysis and for different research purposes according to the articles analyzed.

Future research topics are the integration of the four elements of the SCR. First, external and internal stimuli. Second, adaptation of the activities of the SC. Third, fulfillment of the objectives of the SC and those of the response. Fourth, evaluation of the answer. Also, the characterization and planning of the SCR system. To explore how the analysis of mass data contributes to improve the SCR. In addition, investigate the differences between SC and SCR management.

Acknowledgements This research is supported by the "Conceptos, Instrumentos, Modelos y Algoritmos para el Diseño de la Supply Chain (CIMADISC)" project, funded by Ministerio de Economía y Competitividad and Fondo Europeo de Desarrollo Regional (FEDER) (Grant Number: DPI2015-67740-P).

References

1. Arbeláez Gómez MC, Onrubia Goñi J (2014) 'Análisis bibliométrico y de contenido. Dos metodologías complementarias para el análisis de la revista colombiana Educación y Cultura'. Revista de Investigaciones, UCM 14(1):14. https://doi.org/10.22383/ri.v14i1.5
2. Bernardes ES, Hanna MD (2009) A theoretical review of flexibility, agility and responsiveness in the operations management literature: toward a conceptual definition of customer

responsiveness. Int J Oper Prod Manag 29(1):30–53. https://doi.org/10.1108/014435709109 25352
3. Bourlakis M, Maglaras G, Aktas E, Gallear D, Fotopoulos C (2014) 'Firm size and sustainable performance in food supply chains: insights from Greek SMEs. Int J Prod Econ Elsevier 152:112–130. https://doi.org/10.1016/j.ijpe.2013.12.029
4. Cannella S, Dominguez R, Ponte B, Framinan JM (2018) 'Capacity restrictions and supply chain performance: modelling and analysing load-dependent lead times'. Intern. J Prod Econ, Elsevier BV 204:264–277. https://doi.org/10.1016/j.ijpe.2018.08.008.
5. Catalan M, Kotzab H (2003) Assessing the responsiveness in the Danish mobile phone supply chain. Int J Phys Distrib Logist Manag 33(8):668–685. https://doi.org/10.1108/096000303105 02867
6. Dreyer HC, Swahn N, Kiil K, Strandhagen JO, Romsdal A (2015) the responsiveness of food retail supply chains: a norwegian case study. IFIP Adv Inf Commun Technol 459:152–160. https://doi.org/10.1007/978-3-319-22756-6
7. Fattahi M, Govindan K, Keyvanshokooh E (2017) Responsive and resilient supply chain network design under operational and disruption risks with delivery lead-time sensitive customers. Transp Res Part E Elsevier Ltd 101:176–200. https://doi.org/10.1016/j.tre.2017. 02.004
8. Gubi E, Arlbjørn JS, Johansen J (2003) Doctoral dissertations in logistics and supply chain management. Int J Phys Distrib Logist Manag 33(10):854–885
9. Gunasekaran A, Lai K, Hung, Cheng TCE (2008) 'Responsive supply chain: a competitive strategy in a networked economy'. Omega 36(4):549–564. https://doi.org/10.1016/j.omega. 2006.12.002
10. Halldórsson Á, Arlbjorn JS (2005) 'Research methodologies in supply chain management-What do we know?'. Research methodologies in supply chain management: in collaboration with magnus westhaus, pp 107–122. https://doi.org/10.1007/3-7908-1636-1_8.
11. Hum SH, Parlar M (2014) Measurement and optimization of supply chain responsiveness. IIE Trans (Inst Indus Eng) 46(1):1–22
12. Hum S, Parlar M, Zhou Y (2018) 'Measurement and optimization of responsiveness in supply chain networks with queueing structures R'. Euro J Oper Res Elsevier BV 264(1):106–118. https://doi.org/10.1016/j.ejor.2017.05.009
13. Jahre M, Fabbe-Costes N (2015) How standards and modularity can improve humanitarian supply chain responsiveness: the case of emergency response units. J Human Logistics Supply Chain Manage 5(3):348–386. https://doi.org/10.1108/JHLSCM-06-2015-0026
14. Jajja MSS, Asif M, Shah SAA, Chatha KA (2019) Supply chain innovation research: content analysis based review. Benchmarking 27(2):666–694. https://doi.org/10.1108/BIJ-09-2018-0297
15. Kim D, Cavusgil ST, Calantone RJ (2006) Information system innovations and supply chain management: channel relationships and firm performance. J Acad Mark Sci 34(1):40–54. https://doi.org/10.1177/0092070305281619
16. Klibi W, Martel A, Guitouni A (2010) 'The design of robust value-creating supply chain networks: a critical review'. Euro J Oper Res Elsevier BV 203(2):283–293. https://doi.org/10. 1016/j.ejor.2009.06.011
17. Krippendorff Klaus (2018) Content Analysis: An Introduction to Its Methodology. Edited by I SAGE
18. Kristianto Y, Gunasekaran A, Helo P (2017) Building the "Triple R" in global manufacturing. Int J Prod Econ Elsevier 183:607–619. https://doi.org/10.1016/j.ijpe.2015.12.011
19. Kritchanchai D, MacCarthy BL (1999) Responsiveness of the order fulfilment process. Int J Oper Prod Manag 19(8):812–833. https://doi.org/10.1108/01443579910274419
20. L'Hermitte C, Bowles M, Tatham P, Brooks B (2015) An integrated approach to agility in humanitarian logistics. J Human Logistics Supply Chain Manage 5(2):209–233. https://doi. org/10.1108/JHLSCM-04-2014-0016
21. Mayring P (2014) 'Qualitative content analysis: theoretical foundation, basic procedures and software solution'. The SAGE handbook of qualitative data analysis, 143

22. Merminod N, Nollet J, Pache G (2014) Streamlining humanitarian and peacekeeping supply chains. Soc Bus Rev 9(1):4–22. https://doi.org/10.1108/sbr-06-2013-0048
23. Mir S, Lu S-H, Cantor D, Hofer C (2018) Content analysis in SCM research: past uses and future research opportunities. Int J Logistics Manage 29(1):152–190
24. Rajagopal P, Azar NAZ, Bahrin AS, Appasamy G, Sundram VPK (2016) Determinants of supply chain responsiveness among firms in the manufacturing industry in Malaysia. Int J Supply Chain Manage 5(3):18–24
25. Reichart A, Holweg M (2007) Creating the customer-responsive supply chain: a reconciliation of concepts. Int J Oper Prod Manag 27(7):1144–1172. https://doi.org/10.1108/01443570710830575
26. Squire B, Cousins PD, Lawson B, Brown S (2009) The effect of supplier manufacturing capabilities on buyer responsiveness: the role of collaboration. Int J Oper Prod Manag 29(8):766–788
27. Sundram VPK, Rajagopal P, Azar NAZ, Bahrin AS, Appasamy G (2018) Supply chain responsiveness in an asian global electronic manufacturing firm: ABX Energy (M). Int J Supply Chain Manage 7(2):23–31
28. Tangpong C (2011) Content analytic approach to measuring constructs in operations and supply chain management. J Oper Manage Elsevier BV 29(6):627–638. https://doi.org/10.1016/j.jom.2010.08.001
29. Williams BD et al (2013) Leveraging supply chain visibility for responsiveness: the moderating role of internal integration. J Oper Manag 31(7–8):543–554. https://doi.org/10.1016/j.jom.2013.09.003
30. Wong CY, Hvolby HH (2007) Coordinated responsiveness for volatile toy supply chains. Prod Plan Control 18(5):407–419. https://doi.org/10.1080/09537280701425564

Chapter 7
A Blockchain Applications Overview

E. Ponce, J. Mula, and D. Peidro

Abstract This paper presents a general review of blockchain applications. It addresses financial applications that are the origin of this recent technology, Internet of things (IoT) applications and also logistics and supply chain areas. Additionally, other application areas are reviewed as reference models for further research. The most widely employed modelling approaches and software tools are also identified. Here HyperLedger, which is a private blockchain, seems to be the most convenient one for supply chains. Finally, a critical discussion about the reviewed articles is carried out.

Keywords Blockchain · Applications · Supply chain management · Industry 4.0

7.1 Introduction

The blockchain concept appeared in 2008 and was applied to a peer-to-peer network of electronic payments between two parties without depending on a financial institution [36]. Blockchain can be defined as a database that is distributed among different users, is cryptographically protected and comes in transactional blocks of information that are mathematically related. Its main use is framed within a system with several interacting parts, but without "trust" among them. This technology allows the parts that participate in a blockchain to trust registered information because it is consensus-based information [42]. A blockchain presents numerous applications and uses, thus, despite this being an emerging research line, it is expected to

E. Ponce · J. Mula (✉) · D. Peidro
Research Centre on Production Management and Engineering (CIGIP), Universitat Politècnica de València, Alarcón, 1, 03801 Alcoy, Alicante, Spain
e-mail: fmula@cigip.upv.es

E. Ponce
e-mail: erpon@epsa.upv.es

D. Peidro
e-mail: dapeipa@cigip.upv.es

C. Avilés-Palacios and M. Gutierrez (eds.), *Ensuring Sustainability*,
Lecture Notes in Management and Industrial Engineering,
https://doi.org/10.1007/978-3-030-95967-8_7

be significant in the future and in several domains [16]. Zheng et al. [58] identify blockchain as an enabling technology for implementing Industry 4.0. In supply chain 4.0 management, [25] propose blockchain as a digital technology to integrate data into the information used by artificial intelligence algorithms in the cyber supply chain and managers in the physical supply chain. Supply chain management will be transformed by blockchain in terms of facilitating, validating and measuring key objectives, such as cost, quality, speed, dependability, risk reduction, sustainability and flexibility [29].

This article proposes a preliminary summary of a classification of articles from the scientific literature that have applied the blockchain technology to different sectors. The present article addresses the finance sector, where this technology originated from, Internet of things (IoT) applications, logistics and supply chains, but other applications from or other sectors are considered reference models for future research lines. Readers are referred to [54] and Zibin et al. [59] for other literature reviews on blockchain, although both are based on technical aspects. The present review is based mainly on blockchain applications.

The rest of the article is as follows. Section 7.2 presents the review methodology. Section 7.3 offers the classification taxonomy. Section 7.4 critically deals with the reviewed articles. Section 7.5 identifies the conclusions and future research lines.

7.2 Review Methodology

The review methodology was based on exploring articles by mainly employing the Scopus technical and scientific database. Scientific articles published during the 2008–2020 period were considered because the first blockchain came about in 2008, and the use of this technology has rapidly risen in the last 4 years. The search for and observation of articles were carried out by combining several key words according to the environment in which blockchain technology has been used and its name. The finally 45 selected articles were those based on contents of applications that can act as references for new digital planning proposals in supply chain and logistics.

7.3 Literature Review

Blockchains provide a distributed, immutable and ordered database that is based on a growing sequence of blocks. This technology can be open so that anyone can participate and may wish to belong to a blockchain. This technology, known as an open or public blockchain, allows participants to read, register transactions in a ledger and even can participate in the validation of the processes of consensus to be able to add blocks to the chain, and blockchains of this type are considered decentralised [9]. They are applied to the cryptoeconomy, the combination of economic incentives and verification mechanisms that use cryptography. Information can only be added to the blockchain if an agreement has been reached by most parties. It is worth stressing that after a certain time, any information added to the block cannot be modified [16]. Participation in a blockchain may also be limited; that is, it is a private blockchain in which not all the registered data are publicly diffused, but may only be consulted at any time by a user belonging to the network [42]. In a private blockchain the writing permissions are maintained centralised, and the autorisation of consensus is validated by a number limited of participants. However, the reading permissions can be arbitrarily public or restricted [9]. A private blockchain is applied in the transfers of monetary or non-monetary actives and as a register in intelligent agreements by executing automatically the terms and conditions of them without the participation of a person, and this type of blockchain does not necessarily use mechanisms based on the cryptography [21]. Moreover, a private blockchain is not immutable because authorised network parties can amend information according to their considerations. We ought to bear in mind that some blockchains use a system with a white list on which only listed nodes can participate [13], and a blockchain would be a consortium blockchain when these nodes are from different organisations [57].

Blockchain has an extremely high potential, a substantial development capacity based on computer security by means of cryptography and a legal capacity based on contractual agreements [42]. In this way, this system allows "trustless" networks because parties can make transactions in blockchains, even though users do not trust one another [45]. Some of this system's main characteristics are no trustworthy go-between; certification of consensus-based information; transparent information.

Garay et al. [18] consider a blockchain to be robust if it guarantees two properties: availability and consistency. Availability allows a reliable or trustworthy transaction to be added to a blockchain and avoids it being rejected. Consistency guarantees that, if a node provides a stable transaction, the other nodes that are considered honest validate it by making it immutable. Inserting blocks into a blockchain by the peer-to-peer network avoids a single participant or group controlling the system because all the network members agree to work with the same protocols [16]. Different states are distinguished for the block's information that is being processed: candidate information to be added, which is the information that nodes have sent to other nodes, but has not yet been validated in any block, confirmed information, which is the information validated by the network and is then added to the next block; stable information, which immutably forms part of the blockchain. In summary, key blockchain characteristics are decentralisation, persistency, anonymity and auditability [57].

7.3.1 Financial Applications

The first and main blockchain application took place in the finance system by the creation of the so-called cryptocurrency. The first cryptocurrency was Bitcoin, which appeared in 2008 as a virtual intangible currency to be used with Internet purchases [36]. In this context, the digital currency employed to make exchanges is known as cryptocurrencies [37]. Finally, it is worth stressing the storage of smart contracts that can be viewed by any blockchain user, including coded clauses and information that are impossible to amend [10]. This protocol is interpreted and run automatically in the terms set out and describes all the likely outcomes of contracts if they are well written [13]. Some banks and financial entities use this blockchain technology because it allows to speed up companies and reduce risks and costs in transactions among the participants [46]. One of the profits of the blockchains in the financial field is reflected in the transnational processes of payments, and other profits are regarding the digital identity, emission of primary values, compensations and reports, among others [12].

7.3.2 IoT Applications

The principal use of blockchain technology in an IoT environment is to keep devices up to date because many manufacturers find that acquiring the last version of their programmes is expensive [13]. Cryptocurrency can be added to this application. When an update is employed, payment for the rendered service is sent [16].

Another use related to the IoT is security owing to users' mistrust in communicating these devices with their manufacturer if no supervision takes place. In line with this, [27] address how blockchain can be used to solve some IoT security problems.

Combining blockchain, the IoT and machine-to-machine (M2M) comprises part of the digitalisation of factories or Industry 4.0. Blockchain plays the role of a decentralised organisation to: receive information from devices installed in factories; register industrial network activity; guarantee the integrity of this information; ensure that this information cannot be amended.

In the domestic domain, [17] use a local private blockchainto provide secure access control to IoT devices and smart home data.

Another area it can be applied in is the market with different kinds of insurance companies, e.g. car insurance which, by employing blockchain and the IoT, will acquire information by means of the vehicle's sensors: on the one hand, studying driving habits and, by contemplating other factors, the insurance company will determine the price of the insurance policy; on the other hand, the driver can use these data to send them to several insurance companies to receive smart contracts and offers [42]. Finally, it is worth highlighting IoT devices [3, 30, 44, 52, 55, 61].

7.3.3 Logistics and Supply Chain Applications

In a supply chain, blockchain can determine who is performing an action, e.g. shipments, deliveries, quality of transportation and progress, among others, and the time and location of such action [14]. Based on multiple case studies, [29] illustrates the mechanisms through which blockchain will help supply chains to accomplish their key objectives. The blockchain technology has drawn the attention of the agriculture supply chain [8, 34, 36, 41]. In order to once again win consumers' trust after food security warnings, measures are being taken that are related to the agrofood supply chain's transparency [1]. Arena et al. [4] propose BRUSCHETTA, a blockchain-based system to enforce certifications of extra virgin olive oil (EVOO) by tracing its entire supply chain, e.g. from plantations to shops. Bechtsis et al. [6] establish a blockchain framework and infrastructure for containerised food supply chains that uses smart interconnected devices, shares information, enhances process control and traceability and prevents potential risks. Guo et al. [22] propose a SafeFood solution using blockchain and the IoT to track a beef supply chain from farms to storage shelves.

Tian [48] presents an RFID-based agrofood supply chain by means of which the information found on labels and sent to a blockchain continues to inform the supply chain about the product's origin, the kind of chemicals employed to produce it, among other details. With meat, the food type that the animal source ate is registered, along with the personnel who handled the animal in the slaughterhouse, etc. By means of wireless sensors, monitoring systems and RFID reader devices, it is possible to determine inventories, delivery times, storage temperatures, a cargo's value, etc., in transport vehicles minute by minute. Similarly, [11] propose a model by considering blockchain, smart contract and multi-agent systems to coordinate the tracking of food in an agricultural supply chain with circular economy criteria. Yusuf et al. [56] use blockchain case studies in the field of perishable vegetable supply chains. They report a reduction in the duration of ledger problems and enhanced privacy and transparency using multichannel functions.

Blockchain and smart contracts offer a real opportunity for small farms and cooperatives to participate more in the market [49]. Nonetheless, some innovations in blockchain architectures and business models related to the agrifood sector are still not very well known [31, 49, 52]. In the logistics field, sending containers as part of international trade can find an ally in blockchain because those participating in this network can share up-to-date information about the state that a certain merchandise is in because all updates are signed using the private keywords of those delivering and picking them up [16].

Using a blockchain in a supply chain has been considered a solution to the shortages shown by its conventional logistics system, e.g. in iron and steel firms. Here the information generated by a given operation is dispersed and stored in several of the system's links. This means that this information is not transparent and is no longer traceable, which leads to high transaction costs. To make further improvements in this field, devices and tools are used along with blockchain, such as RFID, bar codes, voice detection and video. All this makes the transport process more efficient and more transparent [53]. Kouhizadeh and Sarkis [28] deal with the potential applications of the blockchain technology to facilitate ecological practices in sustainable supply chains.

7.3.4 Other Application Areas

Blockchains are also used in the health area to keep all patient information centralised, in which doctors, pharmacists, insurance companies and laboratories participate, among others. This allows a patient's complete medical record to be located at a single point [5, 35]. Applying blockchains in medicine also reaches drug production and avoids counterfeit medicines. In this field, a research network launched a counterfeit medicines project in an attempt to establish when and where medicine was prepared and to register the materials employed to produce it [43, 47]. The medicine distribution field is highly regulated given the EU regulation that obliges any change in the temperature at which the merchandise is being transported to be informed to both the distributor of medicines and the addressee. In this context, a project called MONDUM.IO AG employs the blockchain technology associated with sensors (IoT) to guarantee data integrity and security by means of smart contracts [7].

The blockchain technology can be applied in many fields, like companies' operations and business processes for producing and distributing energy, by favouring issues like turnover, using smart devices in the network, benefitting sales by identifying customer consumer patterns, improving network management as regards the transparency of records, viewing used energy levels and paying for a service with cryptocurrency [2]. A case study was conducted by the Brooklyn microgrid, which commercialised energy based on blockchains in a micronetwork where consumers and prosumers sell surplus energy to their neighbours using blockchain by means of smart contracts [33].

The blockchain technology also has considerable potential for commercial applications [39]. For instance, Hassija et al. [23] propose a decentralised blockchain platform for car rental applications with minimum transaction charges and smart contracts. Pallam and Gore [38] put forward Boomerang, a decentralised freelance administration paradigm for users who seek employment opportunities. Zyskind et al. [61] combine blockchain and off-blockchain storage to construct a personal data management platform that focuses on privacy issues, such as data ownership, data transparency and auditability and fine-grained access control. Finally, Wai et al. [51] present a storage approach for student record data to develop a blockchain-based decentralised higher education system.

7.3.5 Modelling Approach and Software Tool

Several blockchain modelling algorithms exist, of which the most frequent is proof of work (PoW) consensus [13], with sha256 of Bitcoin and Ethash of Ethereum, where a node must solve a cryptographic puzzle that confers the right to validate the new block. Other consensus algorithms include practical Byzantine fault tolerance (PBFT) of HyperLedger [50], proof of stake (POS) information [42], delegated proof of stake (DPOS), Ripple and Tendermint [57]. It is also worth highlighting cryptographic issues like elliptic curve cryptography (ECDSA, ECDH, ECIES) [26, 42], hash, zeros that change every given time into Bitcoins to make mining times difficult, smart contracts amend [9], the Solidity language and Chaincode, among others [3].

Some leading blockchain software include: Etherum, which is a public blockchain [5, 7, 20, 33] and Hyperledger, a private blockchain [22, 35, 47, 56] HyperLedger Fabric is a distributed operating system for permissioned blockchain that is well supported and is being used in many different projects. It is employed by Walmart in its pork supply chain [15, 24]. We also find BigChainDB [32], BRUSCHETTA [4], Boomerang [38] and SeaFood [22]. Other blockchain frameworks propose using both Etherum and HyperLedger Fabric [6, 56, 4].

HyperLedger Fabric is a modular distributed system that works with authorised blockchains and separates executing the transaction from consensus, which allows for policy-based support towards strength, flexibility, scalability and confidentiality; it is an Ethereum-based private blockchain. This is the first blockchain to support the execution of distributed applications written in standard programming so they can be executed consistently across many nodes. The HyperLedger architecture follows the execution order validation sequence for the distributed execution of unreliable codes in a trusted working environment. It divides the sequence of the transaction flow into three steps that can be executed in different system entities: (1) execute a transaction and check its correctness; (2) organise with a consensus protocol regardless of the transaction; (3) validate or approve transactions by application-specific assumptions of trust. It combines the two approaches to replication (active and passive) in the Byzantine model. This makes Fabric a scalable system and confers its flexibility to support trust assumptions [3].

7.4 Discussion

Blockchain applications are spreading in many areas, including financial services, the IoT, logistics and supply chains, commercial applications, insurance, health, agriculture/food and power sectors. Nonetheless, very few blockchain applications related to real cases have gone beyond the pilot test phase concept or small-scale pilot testing [19]. Blockchain involves one important aspect to improve, which is scalability, defined by the number of transactions per second at the maximum throughput of the chain. Bitcoin and Ethereum have a speed of 1–5 transactions/second. A private blockchain has more transactions/second than a public one [40].

Other aspects to be addressed are security issues, such as privacy leakage and selfish mining [57]. Additionally, the cost of developing/operating or leasing the required platforms, which is related to the volume of exchanged data, as well as additional costs (sensors, IoTs devices) for collecting the necessary information, will be the key parameters that guarantee this blockchain system's economic feasibility for a supply chain [6].

Based on potential applications to other industrial sectors, blockchain can be introduced into the plastics industry by promoting and facilitating the work associated mainly with the injection companies that work with partners to complete a production process, such as assembling parts, finishing or painting components. Suppliers of the materials needed to inject parts might also belong to this chain. As all these domains are linked, the traceability of the components that an injection company delivers to its assembling partner remains, and it can also control the lots and lead times of the finished product. Plastic injection companies can view the promised lead dates of their suppliers of materials, which allows them to control the lead times of the injected products to their end customers by planning more accurately. Blockchains will offer many benefits because of the globalised and accelerated environment in the industrial field with more personalised production and just in time.

Another industry it can be applied to is the automobile industry by linking blockchains to all suppliers of parts to assemble a car. In turn, these can be linked with their suppliers. This will provide a car assembly plant with a vision of the whole parts supply chain and a register of information about the production process in order to control the state and application of each part. The blockchain application horizon can be extended to car showrooms and car repair garages. Such information in a ledger can be very useful for all the chain nodes, and the automobile user can be considered yet another of its links. Blockchain software has a before and an after with the appearance of HyperLedger, which is a private or "permissioned" blockchain that makes it different for Ethereum and Bitcoin, which are public. This is most convenient for B2B and, therefore, for supply chains. HyperLedger is supported by the Linux Foundation and companies like IBM (HyperLedger Fabric) and Intel (Sawtooth).

7.5 Conclusions

This paper presents a preliminary literature review on blockchain applications and aims to be a starting point for researchers and practitioners to develop and implement blockchain applications. Implementing blockchain into a supply chain could enhance the efficiency of the following: product accuracy and follow-up; making processes more visible; knowledge for planning and forecasting; improving personnel scheduling and transport and/or production capacity; controlling storage conditions. Other technical advantages are track and trace; fight counterfeit and visibility with extended security [43]. Conversely, blockchain is an emerging technology in its initial stage. Some of its limitations are immature digital currency, no regulations available and speed of transactions.

Further research will aim to make prescriptive recommendations for blockchain developments and implementations in supply chain 4.0 contexts.

Acknowledgements This work was supported by the Spanish Ministry of Science, Innovation and Universities Project CADS4.0 (RTI2018-101344-B-I00) and the Valencian Community ERDF Programme 2014-2020 IDIFEDER/2018/025.

References

1. Akkerman R, Farahani P, Grunow M (2010) Quality, safety and sustainability in food distribution: a review of quantitative operations management approaches and challenges. OR Spectr 32:863–904
2. Andoni M, Robu V, Flynn D et al (2019) Blockchain technology in the energy sector: a systematic review of challenges and opportunities. Renew Sustain Energy Rev 100:143–174
3. Androulaki E, Barger A, Bortnikov V et al (2018) Hyperledger fabric: a distributed operating system for permissioned blockchains. Proc 13th Euro Sys Conf Euro Sys 2018 2018-Janua
4. Arena A, Bianchini A, Perazzo P, et al. (2019) BRUSCHETTA: An IoT blockchain-based framework for certifying extra virgin olive oil supply chain. Proc - 2019 IEEE Int Conf Smart Comput SMARTCOMP 2019:173–179
5. Azaria A, Ekblaw A, Vieira T et al (2016) MedRec: using blockchain for medical data access and permission management. Proc—2016 2nd Int Conf Open Big Data OBD 2016:25–30
6. Bechtsis D, Tsolakis N, Bizakis A et al (2019) A blockchain framework for containerized food supply chains. Comput Aided Chem Eng 46:1369–1374
7. Bocek T, Rodrigues BB, Strasser T et al (2017) Blockchains everywhere—a use-case of blockchains in the pharma supply-chain. Proc IM 2017—2017 IFIP/IEEE Int Symp Integr Netw Serv Manag, 772–777
8. Borrero JD (2019) Agri-food supply chain traceability for fruit and vegetable cooperatives using Blockchain technology. CIRIEC-Espana Rev Econ Publica, Soc y Coop:71–94
9. Buterin V (2015) On public and private blockchains. Ethereum foundation blog. Available via DIALOG. https://blog.ethereum.org/2015/08/07/on-public-and-private-blockchains/. Accessed 22 Feb 2020
10. Buterin V (2016) Chain interoperability. R3 Res Pap
11. Casado-Vara R, Prieto J, de La Prieta F et al (2018) How blockchain improves the supply chain: case study alimentary supply chain. Procedia Comput Sci 134:393–398
12. Casey M, Crane J, Gensler G et al (2018) The impact of blockchain technology on finance: a catalyst for change. Geneva Reports World Econ 2018:1–82
13. Christidis K, Devetsikiotis M (2016) Blockchains and smart contracts for the internet of things. IEEE Access 4:2292–2303
14. Da XL, Xu EL, Li L (2018) Industry 4.0: state of the art and future trends. Int J Prod Res 56:2941–2962
15. del Castillo M (2016) Walmart blockchain pilot aims to make China's pork market safer, CoinDesk
16. Dolader C, Bel J, Muñoz JL (2017) La blockchain: fundamentos, aplicaciones y relación con otras tecnologías disruptivas. Econ Ind 405:33–40
17. Dorri A, Kanhere SS, Jurdak R et al (2017) (2017) Blockchain for IoT security and privacy: the case study of a smart home. IEEE Int Conf Pervasive Comput Commun Work PerCom Work 2017:618–623
18. Garay J, Kiayias A, Leonardos N (2015) The Bitcoin backbone protocol: analysis and applica tions. Lect Notes Comput Sci (including Subser Lect Notes Artif Intell Lect Notes Bioinfor) 9057:281–310
19. Ge L, Brewster C, Spek J et al (2017) Blockchain for agriculture and food: findings from the pilot study. Wageningen EconRes
20. Gerring T (2016) Cut and try: building a dream-Ethereum Blog. Ethereum Blog
21. Guegan D (2017) Public Blockchain versus Private blockhain
22. Guo M, Liu XJ, Zhang W (2018) Using blockchain technology in human food chain provenance. WIT Trans Built Environ 179:391–396
23. Hassija V, Zaid M, Singh G et al (2019) Cryptober: A blockchain-based secure and cost-optimal car rental platform. 2019 12th international conference on contemporary computing (IC3) 2019
24. Higgins S (2017) Walmart: blockchain food tracking test results are 'very encouraging.' Available via DIALOG. http://www.coindesk.com/walmart-blockchain-foodtrackingtest-results-encouraging/. Accessed 25 Feb 2020

25. Ivanov D, Dolgui A, Sokolov B (2019) The impact of digital technology and Industry 4.0 on the ripple effect and supply chain risk analytics. Int J Prod Res 57:829–846
26. Johnson D, Menezes A, Vanstone S (2001) The elliptic curve digital signature algorithm (EC-DSA). Int J Inf Secur 1:36–63
27. Khan MA, Salah K (2018) IoT security: review, blockchain solutions, and open challenges. Futur Gener Comput Syst 82:395–411
28. Kouhizadeh M, Sarkis J (2018) Blockchain practices, potentials, and perspectives in greening sup-ply chains. Sustain 10
29. Kshetri N (2018) 1 Blockchain's roles in meeting key supply chain management objectives. Int J Inf Manage 39:80–89
30. Lin C, He D, Huang X et al (2018) BSeIn: a blockchain-based secure mutual authentication with fine-grained access control system for industry 4.0. J Netw Comput Appl 116:42–45
31. Lin Y-P, Petway JR, Anthony J et al (2017) Blockchain: the evolutionary next step for ICT e-agriculture. Environments 4:50
32. McConaghy T (2017) BigchainDB 2017 roadmap. Available via DIALOG. https://blog.bigchaindb.com/bigchaindb-2017-roadmap-d2e7123f9874. Accessed 22 Feb 2020
33. Mengelkamp E, Gärttner J, Rock K et al (2018) Designing microgrid energy markets: a case study: the Brooklyn Microgrid. Appl Energy 210:870–880
34. Morris D (2016) Leaderless, blockchain-based venture capital fund raises $100 million, and counting. Available via DIALOG. https://fortune.com/2016/05/15/leaderless-blockchain-vc-fund/. Accessed 25 Feb 2020
35. Murugan A, Chechare T, Muruganantham B et al (2020) Healthcare information exchange using blockchain technology. Int J Electr Comput Eng 10:421–442
36. Nakamoto S (2008) Bitcoin: un sistema de dinero en efectivo electrónico peer-to-peer. Available via DIALOG. https://bitcoin.org/files/bitcoin-paper/bitcoin_es.pdf. Accessed 15 Jan 2020
37. Pacheco M (2016) Criptodivisas: del bitcoin al MUFG. El potencial de la tecnología blockchain. Rev CESCO Derecho Consum, pp 6–15
38. Pallam B, Gore MM (2019) Boomerang: Blockchain-based freelance paradigm on hyperledger. 2019 10th Int Conf Comput Commun Netw Technol ICCCNT 2019
39. Panetta K (2017) Top trends in the gartner hype cycle for emerging technologies, 2017. Gartner, Stamford, US
40. Perboli G, Musso S, Rosano M (2018) Blockchain in logistics and supply chain: a lean approach for designing real-world use cases. IEEE Access 6:62018–62028
41. Popper N (2016) A venture fund with plenty of virtual capital, but no capitalist. Available via DIALOG. https://www.nytimes.com/2016/05/22/business/dealbook/crypto-ether-bitcoin-currency.html. Accessed 25 Feb 2020
42. Preukschat A (2017) Blockchain: la revolución industrial de internet. Gestión 2000
43. Raj R, Rai N, Agarwal S (2019) Anticounterfeiting in pharmaceutical supply chain by establishing proof of ownership. IEEE region 10 annual international conference proceedings/TENCON 2019-October, pp 1572–1577
44. Shen M, Deng Y, Zhu L et al (2019) Privacy-preserving image retrieval for medical iot systems: a blockchain-based approach. IEEE Netw 33:27–33
45. Tapscot D, Tapscott A (2016) Blockchain revolution: how the technology behind bitcoin is changing money, business, and the world. Penguin
46. Tapscott A, Tapscott D (2017) How blockchain is changing finance. Harv Bus Rev 1:2–5
47. Taylor P (2016) Applying blockchain technology to medicine traceability. Available via DIALOG. https://www.securingindustry.com/pharmaceuticals/applying-blockchain-technology-to-medicine-traceability/s40/a2766/. Accessed 15 Jan 2020
48. Tian F (2016) An agri-food supply chain traceability system for China based on RFID and blockchain technology. 2016 13th international conference on service systems and service management ICSSSM 2016
49. Tripoli M, Schmidhuber J (2018) Emerging opportunities for the application of blockchain in the agri-food industry agriculture. Food Agric Organ United Nations

50. Vukolić M (2016) Hyperledger fabric: towards scalable blockchain for business. Trust Digit Life
51. Wai KSS, Htoon EC, Thein NNM (2019) Storage structure of student record based on hyperledger fabric blockchain. 2019 international conference on advanced information technologies ICAIT 2019, pp 1–108
52. Yang Y, Jia Z (2017) Application and challenge of blockchain technology in the field of agricultural internet of things. Inf Technol 258:24–26
53. Yang A, Li Y, Liu C et al (2019) Research on logistics supply chain of iron and steel enterprises based on block chain technology. Futur Gener Comput Syst 101:635–645
54. Yli-Huumo J, Ko D, Choi S et al (2016) Where is current research on blockchain technology?—a systematic review. PLoS ONE 11:e0163477
55. Yong B, Shen J, Liu X, et al. (2019) An intelligent blockchain-based system for safe vaccine supply and supervision. Int J Inf Manage
56. Yusuf H, Surjandari I, Rus AMM (2019) Multiple channel with crash fault tolerant consensus blockchain network: a case study of vegetables supplier supply chain. 2019 16th international conference on service systems and service management, 1–4
57. Zheng Z, Xie S, Dai H et al (2017) An overview of blockchain technology: architecture, consensus, and future trends. Proceedings—2017 IEEE 6th international congress big data bigdata congress 2017, pp 557–564
58. Zheng Z, Xie S, Dai HN et al (2018) Blockchain challenges and opportunities: a survey. Int J Web Grid Serv 14:352–375
59. Zibin Z, Shaoan X, Hongning D, Xiangping C, Huaimin W (2017) An overview of blockchain technology: architecture, consensus, and future trends. In: IEEE 6th international congress on big data, pp. 557–564
60. Zinas N, Kontogiannis S, Kokkonis G et al. (2017) Proposed open source architecture for Long Range monitoring. The case study of cattle tracking at Pogoniani. Proceedings 21st pan-hellenic conference informatics, 1–6
61. Zyskind G, Nathan O, Pentland AS (2015) Decentralizing privacy: using blockchain to protect personal data. Proceedings—2015 IEEE securty privacy work SPW 2015, 180–184

Chapter 8
Blockchain Application for a Sustainable Supply Chain Management

S. Fernández-Vázquez, R. Rosillo, P. Priore, A. Gomez, and J. Parreño

Abstract Today, complex and challenging industrial requirements face challenges related to distribution and production necessities. This changing environment also creates opportunities, which with the introduction of a disruptive technology such as blockchain will enhance the emergence of the connected supply chain as an essential element in the industrial world. By using blockchain, firms will have faster access to data, such as sales patterns on newly commercialized products or interruptions in the upstream supply, with the guarantee that the data is verifiable, transparent and perfectly traceable. This paper offers an analysis of blockchain technology and smart contracts exploring the capabilities and advantages of this technology in making traditional supply chain management a new and sustainable digital process.

Keywords Blockchain · Smart contracts · Supply chain · Sustainability

8.1 Introduction

Blockchain technology has grown massively in recent years. It has revolutionized the way financial systems operate and is currently penetrating into different industries.

S. Fernández-Vázquez (✉) · R. Rosillo · P. Priore · A. Gomez · J. Parreño
Business Management Department Edificio Departamental Este, 1 Planta. Campus de Viesques s/n. 33204, Gijón, Asturias, Spain
e-mail: simonfe@uniovi.es

R. Rosillo
e-mail: rosillo@uniovi.es

P. Priore
e-mail: priore@uniovi.es

A. Gomez
e-mail: albertogomez@uniovi.es

J. Parreño
e-mail: parreno@uniovi.es

C. Avilés-Palacios and M. Gutierrez (eds.), *Ensuring Sustainability*,
Lecture Notes in Management and Industrial Engineering,
https://doi.org/10.1007/978-3-030-95967-8_8

Supply chain management is no exception, and blockchain can help the integration of sustainable distribution, communication and management functions [14].

The chain of blocks can bring many advantages to supply chain issues, such as an improvement of real-time communications, guaranteeing transactions take place in a trustworthy environment, improvement in payment processing speed, smaller transaction fees, minor transaction costs, a reduction on bottlenecks and better forecasting [6].

However, there are also some issues that need to be addressed to improve the widespread of this technology. In the first place, those regarding the technological aspects of blockchain, such as performance, scalability, security, capacity and robustness. Since blockchain technology is rather new, its adoption becomes difficult to predict. Transparency in processes, products and information has still not been used as an evaluation technique; therefore, there are difficulties in its measurement. Some companies are reluctant to the application of blockchain regarding its transparency and the traceability of information [3].

8.2 Blockchain

Blockchain can be defined as a distributed, decentralized technology, which allows trust between third parties solving traditional system architectures. This technology was developed in 2008 by Nakamoto [19], as an electronic peer-to-peer payment system that functioned without the intermediation of third parties, such as financial institutions. The double-spending problem was solved through this mechanism.

Blockchain works as a peer-to-peer network that is linked by the nodes that form the system. The idea behind this technology is that copies of files are stored in different locations around the world, known as nodes, as shown in Fig. 8.1. Any user with a server which provides the ability to update the blockchain files can participate in the system. The investment in hardware is significant, as well as in electricity; therefore, the system rewards those nodes involved. These users, known as miners, ensure the validity of the transactions. Miners usually gather in large communities, or mining pools, to reduce costs and share the rewards [21].

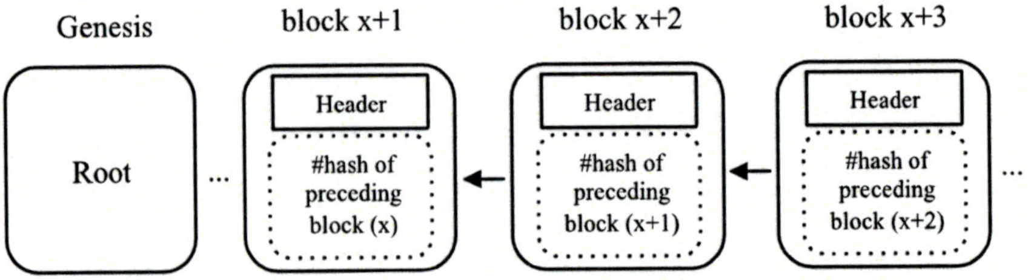

Fig. 8.1 Blockchain network

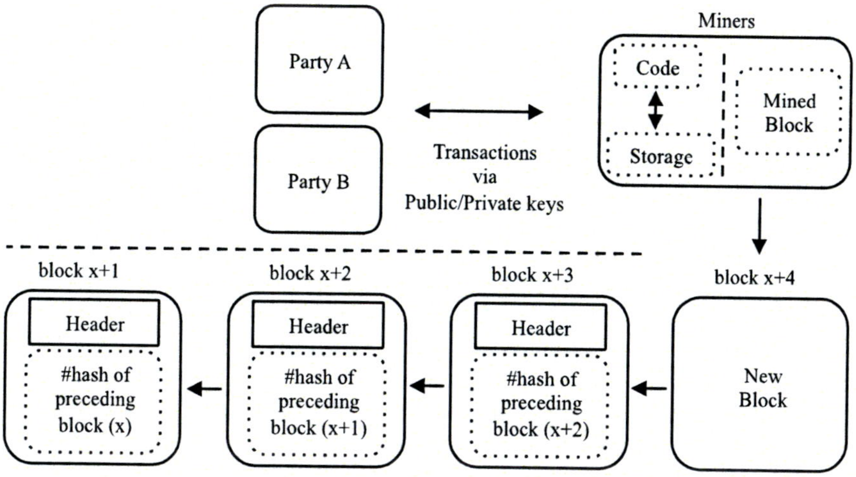

Fig. 8.2 Smart contract system

8.3 Smart Contracts

Smart contracts are computing protocols, envisioned to enable, verify and enforce the implementation of an agreement [9]. They allow users to achieve consensus in the implementation of transactions without the intervention of third parties [7]. They can be externally invoked through the use of function modifiers that are attached to different functions, for example, in data validation [10].

There are multiple blockchain platforms that allow the development of smart contracts, as shown in Fig. 8.2. One of the most widely used is Ethereum as it supports turing features that outstand from other languages as it can be easily customized for the creation of contracts. These smart contracts can be used in different industries, such as e-commerce or housing properties [1].

8.4 Blockchain-Based Supply Chain

Supply chain can be defined as a collection of agents that are directly involved in both upstream and downstream flow of goods, services or information, from the source to the final client; information flow, therefore, is key, mainly at process levels [18]. Communication is increased by the use of strategic information exchange among suppliers (commercial, fabrication or design). This enables the automatization of processes and digital processing between customers and suppliers [20].

There are multiple decisions that need to be taken at different levels, as the management of the supply chain resources is a complicated endeavor. One of the ways in which sustainable competitive advantage can be obtained is through a tight control over the main supply chain resources. These can be tangible, such as physical resources, or intangible, such as knowledge management [11].

The competitive advantage of a company is a mixture of the efficiency in the management of the supply chain resources. Extraordinary organization is required between the activities within the company and the capacity to align information [12].

Nevertheless, in order to obtain a competitive advantage, the degree of acquisition and development of the resources is extremely important. These resources can be classified in six different types: physical, human, financial, organizational, intangible and technological [4].

One of the most important innovation that blockchain will bring is data-driven supply chains. In order to obtain sustainability, these data chains bring many advantages such as the availability of data that offers multiple chances for information sharing as well as sustainable decision taking. For example, blockchain applied to the Internet of things (IoT) in the agricultural sector will offer social benefits to some underdeveloped communities. The traceability of products will give the final consumer information about pesticides and other dangerous products that could have been used in the process. Due to the particularities of blockchain, the traceability of agricultural products will improve, as well as its transparency and authenticity, as seen in Fig. 8.3 [12].

The application of this technology to supply chain has the possibility to make the chain responsive to actions and trends, as well as resistant against market disruptions

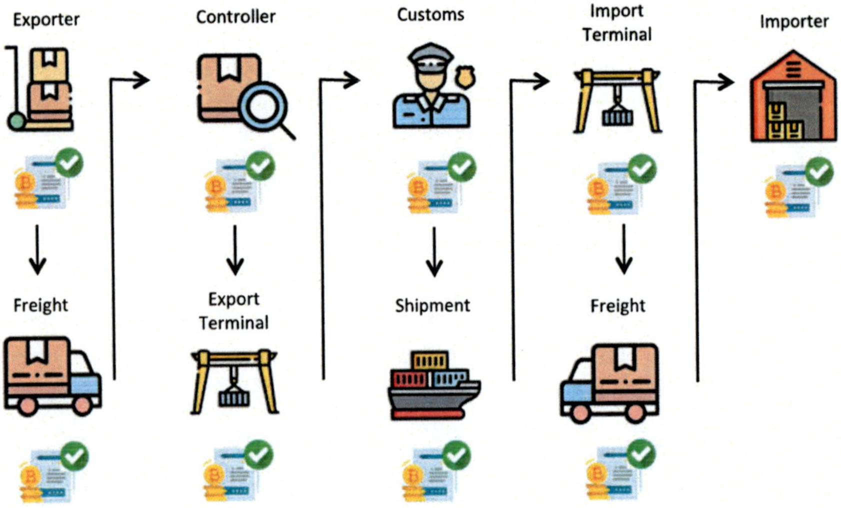

Fig. 8.3 Supply chain logistics using blockchain technology to increase traceability on international shipments

[5]. Partners are connected through the network; therefore, firms can have quick access to data, such as sales patterns on newly commercialized products or interruptions in the upstream supply. The system can be implemented in a way that even in an event that a blockchain node breaks down, it will still be able to carry on with its tasks [2].

Firms willing to know about market movements or trends related to demand in the short or long term increase their market competitiveness through the use or renewable energy, through innovation or new product development (NPD), implement a novel business model or capture the results of technological development, can do so through blockchain-based traceability applications [22]. Companies using this information earlier than their competitors are set to have a competitive advantage [8].

8.5 Advantages in the Use of Blockchain Applications

Blockchain offers multiple advantages in supply chain. In the first place, it helps creating and sharing information in an immutable and transparent way, with the intervention of an expensive centralized information system. This reduces the need for third-party intervention, decreasing the amount of data to be shared among parties. As the information is stored in each and every node, it is easily accessible, and keeping track of previous transactions is as easy as following the chain of previous blocks.

Modifications made to the information stored in the blockchain need to be validated by a consensus reached among all members of the network. When the information is validated, it is stored in every node that comprises the chain, preventing information disruption or security attacks. Due to the heavy use of cryptography, modification of the information stored in the nodes is extremely challenging, which avoids issues such as forgery or fraud. Finally, one of the most important benefits that blockchain can bring to supply chain is the traceability related to the certification of the origin of the materials and end-to-end product traceability [13].

Blockchain can currently be implemented in a digital business ecosystem, establishing what is known as the digital supply chain (DSC). One of the main advantages of this system is the reduction in transaction costs and the increase of activities that create value to firms, providers, employees and clients [17]. This technology can reduce costs, increase the information quality, improve transactional speed and reduce dependability and risk [16].

Regarding sustainability, data related to vendors historical performance could be accessible using blockchain. This helps the firms in different ways; firstly, it creates a system of accurate and secure data. Secondly, it can help companies decide their vendor selection based on certain performance values related to their sustainable environmental implications. For instance, blockchain provides information of the whole supply chain, helping firms select their vendors at different levels of the chain. This can serve as a basis for reducing intermediaries and as an important source for vendor selection based on the company's interests [15].

In relation to other systems such as IoT, sensors that offer real-time information can be encrypted, validated by the nodes and shared through the supply chain with other agents. This could guarantee, for example, that the temperature or the levels or humidity is adequate, or that the quality of the materials is as expected, verifiable through their traceability [5].

8.6 Conclusions

Sustainability is one of the critical factors that many companies face. Social or environmental sustainability is the major concern, and blockchain can help facilitate the transition. Traceability is just one of the many advantages of this new system, although managers have to think if this is the best solution in the short and long run. The supply chain of these firms would need to be redesigned in order to cope with the new demands.

Some companies will prefer to extend their current capabilities instead of betting on the development of blockchain applications that will lead them to further solutions. Other companies are reluctant to the use of this technology, as today there is still limited experience in the benefits of using blockchain related to this area. This limitation also reaches the way in which companies can evaluate in an effective and efficient way the selection of blockchain with respect to its sustainability. In any case, the characteristics of blockchain can only benefit the company and the final consumers in the long term, and it is a matter of time that this technology is implemented in our day-to-day life.

References

1. Alharby M, van Moorsel A (2017) Blockchain based smart contracts: a systematic mapping study. https://doi.org/10.5121/csit.2017.71011
2. Bahga A, Madisetti VK (2016) Blockchain platform for industrial Internet of Things. J Softw Eng Appl 09:533–546. https://doi.org/10.4236/jsea.2016.910036
3. Bai C, Sarkis J (2020) A supply chain transparency and sustainability technology appraisal model for blockchain technology. Int J Prod Res 1–21. https://doi.org/10.1080/00207543.2019.1708989
4. Braganza A, Brooks L, Nepelski D, Ali M, Moro R (2017) Resource management in big data initiatives: processes and dynamic capabilities. J Bus Res 70:328–337. https://doi.org/10.1016/j.jbusres.2016.08.006
5. Calatayud A (2017) The connected supply chain: enhancing risk management in a changing world. Inter-American Development Bank. IDB-DP-508
6. Casado-Vara R, Prieto J, la Prieta FD, Corchado JM (2018) How blockchain improves the supply chain: case study alimentary supply chain. Proc Comput Sci 134:393–398. https://doi.org/10.1016/j.procs.2018.07.193
7. Dai W, Daia C, Choo KKR, Cui C, Zou D, Jin H (2019) SDTE: a secure blockchain-based data trading ecosystem. IEEE Trans Inf Forensics Secur 1–1. https://doi.org/10.1109/tifs.2019.2928256

8. Edwards N (2017) Blockchain meets the supply chain. MHD Supply Chain Solutions 47(4):48–50
9. Fernandez-Vazquez S, Rosillo R, De La Fuente D, Priore P (2019) Blockchain in FinTech: a mapping study. Sustainability 11(22). https://doi.org/10.3390/su11226366
10. Frantz C, Nowostawski M (2016) From institutions to code: towards automated generation of smart contracts. https://doi.org/10.1109/FAS-W.2016.53
11. Gupta M, George JF (2016) Toward the development of a big data analytics capability. Inf Manage 53(8):1049–1064. https://doi.org/10.1016/j.im.2016.07.004
12. Kamble SS, Gunasekaran A, Gawankar SA (2019) Achieving sustainable performance in a data-driven agriculture supply chain: a review for research and applications. Int J Prod Econ. https://doi.org/10.1016/j.ijpe.2019.05.022
13. Kim H, Laskowski M (2016) Towards an ontology-driven blockchain: design for supply chain provenance
14. Korpela K, Hallikas J, Dahlberg T (2017) Digital supply chain transformation toward blockchain integration. In: International conference on system sciences. Waikoloa Beach, USA
15. Kouhizadeh M, Sarkis J (2018) Blockchain practices, potentials, and perspectives in greening supply chains. Sustainability 10:3652. https://doi.org/10.3390/su10103652
16. Kshetri N (2018) Blockchain's roles in meeting key supply chain management objectives. Int J Inf Manage 39:80–89. https://doi.org/10.1016/j.ijinfomgt.2017.12.005
17. Maier R, Passiante G, Zhang S (2011) Creating value in networks. Int J Innov Technol Manag 8(03):357–371
18. Mentzer JT, DeWitt W, Keebler JS, Min S, Nix NW, Smith CD, Zacharia ZG (2001) Defining supply chain management. J Bus Logist 22(2):1–25. https://doi.org/10.1002/j.2158-1592.2001.tb00001.x
19. Nakamoto S (2008) Bitcoin: a peer-to-peer electronic cash system. https://bitcoin.org/bitcoin.pdf. Accessed 21 Jan 2020
20. Paulraj A, Chen IJ, Flynn J (2006) Levels of strategic purchasing: impact on supply integration and performance. J Purch Supply Manag 12(3):107–122. https://doi.org/10.1016/j.pursup.2006.08.002
21. Stupar S, Bičo Ćar M, Šahić E (2019) Challenges of applying blockchain technology. Lecture Notes Comput Sci 355–364. https://doi.org/10.1007/978-3-030-18072-0_41
22. Van Bockstael S (2018) The emergence of conflict-free, ethical, and fair trade mineral supply chain certification systems: a brief introduction. Extract Indus Soc 5. https://doi.org/10.1016/j.exis.2017.12.014

Chapter 9
Methodology to Treat Synergies in the Distribution to Multiple Kinds of Clients

N. Anich and M. Mateo

Abstract Many businesses distribute their goods though several supply chains in order to cover different kinds of clients. A methodology is proposed to evaluate if the best solution supposes a separated flow for each one or some common activities along the supply chain may be shared and obtain scale economies. The objective is to evaluate the level of synergies among different supply chains, each for a different set of clients. The evaluation for the distribution of pharmaceutical products is given as example. In this case, three types of clients are provided: pharmacies, public health centres and public health companies.

Keywords Supply chain management · Distribution · Risk · Synergies · Pharmaceutical sector

9.1 Introduction

Supply chain management (SCM) involves all activities to satisfy the needs of a customer in the supply, which includes processes such as procurement, manufacturing, storage, distribution and reverse logistics [18]. However, these processes must be carried out by one or several units, which implies the existence of ties through flows of information, materials or financial resources [32]. In the supply chain (SC), those entities are involved directly or indirectly to satisfy a client requirement; this is part of a wider concept known as configuration or design of the SC [21].

The SC can be considered as this network of organizations involved, through up and down links, in the different processes and activities that produce value in the provided products and services [7]. A SC has been usually designed for a single kind of products or a single sales channel, but later some additional products, clients

N. Anich · M. Mateo (✉)
Department of Management, Institut Organització I Control, ETSEIB Universitat Politècnica de Catalunya, Av. Diagonal 647, 7th floor, 08028 Barcelona, Spain
e-mail: manel.mateo@upc.edu

N. Anich
e-mail: nicolas.anich@upc.edu

C. Avilés-Palacios and M. Gutierrez (eds.), *Ensuring Sustainability*, Lecture Notes in Management and Industrial Engineering,
https://doi.org/10.1007/978-3-030-95967-8_9

or sales channels have been added independently, without taking into account the existent structure. We ask themselves if a redesign is suitable in mixing independent SCs in some processes. We propose a methodology and apply it to a pharmaceutical distribution in Chile. This methodology is integrated in a more general one, whose objective is the redesign of a supply chain considering risks [3]. Therefore, the results obtained here will be used for next steps in the search of the best options for the redesign, using complementary tools such as simulation and optimization.

The paper is organized as follows. Some definitions on SC and SCM are given in Sect. 9.2. Section 9.3 introduces the typology of supply chains, while Sect. 9.4 defines the synergic supply chains. In Sect. 9.5, the methodology is proposed, and in Sect. 9.6, it is evaluated in a pharmaceutical SC. Finally, Sect. 9.7 provides the conclusions and some future work to be done.

9.2 Definitions on Supply Chain Management

The several dimensions in the SCM lead to a lack in a unique definition [26]. Some different points of view are shown in Table 9.1 and later translated into different types of chains.

According to [28], the SC covers all activities involved in the delivery of a product from raw materials to the customer, including, among others, order entry and order management, distribution in all channels, delivery to the client and information systems. Lambert and Enz [24] emphasize the mainstreaming of supply chain activities and define them: design, demand planning, procurement, manufacturing, storage, distribution, returns and reverse logistics. This coincides with the definition given by the Council of Supply Chain Management Professionals [9]: "Planning and management of all activities related to procurement and acquisition, conversion and all logistics management activities. In essence, supply chain management integrates supply and demand management within and between companies".

The SC is also a sequential network of business partners involved in production. It includes not only the manufacturer and the supplier, but also transporters, retailers, distributors and the customers [6]. Flynnet al. [13] describe the SC as upstream, downstream and the organizational unit in the centre. In downstream, the greatest concern is the quality of service, lead times and the environmental decisions nowadays.

The SCM integrates the demand and the offer (inter and intra organizations), to reach distinctive capacities with the target of achieving competitive advantage and value generation [15]. Based on the SC integration, each chain can be defined through efficiency or effective differentiation.

Vitaseck et al. [34] define the SC "as a set of organizations directly linked by upward and downward flows of products, services, information and funds that work collaboratively to reduce costs and waste by efficiently extracting what is needed to meet the needs of individual clients". The architecture of this network includes the way in which the activity is carried out, including the types of human and physical assets, as well as associated organizational structures and capabilities. In turn, this

Table 9.1 Definitions about supply chain management

	Authors	Definitions
1	Lummus and Vokurka [28]	All activities involved in the delivery of a product from the raw material to the customer, including the supply of raw materials and parts, manufacturing and assembly, storage and inventory, order entry and management, distribution, delivery to the client and information systems
2	Vitaseck et al. [34]	Group of organizations directly linked by up and down flows of products, services, information and funds that work collaboratively to reduce costs and waste by efficiently extracting what is needed to meet the needs of individual customers
3	Gunasekaran et al. [15]	Resource coordination and optimization of activities throughout the value chain to obtain competitive advantages
4	Flynn et al. [13]	Degree in which a manufacturer collaborates strategically with its partners and collaboratively manages intra- and inter-organizational processes
5	Christopher and Holweg [7]	Network of organizations involved, through up and down links, in the different processes and activities that produce value in the form of products and services in the hands of the final consumer
6	Wieland et al. [36]	Active management of supply chain activities to maximize value and achieve a sustainable competitive advantage
7	Chopra and Meindl [6]	All the parties involved, directly or indirectly, in the fulfilment of a client request
8	Lambert and Enz [24]	Integration of key business processes from the end user to the original suppliers that offer products, services and information that add value to customers and other stakeholders
9	LeMay et al. [26]	Design and coordination of a network through which organizations and individuals obtain, use, deliver and dispose of goods; acquire and distribute services; and are made available to markets and customers

configuration must respond to the definition of the business, which seeks to generate value in its broadest sense to customers. Currently, all this is built in an environment of uncertainty and sustainability [7]. Strategic guidelines of the SC will lead to some typologies in search of competitive advantage [36].

Concluding, we can realize that, regardless of the set of presented definitions, we believe that the SC is mainly defined by its processes, which flow from the supplier–supplier to the client–client [24], but the design is only taken into account by [26]. In this sense, Table 9.2 shows in columns if a set of elements appears or not in the definitions given in Table 9.1: activities, entities, flows, reverse (green logistics), coordination, integration, performance, value and design or redesign.

Table 9.2 Elements considered in the definitions about supply chain management

Definition	Activities	Entities	Flows	Reverse	Coordinate	Integration	Perform	Value	Design and redesign
1	*				*	*			
2		*	*	*	*	*	*		
3	*	*			*				
4		*	*		*				
5	*	*	*		*			*	
6	*							*	
7	*	*					*		
8	*		*	*		*		*	
9		*	*	*					*

9.3 Types of Supply Chains

Twenty-six papers were reviewed that emphasize different typologies to define a SC. A selection of them, starting from [12], is included in Table 9.3. In columns, there are different characteristics in the SC (in bold fonts in this section) according to its orientation and objectives: efficient (efnt), effective (efve), resilient (res), green, sustainable (sust), related (relt), quality (qual) and synergic (syng).

An **efficient SC** will be linked to using the means available to meet an objective, that is, to optimize the used resources. An **effective SC** is the ability to achieve what is expected or desired after the performance of an action, in many cases related to response times [11]. Achieving the ideal combination of these two components is one of the purposes within a SC, for which its degree of complexity must be managed. Both variables will depend how the company defines its strategy [11].

A SC depends on the location of the nodes, since there are risks from the infrastructure or even catastrophic risks, such as epidemics, natural hazards or sociopolitical instability [35]. A complex SC is prone to interruptions and its propagation, with a degree of vulnerability as a result from the exposure to a disruptive event [4]. For a network, the capacity to return to a stable state after a disturbance is necessary. A **resilient** SC has also the ability to respond, recover and transform after an interruption [2, 10]. A SC must increasingly consider environmental concerns in their activities.

Supply chains must increasingly consider environmental concerns in their activities. A **green** SC links performance with ecological capabilities, such as pollution prevention [14, 33]. This involves preventing waste and emissions, allowing recycling activities, reducing CO_2 emissions, etc. [17]. A more up-to-date view is incorporated in a **sustainable** SC, which searches for the maximum profitability and the minimum negative environmental impacts, while improving the social welfare of its stakeholders [19, 23, 31].

Table 9.3 Summarized classification of types of SC according to some outstanding papers

Papers	efnt	efve	res	green	sust	relt	qual	syng
Fisher [12]	*	*						
Seuring and Müller [31]					*			
Flynn et al. [13]	*	*						
Gavronski et al. [14]				*				
Elleuch et al. [10]			*			*		
Fahimnia et al. [11]	*	*				*		
Rozen-Bakher [30]								*
Abdel-Basset et al. [1]						*		
Kusi-Sarpong and Sarkis [23]					*			
Tseng et al. [33]				*				
Kaur et al. [20]							*	

As a way to increase the degree of resilience, **related** SCs appear. They define the organizational arrangements that use resources and/or governance structures between two partners in the SC [1, 16]. Following the same line, total quality management (TQM) takes on the concept of synergy, which gives higher relevance to the performance of the total system than to the sum of each individual component. In this sense, it is defined a **quality** SC [20].

However, this has not been enough, since multichannel has forced to manage the network with great flexibility. A network coexists with a combination of typologies, depending on the characteristics of each different business absorbed by the SC. In this context, a new typology is proposed to help address these complexities, a **synergic** SC [30].

9.4 Synergic Supply Chains

Contemporary researchers on operations management have discussed synergies and problems in the integration of operations [22]. Horizontal and vertical integration has been the focus of research in management synergy [38]. However, such integration has received little attention related to operations considering the SC types and the decision levels in the SC.

In a vertical integration, effective coordination is essential. Companies have managed to establish different types of relationships to increase synergies, which are materialized in contracts and actions: joint reimbursement contracts, wholesale discount and cost-sharing contracts [25]. However, the best results are produced from the operational point of view, which improves the response capacity to customers [25]. The horizontal integration consists of two or more companies producing goods or services that perform the same task or produce the same good, integrated into a single organization. The greatest amount of research has been done here [5]. However, this result is only met when a merger occurs at the SC level that the business acting as leader [5].

Zhang and Burke [37] focus on the symptoms that occur when horizontally demand points are integrated, with the aim of a higher accumulative demand and a lower variability. This implies low levels of risk, improves planning or decreases the bullwhip effect on inventories. In the same line, two companies produce operational synergies and as a consequence increase flexibility and robustness, by merging horizontally in different vertical stages of the network. [29] redefine material flows along a horizontally merged network in different stages, based on cost optimization, but in terms of risk they only consider variations in demand.

A set of works has focused on models to quantify operational synergies in mergers. Managers need to evaluate investment in new product lines, demand volume and volatility [27]. The retail has evolved towards marketplaces, which has put the challenge in managing multi-business channel systems. This has increased the complexity of the network: different channels with different and varied objectives. In this context, the synergic chains appear, as a measure of controlling complexity.

9.5 Methodology for the Degree of Synergic Supply Chains

A general methodology, composed of seven steps, for the redesign in the supply chain managing risks is described in [3]

1. Definition of the object of the SC, environment analysis and formulation of objectives.
2. Definition of the existing SC, as a priori design, at several levels (macro, meso and micro).
3. Definition and/or evaluation of the model of SCM and the strategic coherence of the SC (reality versus objectives).
4. Study of customers and possible scenarios of demand.
5. Identification and evaluation of risks in the supply chain, by determining current and latent threats.
6. Analysis of the redesigned chain, final state, using optimization if possible. The study of the synergic supply chains is carried out in this step.
7. Evaluation of the objective fulfilment. If requirements and objectives fixed step 1 are not satisfied, go back to 5 (change policies for risk detection and mitigation) or 1 (change objectives).

From a static point of view, the members of a SC are selected and deselected according to the degree of symmetry between businesses, which will allow them to move to new markets and adopt integrated processes. However, the nature of the relationships between these members is holistic, since it will depend at all levels of a SC (strategic, tactical, operational). Therefore, suppliers vary according to the complementarity of the members in the flow of products, which is reflected in the kinds of contract. This same concept is applied, in the tactical aspects, planning the different product/business lines, to build a synergistic network. The companies try to increase performance through some dimensions such as delivery, cost, quality and flexibility; therefore, it is essential to define their inbound, outbound and transport operations with high degrees of integration, with the objective of share processes and capacities, which allows to achieve these dimensions. The advantage of synergic chains is the high degree of linkage between the risks associated to the SC of each business, since the same uncertainty events are shared and redundancy pathways may be defined. This can help to achieve lower degrees of vulnerability and high **resilience**.

A company with n businesses has $C(n,2)$ combinations of complementarities between chains, where n is the number of supply chains and 2 is the number of compared chains. For example, a company has five businesses: A, B, C, D and E. Each of them has its own strategic objectives (offer, differentiation and segment of business), which leads to differences between them. There should be five logistics chains that operate independently; however, it would be convenient to study a fusion of chains to allow a greater efficiency, increase flexibility and reduce the total complexity.

As a way to rationalize the number of evaluations, first, the pivot chain must be defined. It will be the one with the greatest impact in terms of the flow of products or benefit among the alternatives. In case of two or more pivot candidates, the one with the greatest strategic importance will be chosen. In our example, let us suppose that the pivot chain is A.

Once determined the pivot chain, the rest of alternatives must be compared with this chain. This implies a set of $n-1$ pairs. The set Z includes the $n-1$ synergistic chains to be evaluated Γ (τ, υ), being

τ = number of the pivot chain $1 \leq \tau \leq n$.
υ = number of the complementary chain $1 \leq \upsilon \leq n$; $\upsilon \neq \tau$.

In the example, the set Z is defined as $Z = \{(A, B), (A, C), (A, D), (A, E)\}$.

Once the ordered pairs are selected, the horizontal synergy degree of each pair of chains is sequentially evaluated by calculating a global synergic factor GSF($\Gamma(\tau,\upsilon)$). In the example, the first element in Z is the pair (A, B).

GSF($\Gamma(\tau,\upsilon)$) is calculated by the local horizontal synergic factor of each decision level ($1 \leq \xi \leq 3$) (operational, tactical and strategic) in a possible synergistic network formed from the pair $\Gamma(\tau,\upsilon)$. This involves evaluating the synergistic sources of their respective decisions (λ_ξ). Initially, four operational decisions are considered (**procurement, inbound, outbound** and **transport**), 1 at the tactical level (**planning**) and 2 at the strategic level (business **strategy** and **risk**). That is, $\lambda_1 = 4$; $\lambda_2 = 1$; $\lambda_3 = 2$. The initial weight associated with each of the local sources is 1. In future studies, it is expected that the weights will be more precisely obtained.

The decision to merge two chains will depend on GSF($\Gamma(\tau, \upsilon)$) and can reach a maximum value of 7. Therefore, a synergic network with GSF($\Gamma(\tau,\upsilon)$) ≤ 3 shows few common characteristics, and it is advisable to operate them individually. In the case $3 <$ GSF($\Gamma(\tau,\upsilon)$) ≤ 5, although there are synergistic degrees, it is not a conclusive scenario; other tools (simulation, mathematical programming) should be added to define the design of these chains. Finally, if $5 <$ GSF($\Gamma(\tau,\upsilon)$) ≤ 7, the network has a high degree of synergy, which determines the fusion of the common links and nodes in most of the stages and defines a synergistic design for two businesses. This methodology can be used as a priori design for several chains or a redesign between supply chains and a previous work to the SCOP methodology [8]. The GSF($\Gamma(\tau,\upsilon)$) is obtained for each pair of businesses, if there are more than two, and always uses the same pivot chain, until the total number of elements in the Z set is evaluated.

Additionally, if there are two pairs that demonstrate a high degree of synergy (GSF($\Gamma(\tau,\upsilon)$) > 5), a single solution is finally evaluated whether is better that all the involved chains are joined in a single network, whose common axis is the pivot chain (τ). In our example, if GSF($\Gamma(A,B)$) > 5 and GSF($\Gamma(A,C)$) > 5, the network formed by the chains A, B and C must be evaluated considering the local sources.

9.6 Application to a SC of Pharmaceutical Distribution

The supply chain to be evaluated corresponds to a distributor of pharmaceutical products in Chile [3]. More than 750 active ingredients are sold, provided by more than 80 suppliers, but currently only 65 are active suppliers. The products must reach points of sale, for three different kinds of clients: 650 public health locations (*A*), 400 pharmacies (*B*) and 60 private health institutions (*C*). The respective objectives are low cost for *A*; availability and response times for *B*; and right delivery quantities in short time for *C*. In this case, the evaluation made looking for synergies between two of its business, the public health (*A*) and the pharmacies (*B*), has led to the following results:

- Procurement. As the same supplier can provide similar products, an order may contain products for different kinds of clients. Score = 0.84 (84% of times the products, main active ingredients or therapeutic classes are the same).
- Inbound. This process is equivalent for any kind of client. Score = 1 (always).
- Outbound. There are some specificities (characteristics of the products, types of orders…) for any kind of client. Score = 0 (never common).
- Transport. The transport for both chains coincides in date and lead time in a 64% of the times, and it can be shared. Score = 0.64 (simultaneous delivering for both kinds of clients in a single transport).
- Planning. The existence of seasonality has influence on the procurement orders. Score = 0.46 (46% of times the EAN coincides in both).
- Strategy. About the three criteria that describe the strategy (differentiation, segmentation and offer). Score = 0.33 (only offer is common for both business).
- Risk. The number of common risks forms a list of potential risks. Score = 1 (in this case, 17 in the pharmacies out of 17 risks in the public health institutions).

The final score is GSF $GSF(\Gamma(A,B)) = 4.27$; therefore, a more detailed analysis is recommended. Moreover, if a similar evaluation is done taking *A* as pivot and *C* the other member of the pair, the final score is $GSF(\Gamma(A,C)) = 3.47$. Finally, the network formed by chains *A*, *B* and *C* is evaluated considering the local synergic sources. As a summary of the results obtained, Table 9.4 shows some representative data (number of suppliers, references, clients, warehouses and kinds of transport) for the three individual supply chains, and in the last column, a potential synergic

Table 9.4 Comparison of three individual supply chains versus the synergic network adding the three

	(A) Public health	(B) Pharmacies	(C) Private health	(A + B + C) Synergic network
Number of active suppliers	65	60	60	65
Number of references	1280	1027	665	1499
Number of clients	650	396	61	1107
Number of warehouses	2	2	2	1
Number of different transports	3	3	3	5 (A + B); 3(C)

network is formed by the addition of chains *A*, *B* and *C*. In this case, five kinds of transport can be shared by businesses *A* and *b*, while the three kinds of transport for the business *C* cannot be shared with chains *A* and *B*.

9.7 Conclusions

A methodology to deal with different kinds of clients is proposed. First, an analysis on different views of SCM and a classification of types of SC are developed. This shows that synergic SCs have been little studied. In order to design a global SC with synergies between each one considered independently, we propose a methodology that addresses the complementarity of multi-business chains in several aspects, scopes and components. Seven types of decisions are evaluated. Applying the proposed methodology, an index for the degree of synergy and complementarity between pairs of chains is calculated, which will allow to conclude their degree of symbiosis. Very low values indicate that the reduced number of synergies leads to the independent management for both chains. On the other hand, very high values point to share most of the steps in both chains. Finally, intermediate values of synergy require the use of other tools to evaluate the operations to be shared. Such tools, as simulation or mathematical programming, can help to achieve better results in the indicators for the objectives of each chain. Later the methodology is applied to a case in the pharmaceutical sector. The results show an intermediate degree of synergy, and, therefore, other tools must be applied starting from the results obtained.

In future, the degree of synergy must be better defined. The methodology can be enriched by considering a greater number of synergistic sources not considered up to now, such as costs and configuration issues. Moreover, the connections between the results in this methodology and the tools to use in case of a mean synergy should be added. They must take into account as many risks as possible and the optimization according to the defined strategic objective for each SC.

Acknowledgements This work was supported by project DPI2015-67740-P (MINECO/FEDER).

References

1. Abdel-Basset M, Gunasekaran M, Mohamed M, Chilamkurti N (2019) A framework for risk assessment, management and evaluation: economic tool for quantifying risks in supply chain. Futur Gener Comput Syst 90:489–502
2. Adobor H, McMullen RS (2018) Supply chain resilience: a dynamic and multidimensional approach. Int J Logist Manag 29:1451–1471
3. Anich N, Mateo M (2019) Development of a methodology for the management of risk in a supply chain. Application to the pharmaceutical sector. In: 13th International conference on industrial engineering and industrial management

4. Blackhurst J, Rungtusanatham MJ, Scheibe K, Ambulkar S (2018) Supply chain vulnerability assessment: a network based visualization and clustering analysis approach. J Purch Supply Manag 24:21–30
5. Cho SH (2014) Horizontal mergers in multitier decentralized supply chains. Manage Sci 60:356–379
6. Chopra S, Meindl P (2016) Supply chain management. Strategy, planning and operation, 6th ed. Pearson, Prentice-Hall, Upper Saddle River
7. Christopher M, Holweg M (2011) Supply chain 2.0: managing supply chains in the era of turbulence. Int J Phys Distrib Logist Manag 41:63–82
8. Corominas A, Mateo M, Ribas I, Rubio S (2015) Methodological elements of supply chain design. Int J Prod Res 53(16):5017–5030. https://doi.org/10.1080/00207543.2015.1013641
9. Council of Supply Chain Management Professionals (2013) Supply chain management terms and glossary. https://cscmp.org/CSCMP/Educate/SCM_Definitions_and_Glossary_of_Terms.aspx
10. Elleuch H, Dafaoui E, Elmhamedi A, Chabchoub H (2016) Resilience and vulnerability in supply chain: literature review. IFAC-Papers OnLine 49:1448–1453
11. Fahimnia B, Jabbarzadeh A, Ghavamifar A, Bell M (2017) Supply chain design for efficient and effective blood supply in disasters. Int J Prod Econ 183:700–709
12. Fisher ML (1997) What is the right supply chain for your products? Harvard Business School Publishing—HBR
13. Flynn BB, Huo B, Zhao X (2010) The impact of supply chain integration on performance: a contingency and configuration approach. J Oper Manag 28:58–71
14. Gavronski I, Klassen RD, Vachon S, Nascimento LFM (2011) A resource-based view of green supply management. Transp Res Part E: Logist Transp Rev 47:872–885
15. Gunasekaran A, Lai KH, Cheng TCE (2008) Responsive supply chain: a competitive strategy in a networked economy. Omega 36:549–564
16. Hammervoll T, Toften K (2010) How can supply chain managers classify and differentiate supply chain partners? Int J Netw Virtual Organ 7:112–131
17. Helo P, Shamsuzzoha AHM (2020). Real-time supply chain—a blockchain architecture for project deliveries. Robot Comput Integr Manuf 63
18. Holweg M, Helo P (2014) Defining value chain architectures: linking strategic value creation to operational supply chain design. Int J Prod Econ 147:230–238
19. Jia F, Gong Y, Brown S (2019) Multi-tier sustainable supply chain management: the role of supply chain leadership. Int J Prod Econ 217:44–63
20. Kaur M, Singh K, Singh D (2019) Synergetic success factors of total quality management (TQM) and supply chain management (SCM): a literature review. Int J Qual Reliab Manage 36:842–863
21. Klibi W, Martel A, Guitouni A (2010) The design of robust value-creating supply chain networks: a critical review. Eur J Oper Res 203:283–293
22. Kumar M, Rodrigues VS (2020) Synergetic effect of lean and green on innovation: a resource-based perspective. Int J Prod Econ 219:469–479
23. Kusi-Sarpong S, Sarkis J (2019) Sustainable supply chains and emerging economies. Resour Conserv Recycl 143:238–243
24. Lambert DM, Enz MG (2017) Issues in supply chain management: progress and potential. Ind Mark Manage 62:1–16
25. Lan Y, Yan H, Ren D, Guo R (2019) Merger strategies in a supply chain with asymmetric capital-constrained retailers upon market power dependent trade credit. Omega 83:299–318
26. LeMay S, Helms MM, Kimball B, McMahon D (2017) Supply chain management: the elusive concept and definition. Int J Logist Manage 28:1425–1453
27. Liang B, Shi C (2012) Quantifying operational synergies of M&A: an stochastic equilibrium approach. Appl Mech Mater 174–177:2830–2835
28. Lummus RR, Vokurka RJ (1999) Defining supply chain management: a historical perspective. Ind Manag Data Syst 99:11–17

29. Nagurneya A, Qiang Q (2012) Fragile networks: Identifying vulnerabilities and synergies in an uncertain age. Int Trans Oper Res 19:123–160
30. Rozen-Bakher Z (2018) The trade-off between synergy success and efficiency gains in M&A strategy. EuroMed J Bus 13:163–184
31. Seuring S, Müller M (2008) From a literature review to a conceptual framework for sustainable supply chain management. J Clean Prod 16:1699–1710
32. Simatupang TM, Sridharan R (2002) The collaborative supply chain. Int J Logist Manage 13:15–30
33. Tseng ML, Islam MS, Karia N, Fauzi FA, Afrin S (2019) A literature review on green supply chain management: trends and future challenges. Resour Conserv Recycl 141:145–162
34. Vitasek K, Manrodt KB, Abbott J (2005) What makes a lean supply chain. Supply Chain Manage Rev 9:39–45
35. Wagner SM, Bode C (2008) An empirical examination of supply chain performance along several dimensions of risk. J Bus Logist 9:307–325
36. Wieland A, Handfield RB, Durach CF (2016) Mapping the landscape of future research themes in supply chain management. J Bus Logist 37:205–212
37. Zhang X, Burke GJ (2011) Analysis of compound bullwhip effect causes. Eur J Oper Res 210:514–526
38. Zhu J, Boyaci T, Ray S (2016) Effects of upstream and downstream mergers on supply chain profitability. Eur J Oper Res 249:131–143

Part III
Production Planning and Control

Chapter 10
Conceptual Framework for the Integration of Tactical and Operational Decisional Levels

D. Pérez-Perales, F. Alarcón, P. Gómez-Gasquet, and M. M. E. Alemany

Abstract The need to adapt to increasingly competitive markets, adapting to new organizational forms and pursuing greater flexibility, forces companies to make decisions more agile. To face current dynamism, it is necessary to provide information systems for planning with sufficient flexibility to achieve the proposals established in the traditional operations planning and control system (OPCS) scheme. This is possible due to the introduction of new Industry 4.0-based production technologies that give decision-making more flexibility and efficiency. In this paper, a conceptual framework for the integration of the tactical and operational planning is proposed, doing more emphasis in the expert system that integrates and coordinates the specific decisions of both levels.

Keywords Conceptual framework · Tactical planning · Operational planning · Integrated planning

10.1 Introduction

It is a fact in many companies the need to properly coordinate decisions at different time levels, and more particularly between tactical an operational ones.

To achieve this, two clearly differentiated visions are used in practice: the hierarchical planning of the levels or their simultaneous planning.

D. Pérez-Perales (✉) · F. Alarcón · P. Gómez-Gasquet · M. M. E. Alemany
Research Centre On Production Management and Engineering (CIGIP), Universitat Politècnica de València, Camino de Vera, s/n, 46022 Valencia, Spain
e-mail: dapepe@omp.upv.es

F. Alarcón
e-mail: faualva@omp.upv.es

P. Gómez-Gasquet
e-mail: pgomez@omp.upv.es

M. M. E. Alemany
e-mail: mareva@omp.upv.es

C. Avilés-Palacios and M. Gutierrez (eds.), *Ensuring Sustainability*,
Lecture Notes in Management and Industrial Engineering,
https://doi.org/10.1007/978-3-030-95967-8_10

There are several authors [6, 17] who point out different reasons to address the previous question according to the hierarchical approach.

Other authors such as [20] or [22] advocate for the simultaneous planning of the two decision levels.

The analyses of both approaches have allowed to address some shortcomings and identify a set of elements that could be enhanced. Moreover, the new paradigm of Industry 4.0 and its related technologies have changed the way this tactical-operational planning process is carried out [4, 15, 11].

This paper proposes a conceptual framework to support the integration and coordination of the tactical-operational planning process. One of the key aspects is the consideration of the new advances due to Industry 4.0-based technologies which allow to obtain more flexible and efficient integrated solutions.

The paper is structured as follows: In Chap. 2, a brief review about different visions for the tactical-operational planning process is conducted, basically those ones based on hierarchical and simultaneous planning of the decision levels. From the previous analysis, a conceptual framework is proposed in Chap. 3. Finally, in Chap. 4, some conclusions are drawn.

10.2 Review of Visions for Tactical-Operational Planning

As aforementioned in the introduction, two clearly differentiated visions for tactical-operational planning are used: the hierarchical planning of the levels and their simultaneous planning.

The **hierarchical vision** of tactical and operational planning splits the problem into two subproblems, which require "mechanisms capable of coordinating the different decisions of each level together with the information exchanged" [13].

Many authors have addressed the advantages and obstacles of using this vision. One of the main advantages concerns to the compatibility with the organizational structure of the company and consistency among various planning activities in the different levels of organization's hierarchy [18]. Infeasibility and suboptimality among the decisions made at the different hierarchical levels of the firm are among the main obstacles [16].

Due to space restrictions, just other consulted works from literature are quoted, such as those from [1, 5, 8, 9, 12, 19, 23] or [21].

On the other hand, the **simultaneous planning** of the tactical and operational decisional levels. As in the case of the previous one, some advantages and obstacles exist. Bitran and Tirupati [3] expose that although developing integrated decision models capable of dealing with all decisions at once may seem desirable at first, the integrated models have several drawbacks. First, these models are so complex that their optimality is just guaranteed in a few practical cases and with excessively high computational times. Second, even if computational power does not restrict obtaining their solution, these models do not respond to the hierarchical structure of many companies, since their monolithic approach does not allow the interactions between those responsible for each level of the hierarchy.

Other consulted works are those from [7, 10] or Almada and James [2].

10.3 Proposal of a Conceptual Framework for Tactical-Operational Planning

In this third section, a conceptual framework for tactical-operational planning is proposed. As stated in the acknowledgements, this proposal is developed as part of the funding project NIOTOME [14], where some deficiencies of tactical-operational production planning systems were first identified in order to contribute with some improvements and therefore with an innovative proposal.

The deficiencies identified are:

- The decomposition of the decision process into subproblems facilitates having models approvable by the decision-maker; however, these models have a partial vision of the problem.
- There are no check procedures to determine consistency between levels.
- The traditional operations planning and control system (OPCS) scheme is a common and widely used proposal. New proposals must face the inertia established in this traditional vision.
- Information systems for OPCS are complex and rigid, following the theoretical approach and leaving out practical reality. In many cases, the hierarchical planning system faces alterations in the production process, urgent orders or breakdowns that would force altering the plans.
- The cycles of decision-making are shortened, which force decision-makers to alter the proposals of the information systems in order to greater adapt to that reality and improve the proposals.
- It is key to identify obsolete models. The decision models made by the model designer may not correspond to reality (change in the processes over time, misinformation of the model designer, …).
- The OPCS information system is not ready to absorb and treat the huge amount of available data in the company thanks to Industry 4.0-based technologies.

This paper proposes a conceptual framework for the adaptation of the instruments available in the tactical-operational scope of the OPCS so that they are able to take advantage of the information of its environment to provide more adjusted solutions to the reality of each moment, improving efficiency throughout the production system.

This must be specified in two main objectives, which are aligned with part of the aforementioned project [14].

1. The tactical and operational levels must have decisional independence, as well as a high degree of coordination by extending the decision process that closes

the tactical-operational loop (flow from the operational to the tactical) and improving the efficiency of the joint tactical-operational planning process.

2. Information and its obsolescence must be considered. The data must be processed and converted into relevant information that must reach all decision levels for effective decision-making. The simple updating of the values of the models with the data already available, the elaboration of values based on machine learning systems or the transformation of the constraints or objectives of the models under the tutelage of a manager that establishes the most appropriate time to do so.

Therefore, a change in the traditional OPCS framework has been proposed, which has been presented towards a system with greater integration and better use of information by defining an intermediate layer between the tactical and operational level called tactical-operational objectives integration system (TODIS), as it can be seen above in Fig. 10.1. TODIS nomenclature is defined in Table 10.1.

Therefore, a change in the traditional OPCS framework has been proposed, which has been presented towards a system with greater integration and better use of information by defining an intermediate layer between the tactical and operational level called tactical-operational objectives integration system (TODIS), as it can be seen above in Fig. 10.1. TODIS nomenclature is defined in Table 10.1.

First, an initial plan is generated (optimally or heuristically). Then, an initial schedule is also generated (optimally or heuristically). This schedule is constrained by the IN sent from the initial plan. If this initial schedule is infeasible, the IN must be changed to result in a feasible programme. If feasible, the efficiency of the joint initial plan-schedule is computed.

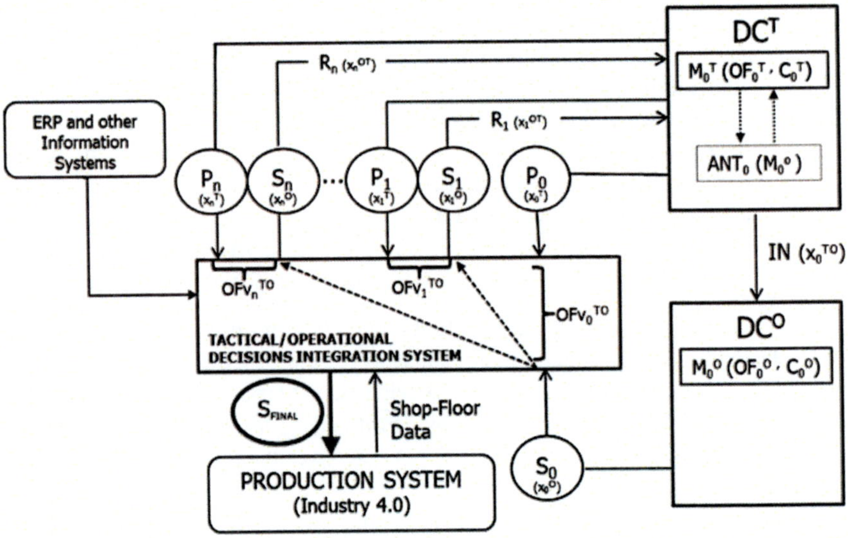

Fig. 10.1 Framework for the integration of tactical and

Table 10.1 Nomenclature

DC^T	It corresponds to the tactical decision centre (human or computer resource), which is responsible of the tactical decisions (plan)
DC^O	It corresponds to the operational decision centre (human or computer resource), which is responsible of the operational decisions (schedule)
M_0^T (OF_0^T, C_0^T)	DC^T makes initial decisions (initial plan) by means of a tactical decisional model M_0^T. This model is made up of an objective function OF_0^T and some constraints C_0^T
M_0^O (OF_0^O, C_0^O)	DC^O makes initial decisions (initial schedule) by means of an initial operational decisional model M_0^O. This model is made up of an objective function OF_0^O and some constraints C_0^O
ANT_0 (M_0^o)	M_0^T can anticipate to a lesser or greater extent M_0^O. This anticipation may affect its objective function OF_0^T, its constraints C_0^T or both
P_0 (x_0^T)	It corresponds to the initial plan P_0 generated by DC^T. This plan is the result of solving M_0^T and obtaining the value of the tactical decision variables x_0^T throughout its planning horizon (made up of a set of planning periods)
IN (x_0^{TO})	Once M_0^T is solved and obtained the value of the tactical decision variables P_0 (x_0^T), some of them (x_0^{TO}) are sent to DC^O within an instruction IN that constrains M_0^O and therefore the value of the operational decision variables (initial schedule) obtained when M_0^O is solved
S_0 (x_0^O)	It corresponds to the initial schedule S_0 generated by DC^O. This schedule is the result of solving M_0^O and obtaining the value of the operational decision variables x_0^O throughout its planning horizon (made up of a set of tactical planning periods)
OFv_0^{TO}	Once M_0^T and M_0^O are solved, the TODIS evaluates which is the efficiency of the joint tactical-operational planning process. This efficiency is the sum of the objective functions values of both models (OFv_0^{TO}), computed only for the schedule horizon. Both values (OFv_0^T and OFv_0^O) must be expressed in monetary units; otherwise, they must be converted
S_n (x_n^O)	The TODIS feeds on the initial plan P_0 (x_0^T) and schedule S_0 (x_0^O), as well as the real data of the production system and proposes a set of alternative schedules S_1 (x_1^O), S_2 (x_2^O), …, S_n (x_n^O) based on possible improvements with respect S_0 (x_0^O)
R_n (x_n^{OT})	Each alternative S_n (x_n^O) generated by the TODIS results in a reaction R_n (x_n^{OT}) sent to DC^T which constraints M_0^T and therefore the new values of the tactical decision variables P_n (x_n^T) obtained when M_0^O is solved
P_n (x_n^T)	The TODIS generates a set of plans P_1 (x_1^T), P_2 (x_2^T), …, P_n (x_n^T) which are the result of solving again M_0^T, but in this case, considering the reactions R_1 (x_1^{OT}), R_2 (x_2^{OT}), R_n (x_n^{OT})
OFv_n^{TO}	The TODIS calculates the efficiency of the joint tactical-operational planning process for each couple of S_n (x_n^O) / P_n (x_n^T), by obtaining the sum of their objective functions values (OFv_n^{TO})
S_{FINAL}	The best alternative S_n (x_n^O) / P_n (x_n^T) is selected based on the "efficiency" of the joint tactical-operational planning process, that is that with the best OFv_n^{TO}

So far, everything is run as it is carried out in the company. Actually, this proposal does not require an integration of complex information systems and is based on a basic data exchange, since the logic of the process is completely located in the new system (TODIS), so its implementation in companies is inexpensive.

At this point, TODIS, an expert system that draws on all the information from the tactical and operational levels, collects this initial schedule and generates different alternatives, relaxing the constraints from the IN. This will mean an improvement to a greater or lesser extent in the efficiency (OFv_n^O) of these schedules. Then, TODIS evaluates the plans that best fit to these new alternative schedules. Basically, each plan is recalculated taking into account the different types of reactions from these new alternative schedules. These plans will have a penalty in a greater or lesser extent in its efficiency (OFv_n^T),

Finally, TODIS will assess which plan-schedule generates the highest joint efficiency (OFv_n^{TO}), that is the highest integration between tactical and operational decisions.

Figure 10.2 shows this closed-loop scheme for the integration and coordination of tactical and operational decisions. Blue-coloured data refers to those specific decisions that link tactical and operational levels.

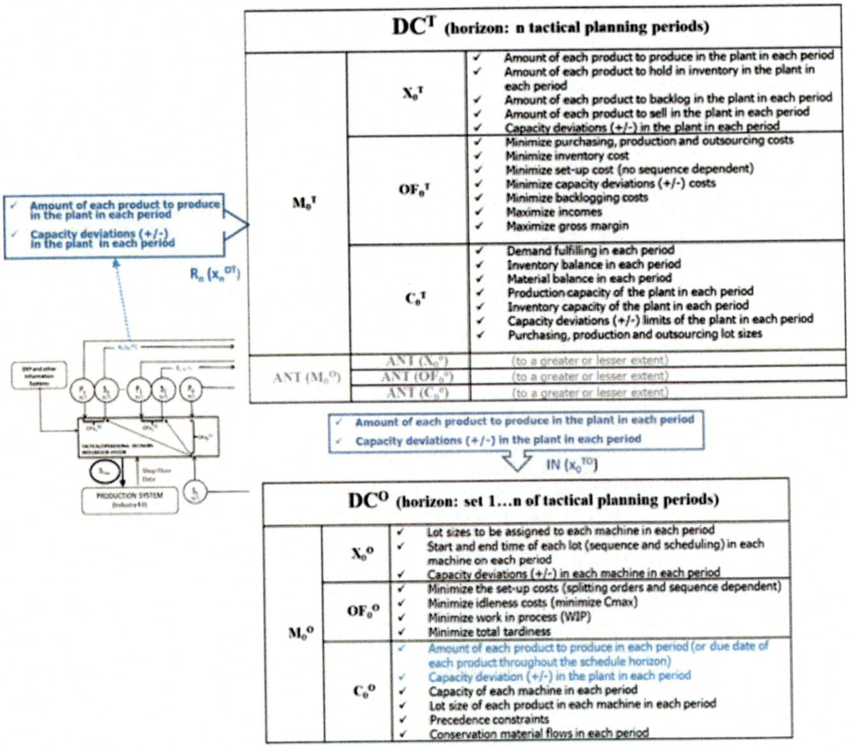

Fig. 10.2 Closed-loop scheme for the integration of tactical and operational decisions

10.4 Conclusions

This paper has proposed a conceptual framework to support the integration and coordination of the tactical-operational planning process. It enhances the traditional hierarchical planning approach by considering the new advances due to Industry 4.0-based technologies which allow to obtain more flexible and efficient integrated solutions.

The framework encompasses an expert system, called TODIS, that integrates and coordinates the specific decisions of both levels, in a closed-loop. It is a layer that feeds on the results (plans and schedules) elaborated in the upper and lower level, respectively, as well as data from ERP and other corporative information systems and real data of the plant (where it is remarkable the use of Industry 4.0-based technologies).

TODIS facilitates integration between plans and schedules, aids in its improvement and proposes alternatives, which if accepted, are managed at the corresponding decision levels, such as consolidated plans or schedules.

It must be remarked that TODIS can be integrated into the current decision-making system, so it is assumed that, on the one hand, the planner generates its definitive plans (tactical level), and these are available for the schedulers who, on the other hand, generate its definitive schedules (operational level). The schedules specify the timed sequence of production orders for a subset of plan periods (schedule horizon).

Acknowledgements Authors of this publication acknowledge the contribution of the project funded by the Ministery of Science, Innovation and Universities entitled "Decision Making Integration of the Tactical-Operative Levels for Improving the Efficiency of the Productive System in Industry 4.0 Environments (NIOTOME)".

References

1. Alemany ME (2003) Metodología y Modelos para el Diseño y Operación de los Sistemas de Planificación Jerárquica de la Producción. Aplicación a una Empresa del Sector Cerámico, Tesis. Universitat Politècnica de València
2. Almada-Lobo B, James RJW (2010) Neighborhood search metaheuristics for capacitated lot-sizing with sequence-dependent setups. Int J Prod Res 48(3):861–878 2
3. Bitran GR, Tirupati D (1993) Hierarchical production planning. In: Graves SC et al (eds) Handbooks in OR&MS, vol 4, Elsevier Science Publishers B.V
4. Boza A, Cortes B, Alemany ME, Vicens E (2015) Event monitoring software application for production planning systems. In: Enhancing synergies in a collaborative environment. Springer, Cham, pp 123–130
5. Boza A, Ortiz A, Vicens E, Poler R (2009) A framework for a decision support system in a hierarchical extended enterprise decision context. IFIP-international workshop on enterprise interoperability. Springer, Berlin, Heidelberg, pp 113–124
6. Fleischmann B, Meyer H (2003) Planning hierarchy, modelling and advanced planning. In: de Kok AG, Graves C (eds) Supply chain management: design, coordination and operation. Elsevier, Amsterdam

7. Gupta D, Magnusson T (2005) The capacitated lot-sizing and scheduling problem with sequence-dependent setup costs and setup times. Comput Oper Res 32:727–747
8. Hax A, Meal HC (1973) Hierarchical integration of production planning and scheduling. Sloan Working Papers, ed. MIT, pp 656–673
9. Hax A, Golovin JJ (1983) Sistemas jerárquicos de planificación de la producción. In: Hax A (ed) Dirección de Operaciones en la empresa, Hispano Europea, pp 513–546
10. Kovacs A, Brown KN, Tarim SA (2009) An efficient MIP model for the capacitated lot-sizing and scheduling problem with sequence-dependent setups. Int J Prod Econ 118:282–291
11. Kumar SPL (2018) Knowledge-based expert system in manufacturing planning: state-of-the-art review. Int J Prod Res 57(15–16):4766–4790
12. MacCarthy B (2006) Organizational, systems and human issues in production planning, scheduling and control. In: Handbook of production scheduling. Springer, pp 59–90
13. Marques M, Agostinho C, Zacharewicz G, Poler R, Jardim-Goncalves R (2018) Responsive production in manufacturing: a modular architecture. In: Practical issues of intelligent innovations. Springer, Cham, pp 231–254
14. NIOTOME, 2020. Available at: https://niotome.blogs.upv.es. Accessed 15 Jul 2020
15. Rossit D.A, Tohmé F, Frutos M (2018) Industry 4.0: smart scheduling. Int J Prod Res 57(12):3802–3813
16. Saad GH (1990) Hierarchical production-planning systems: extensions and modifications. J Oper Res Soc 41(7):609–624
17. Schneweiss C (1999) Hierarchies in distributed decision making. Springer, Berlin
18. Torabi SA, Ebadian M, Tanha R (2010) Fuzzy hierarchical production planning (with a case study). Fuzzy Sets Syst 161(11):1511–1529
19. Tsubone H, Matsuura H, Kimura K (1995) Decision support system for production planning –concept and prototype. Decis Support Syst 13:207–217
20. Urrutia EDG, Aggoune R, Dauzère-Pérès S (2014) Solving the integrated lot-sizing and job-shop scheduling problem. Int J Prod Res 52(17):5236–5254
21. Vargas A, Boza A, Pate S, Patel D, Cuenca L, Ortiz A (2016) Interenterprise architecture as a tool to empower decision-making in hierarchical collaborative production planning. Data Knowl Eng 105:5–22
22. Wolosewicz C, Dauzère-Pérès S, Aggoune R (2015) A Lagrangian heuristic for an integrated lot-sizing and fixed scheduling problem. Eur J Oper Res 244(1):3–12
23. Zhong RY, Pang LY, Pan Y, Qu T, Huang GQ (2012) RAPShell for RFID enabled real-time shop-floor production planning, scheduling and execution. CIE42 Proceedings

Chapter 11
Proposal of a Smart Production Planning and Scheduling in the Industry 4.0 Era

A. Boza, P. Gomez-Gasquet, L. Cuenca, and F. Alarcón

Abstract Production planning and control system (PPCS) proposes a hierarchical system in which decisions follow an (almost) unidirectional flow for coordination between the objectives, plans and activities of the strategic, tactical and operational levels. Information systems have played a fundamental role in the implementation of these systems in organizations. However, these systems also have drawbacks due to their rigidity in decision-making, where coordination between different levels (with frozen plans) is not always possible in a short term, and not so much, by the productive capacity itself. The Industry 4.0 paradigm promotes, among other things, autonomous decision-making, interoperability, agility, flexibility, efficiency and cost reduction. This paper proposes the adaptation of the instruments available in the tactic and operational scope of the PPC system so that they are able to take advantage of the real flexibility and current information in their environment to provide solutions, with and expert system, which are more adjusted to the reality of each moment.

Keywords Production planning and control system (PPCS) · Industry 4.0 · Expert systems

---

A. Boza (✉) · P. Gomez-Gasquet · L. Cuenca · F. Alarcón
Centro de Investigación Gestión e Ingeniería de la Producción, Universitat Politècnica de València, Valencia, Spain
e-mail: aboza@cigip.upv.es

P. Gomez-Gasquet
e-mail: pgomez@cigip.upv.es

L. Cuenca
e-mail: llcuenca@cigip.upv.es

F. Alarcón
e-mail: faualva@cigip.upv.es

C. Avilés-Palacios and M. Gutierrez (eds.), *Ensuring Sustainability*,
Lecture Notes in Management and Industrial Engineering,
https://doi.org/10.1007/978-3-030-95967-8_11

11.1 Introduction

The production planning and control system (PPCS) is a traditional proposal, in which recognized authors [10, 24, 30], establish the decision framework in the production process of most current companies. The PPCS proposes a hierarchical system in which decisions follow an (almost) unidirectional flow for coordination between the objectives, plans and activities of the strategic, tactical and operational levels. Each level pursues its own goals, but always taking into account those of the higher level, on which they depend, and those of the lower level, which they restrict. This hierarchical vision of production planning breaks down decisions into subproblems, which somehow, in the context of an organizational hierarchy, links the highest level of decision with those of the lowest level in an effective way [2]. The development of integrated decision models dealing with all decisions simultaneously may seem attractive, but these models tend to be very large, computationally demanding and inappropriate because it would not respond to the management needs of each level of the organization [3].

Managing the interdependent relationships of these decentralized decision-making systems is complex and requires mechanisms capable of coordinating the different decisions of each level together with the exchanged information [20].

This paper addresses the role that new technologies can play in the PPC systems and specifically in the planning and scheduling processes. For this, the paper has been structured in the following sections: Sect. 11.2 deals with deficiencies in PPCS, Sect. 11.3 presents a proposal to improve the planning and scheduling processes, and finally, Sect. 11.4 presents the main conclusions.

11.2 Deficiencies in PPCS

The importance of the information system that supports the PPCS is shown from the first proposals. The contributions of [2, 5, 11, 16, 17, 19, 28, 29, 34], among others, highlight the role that the information system plays in PPCS. The evolution of these information systems from MRP to ERP systems [9, 12, 25] has placed the information system as a main element in the PPCS.

Despite the advantages of information systems for hierarchical production planning, they also have drawbacks due to their rigidity in decision-making. In many cases, the hierarchical planning system faces alterations in the production process, urgent orders or breakdowns that would force altering the plans.

An example of these alterations is found in the ceramic sector when, due to uncontrollable factors during the manufacturing process, quantities and products different from those planned are obtained, which would cause the replanning of deliveries, the promising quantities or the master plan of production [1], the need to consider rescheduling is also emphasized in the field of hybrid flow shop [15], and something similar occurs in the industries of food processing [33]. Thus, the difficulty of replanning in hierarchical systems means that on many occasions, it is not carried out, and consequently, that companies do not respond adequately to unexpected events. The own structure of the hierarchical planning system establishes complex coordination mechanisms between levels. Decisions made using input data produce frozen plans which are transmitted to inferior levels in the hierarchy to continue making decisions, and these decisions are difficult to back out when new input data arrives. Therefore, the chain of decisions and the information system that supports it makes these changes difficult to do.

Thus, the gap between the theory and practice of PPCS is closely related to the associated information system. How the task is carried out in practice can be distanced of the information system designed [6]. This problem is aggravated when the design of the information system does not adequately collect information about the elements that interact in the domain (reduced or ignored information) [22] and/or the operation of the process itself (the designed system does not meet the real needs) [12]. Some examples are information systems for PPC designed years ago that ignores new information currently available for the production system, such as alterations in the production process, urgent orders or breakdowns that would require new plans. This misalignment between what the business wants (or needs) versus what the information systems allow is addressed from the perspective of business engineering in [8].

The need to adapt toward increasingly competitive markets, new organizational structures and greater flexibility, forces companies to more agile decision-making. To face the current dynamism, it is necessary to provide the information systems for planning with sufficient flexibility to achieve the proposals established in the PPCS framework.

11.2.1 Digital Transformation in PPCS

Technologies that facilitate these processes of change in industrial companies are located under the concept of "Industry 4.0". This new paradigm in the industry promotes, among other things, autonomous decision-making, interoperability, agility, flexibility, efficiency and cost reduction [23]. The digital transformation applied to the processes involves incorporating 4.0 technologies to make them more efficient and flexible [18]. Chavarría-Barrientos et al. [7] also address the areas of change in the scope of S3 (sensing, smart and sustainable—sensitive, intelligent and sustainable), for products, processes or companies, among others. The S3 processes are proposed as a solution to the ever-changing environment of the current digital economy. Continuous monitoring and processing will be provided through the use of emerging technologies, that is, Internet of Things, cloud computing, big data, integrated systems.

The vertical integration of intelligent "production systems" and the acceleration of manufacturing are two of the characteristics of the proposals in Industry 4.0 that add value [13]. Similarly, Moisescu and Sacala [21] present the potential offered by "sensitive" business information systems compared to current ones in terms of adaptability to changes in the environment and where they highlight cyber-physical systems in the production area. Weichhart [31] presents the advantages of what they call S3-ES (sensing, smart and sustainable enterprise systems) where required changes in production (both with respect to the product and services) are quickly identified, the data collected by sensors can be quickly sent to different decision models, and model reengineering can be facilitated. Weichhart [32] indicates the possibility of changes in the decision models through agents that communicate with the models. Shrouf [27] presents it as "new planning methods for the industry" and also the improvement of the management of unexpected events, using 4.0 technologies. In the field of production planning, it is addressed in [4, 26], and in the field of production scheduling, it is addressed in [14]. However, many of these proposals are presented as expected benefits of the set of technologies for Industry 4.0 and only a few specify some part of the set of decisions.

11.2.2 Identification of Deficiencies in the Current PPCS

Thus, the review carried out allows us to identify a set of elements in the field of PPCS where to investigate improvements. The deficiencies identified are described in Table 11.1. The identification of these deficiencies allows working on a significant number of research lines to deal with one or more of these elements in one or more of the activities developed in the PPCS. In our case, we have focused on the planning and scheduling processes.

Table 11.1 Deficiencies in the current PPCSs

Nr.	Description
1	The decomposition of the decision process into subproblems facilitates having models approvable by the decision-maker; however, these models have a partial view of the problem
2	There are no precise assessment procedures to determine coherence between levels
3	The PPCS is a traditional and widely used proposal, so the new proposals face the inertia established in the traditional vision
4	Information systems for PPCS are complex and rigid, following the theoretical approach and leaving out practical reality
5	The decision-making cycles are shortened, which obliges the decision-makers to alter the proposals of the information systems in order to adjust more to that reality and improve the proposals
6	It is key to identify obsolete models. The decision models made by the model designer may not correspond to reality (changes in processes over time, misinformation of the model designer, …)
7	The PPC information system is not prepared to absorb and process the data available in the company thanks to 4.0 technologies
8	The S3 processes in the PPC field are an aspect that needs to be specified

11.3 Proposal to Improve the Planning and Scheduling Processes

The traditional PPCS presents a set of deficiencies, which has been condensed in the previous section. However, at present, there is a vast number of companies, in multiple sectors, managing its production following the PPCS proposal. A representation of the current decision-making systems under the PPCS can be seen in Fig. 11.1. Tactical decision-makers work in search of a planning whose fundamental results are the master production plan (MPP). The decision-makers create models with which they represent their planning policy (levels, inventory, availability dates, lot sizes, periods, horizons, etc.) and look for solutions with tools (optimal or sub-optimal), generating plans periodically. These models and their parameters change very little over time.

The information with which planners define the productive capacity is based on assumptions or calculations that may be unrepresentative of the reality of the plant, either because of the technical difficulty of carrying out the calculation or because of the absence of reliable data. In many cases, they are data based on experience and corrected on the basis of trial and error, which are fixed in the IT solutions remaining unchanged for a long period until a new revision is made. Once the tentative plans are obtained, they are analyzed to include non-modeled aspects (peculiarities of certain orders, etc.), correct proposals that do not conform to reality (unrealistic

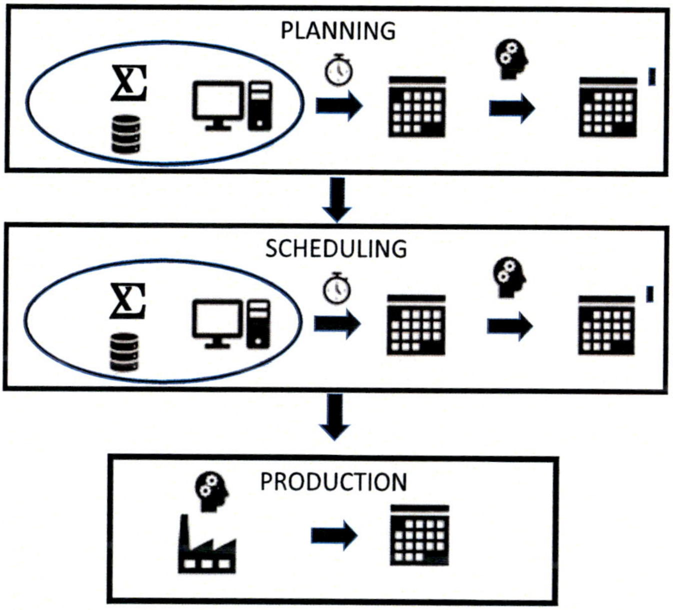

Fig. 11.1 Traditional plan and scheduling process

plant parameters) or simply improve the solution proposed by the algorithm. The final result is the final plan that is sent as input to the operational level.

Decision-makers of the operational level work in a similar way, they also have models to solve periodically where tentative production programs are obtained, and after an analysis, they will become definitive. These models consider the restrictions of the plant in much greater detail and as a consequence with much more complexity. In addition, they usually address a smaller time space (scheduling horizon). The operational level assumes as restrictions the needs of the tactical level and seeks for solutions oriented to sequencing and timing (assignment of orders to machines and/or workers, for example).

The engineering approach provided by the PPCS structure is still valid and provides the advantages indicated above, but we should try to cover the deficiencies presented in Table 11.1. We are in an industrial environment that is incorporating solutions based on the latest existing technology (Industry 4.0 proposals). These solutions deploy the world of Internet of Things (IoT) in the field of production and logistics, allowing companies to have data instantly on what happens in their production plants or in their transport.

A change of the traditional PPCS framework is proposed. Specifically, a system with greater integration and better use of information by defining a complementary layer between the tactical and operational level to obtain a smart production planning and scheduling. Figure 11.2 shows this proposal. This layer for the integration of the tactical-operative levels for improving the efficiency feeds on the plans and programs elaborated in the traditional PPCS. Also, the identification of the possible flexibility

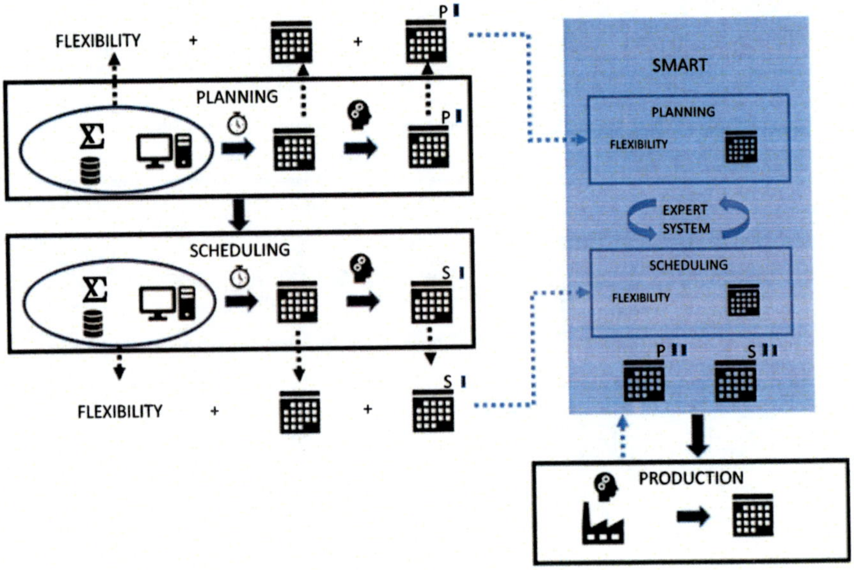

Fig. 11.2 Smart production planning and scheduling

in the system must be identified, as well as real data of the organization which has some kind of impact in plans and programs. These data may be being collected by the manufacturing execution system (MES), warehouse management software (SGA), enterprise resource planning (ERP) or the new systems that arise in the field of Industry 4.0.

The smart planning and scheduling proposed is based in the following hypothesis:

1. Current PPCS is a good starting point
   - The current system generates definitive plans and programs using a top-down approach in the decision-making. First, the plans have been generated and then the programs have been developed.
   - It is assumed that the plans and programs are the best possible for the current (mathematical) model and data used.
2. It is possible to identify real flexibility
   - Hard constraints (conditions that must always be respected) and soft restrictions (conditions that can be changed or have a margin) appear in the planning and scheduling models. The former may correspond to physical or contractual conditions of the plant, for example. The latter may be conditions related to internal policies that company want to use (on average).
   - Decision-makers known a set of alternatives associated with soft restrictions that can be used to adjust plans and programs according to the conditions of each moment. For example, reduce a lot, include more resources.
   - Real flexibility, in addition to constraints, may include the incorporation of new data or the modification of the initial objective.
3. It is possible to introduce flexibility in plans and programs
   - It is assumed that production plans and programs can be improved by considering alternatives thanks to the soft restrictions.

Thus, the smart planning and scheduling will take the set plan-program obtained by the (mathematical) model used in the PPCS; also, it will take the plan-program modified by decision-makers because the introduction of some improvements. Furthermore, the information about the flexibility and new information (in real time) about the situation of some business process will be used in the smart planning and scheduling.

An expert system will identify possible improvements and generate a set of alternative programs, some of which are feasible with respect to the received model and therefore valid, and others are feasible thanks to the flexibility included. Finally, the best alternative (program-plan) will be selected from those existing based on a selection protocol.

This proposal does not require a complex integration in the current information systems and is based on a basic data exchange, since the logic of the process is completely located in the expert system, so its implementation in companies would not be very expensive.

11.4 Conclusions

This paper proposes the adaptation of the instruments available in the tactic and operational scope of the PPC system so that they are able to take advantage of the real flexibility and current information in their environment to provide solutions that are more adjusted to the reality of each moment.

The PPCS and its associate information system are being used in organizations, and the introduction of the Industry 4.0 concept is providing an important range of proposals in this area. The proposal presented in this paper focuses on the validity of the results obtained by the PPCS system as a first approximation, the identification of the flexibility that can be contributed to plans and programs and finally the use of an expert system that offers alternatives improvement of initial plans and programs.

Acknowledgements This research has been funded by the Fondo Europeo de Desarrollo Regional (FEDER)/Ministerio de Ciencia e Innovación (MCI)—Agencia Estatal de Investigación (AEI) of Spain, in the framework of the project entitled "Integración de la Toma de Decisiones de los Niveles Táctico-Operativo para la Mejora de la Eficiencia del Sistema de Productivo en Entornos Industria 4.0 (NIOTOME)" (Ref. RTI2018-102020-B-I00).

References

1. Alarcón F, Alemany MM, Lario FC, Oltra RF (2011) La falta de homogeneidad del producto (FHP) en las empresas cerámicas y su impacto en la reasignación de inventario. Boletín de la Sociedad Española de Cerámica y Vidrio 50(1):49–58
2. Alemany ME (2003) Metodología y modelos para el diseño y operación de los sistemas de planificación jerárquica de la producción. Universitat Politècnica de València, Aplicación a una Empresa del Sector Cerámico, Tesis
3. Bitran GR, Tirupati D (1993) Hierarchical production planning. In: Graves SC et al (eds) Handbooks in OR&MS, vol 4. Elsevier Science Publishers B.V
4. Boza A, Cortes B, Alemany MDME, Vicens E (2015) Event monitoring software application for production planning systems. In: Enhancing Synergies in a Collaborative Environment. Springer, Cham, pp 123–130
5. Boza A, Ortiz A, Vicens E, Poler R (2009) A framework for a decision support system in a hierarchical extended enterprise decision context. In: IFIP-International workshop on enterprise interoperability. Springer, Berlin, Heidelberg, pp 113–124
6. Cegarra J, Van Wezel (2011) A comparison of task analysis methods for planning and scheduling. Behav Oper Plan Schedul, pp 323–338
7. Chavarría-Barrientos D, Camarinha-Matos LM, Molina A (2017) Achieving the sensing, smart and sustainable "everything". In: Working conference on virtual enterprises. Springer, Cham, pp 575–588
8. Cuenca L, Boza A, Ortiz A (2011) An enterprise engineering approach for the alignment of business and information technology strategy. Int J Comput Integr Manuf 24(11):974–992
9. Delgado J, Marín F (2000) Evolución en los sistemas de gestión empresarial. Del MRP al ERP. Economía Industrial 331(1):51–58
10. Dominguez-Machuca JA, Garcia S, Domínguez M, Ruiz A, Alvarez MJ (1995) Dirección de operaciones. McGraw-Hill, Aspectos tácticos y operativos en la producción
11. Doumeingts G, Pun L, Mondain M, Breuil D (1978) Decision-making systems for production control planning and scheduling. Int J Prod Res 16(2):137–152

12. Ellwein C, Elser A, Riedel O (2018) Production planning and control systems—breakage in connectivity make them fit to fit altering conditions. ACM international conference proceeding Series, pp 54–58. https://doi.org/10.1145/3232174.3232176
13. Gilchrist A (2016) Introducing industry 4.0. In: Industry 4.0. Apress, pp 195–215
14. Gomez-Gasquet P (2010) Programación de la producción en un taller de flujo híbrido sujeto a incertidumbre : arquitectura y algoritmos. Universidad Politécnica de Valencia Departamento de Organización de Empresa, Aplicación a La Industria Ceramica
15. Gomez-Gasquet P, Díaz-Madroñero M (2014) Algorithms for reactive production scheduling: an application in the ceramic industry. Boletín de la SE de Cerámica y Vidrio 53(4)
16. Hax A, Golovin JJ (1983) Sistemas jerárquicos de planificación de la producción, en: Hax A (ed) Dirección de Operaciones en la empresa. Hispano Europea, pp 513–546
17. Hax A, Meal HC (1973) Hierarchical integration of production planning and scheduling. en: Sloan Working Papers, ed. MIT, pp 656–73
18. MIET (2015) Industria conectada 4.0: La transformación digital de la industria española, Ministerio de Industria, Energía y Turismo, Madrid
19. MacCarthy B (2006) Organizational, systems and human issues in production planning, scheduling and control. In: Handbook of production scheduling. Springer, pp 59–90
20. Marques M, Agostinho C, Zacharewicz G, Poler R, Jardim-Goncalves R (2018) Responsive production in manufacturing: a modular architecture. In: Practical issues of intelligent innovations. Springer, Cham, pp 231–254
21. Moisescu MA, Sacala IS (2016) Towards the development of interoperable sensing systems for the future enterprise. J Intell Manuf 27(1):33–54
22. Obermayr N (2016) Challenge detailed planning in ERP. In: Piazolo F, Felderer M (eds) Multidimensional views on enterprise information systems. Cham, Springer International Publishing, pp 55–68
23. Perez D, Alarcón F, Boza A (2018) Industry 4.0: a classification scheme. In: Closing the gap between practice and research in industrial engineering. Springer, Cham, pp 343–350
24. Proud J (2007) Master scheduling : a practical guide to competitive manufacturing. Wiley
25. Rondeau PJ, Litteral LA (2001) Evolution of manufacturing planning and control systems: from reorder point to enterprise resource planning. Prod Invent Manag J 42(2):1–7
26. Rossit DA, Tohmé F, Frutos M (2018) Industry 4.0: smart scheduling. Int J Prod Res 1–12
27. Shrouf F, Ordieres J, Miragliotta G (Dec 2014) Smart factories in industry 4.0: a review of the concept and of energy management approached in production based on the internet of things paradigm. In: Industrial engineering and engineering management (IEEM), 2014 IEEE international conference on IEEE, pp 697–701
28. Tsubone H, Matsuura H, Kimura K (1995) Decision support system for production planning—concept and prototype. Decis Support Syst 13:207–217
29. Vargas A, Boza A, Patel S, Patel D, Cuenca L, Ortiz A (2016) Inter-enterprise architecture as a tool to empower decision-making in hierarchical collaborative production planning. Data Knowl Eng 105:5–22
30. Vollmann T, Berry W, Whybark D, Jacobs F (2005) Manufacturing planning and control systems for supply chain management. 5th edn, McGraw-Hill
31. Weichhart G, Molina A, Chen D, Whitman LE, Vernadat F (2016) Challenges and current developments for sensing, smart and sustainable enterprise systems. Comput Indus 79
32. Weichhart G, Stary C (2017) Interoperable process design in production systems. In: OTM confederated international conferences. On the move to meaningful internet systems. Springer, Cham, pp 26–35
33. Van Wezel W, van Donk DP, Gaalman G (2006) The planning flexibility bottleneck in. food processing industries. J Oper Manag 24:287–300
34. Zhong RY, Pang LY, Pan Y, Qu T, Huang GQ (2012) RAPShell for RFID-enabled real-time shopfloor production planning, scheduling and execution. CIE42 proceedings

Chapter 12
Production Scheduling of a Vegetable Packing Machine with Lack of Homogeneity in Raw Material

P. Gomez-Gasquet, P. I. Vidal-Carreras, and S. Liu

Abstract Natural products such as vegetables can be influenced in their cultivation by many variables, some controllable and others not, which affect the homogeneity of the resulting product. This lack of homogeneity must be considered in all processes of the supply chain. Another aspect to consider is the degradation over time of these products, interpreted as shelf-life for inventory policies. In this work, this problem is considered for the production scheduling of a vegetable packaging machine. Specifically, the machine packages broccoli and cauliflower, which are natural raw materials that suffer lack of homogeneity. In addition, the real case described in the paper suggests possible replacement of demand and different combinations of products. The full description of the problem and the framework in which it is located is assumed as one of the contributions of this paper. The other one is the successfully design of two mathematical models which represent the complexity of the problem faced.

Keywords Scheduling · Lack of homogeneity in raw material (LHRM) · Shelf-life · Packing machine · Agri-food sector · Sustainability

12.1 Introduction

Vegetables as others products such as fruits or beverages blood decay or get damaged in course of time as a natural process which has a substantial effect on the inventory

P. Gomez-Gasquet (✉) · P. I. Vidal-Carreras
Dpto. de Organización de Empresas, Universitat Politècnica de València, Camino de Vera S/N, 46022 Valencia, Spain
e-mail: pgomez@cigip.upv.es

P. I. Vidal-Carreras
e-mail: pivicar@omp.upv.es

S. Liu
Plymoth Business School (Faculty of Arts, Humanities and Business), University of Plymouth, 4 8AT Plymouth, England
e-mail: shaofeng.liu@plymouth.ac.uk

C. Avilés-Palacios and M. Gutierrez (eds.), *Ensuring Sustainability*, Lecture Notes in Management and Industrial Engineering,
https://doi.org/10.1007/978-3-030-95967-8_12

strategies [5]. These kinds of items have a shelf-life, that is the length of time that a commodity may be stored without becoming unfit for use, consumption, or sale. Vegetables, as being obtained directly from nature, also suffer of heterogeneity in their intrinsic characteristics. This heterogeneity is interpreted for vegetables production processes as a lack of homogeneity in raw material (LHRM). These two aspects, shelf-life and lack of homogeneity in raw material (LHRM), have to be considered in all the echelons of the supply chain [2]. This work is focused in the stage of scheduling in the agri-food sector, specifically for the casuistic of the broccoli and cauliflower in a vegetable packing machine. In this research line, a complete description of the problem is presented and two mathematical models (mixed integer linear programme—MILP) are developed.

12.2 Problem Description

The study is developed in a British company which is responsible for packaging vegetables. The vegetables considered at this work are broccoli and cauliflower. Before the packaging, the broccoli and the cauliflower have grown in an outdoor proximity field, and they were collected and delivered to the local company. The company includes, among others, the machines required for packaging broccoli and cauliflower. Scheduling the tasks of this vegetable packing machine is one of the aims of the work. The specific factors of this environment that affect the problem are detailed below.

Both, broccoli and cauliflower, suffer from a lack of homogeneity and uncertainty in the quality (size) and quantity of broccoli (raw material) that will be available. According to the literature [4], there are several sources of uncertainty in the agri-food supply chain. At this case, the harvest of the farmer can be affected mainly by inclement weather and pests. These factors can produce different quantity and quality of vegetables collected. Regarding the quality of the broccoli and cauliflower, the packaging machine is affected by its size. There are two different ways of packing the broccoli depending on this weight (size): whole with different grams or chopped into florets. At this point, it has to be remarked that there are other factors related to the quality of the broccoli and cauliflower, but they have been revised and filtered before reaching the entrance of the packaging machine.

The demand for broccoli can be considered substitutable down. Different customers demand different broccoli weights for their packages that can be ordered in a decreasing manner. The substitutability of the demand below implies that if, for example a client will demand a whole broccoli of 350 gram, it would have no problem in being served another integer of 360 gram that corresponds to the weight specification of another client. However, a customer requesting a whole broccoli of 400 gram would not accept one of lower weight. This demand substitutable is also for the cauliflowers. In addition, there is other packaging version called florets, which are chopped vegetables, in mixed trays that include broccoli and cauliflower of any size. For the demand of these mixed flower trays, it is possible to use broccoli or cauliflower of any size. So, requirements of the clients may overlap, such that units packaged for one customer may be used selectively to fill another customer's demand, as usually happens in semiconductor and electronics components sector and petro-chemical processing industries [6]. However, in order to maintain sustainability [1], the most adequate option is not downgrade the demand. To deal with these two different modes of packaging, the company has two different packaging machines, listed below the basic aspects of each type of work centre.

12.2.1 Broccoli/Cauliflower Work Centre

- Family of products associated with the broccoli/cauliflower overwrapped is elaborated.
- Products in a family are differentiated by its weight, e.g. A (400 gram), B (360 gram), or C (350 gram).
- Formed by two conveyor belts in which the broccoli and cauliflower are placed individually, and on each one, they carry out different operations (cleaning, bagging, weighing, and boxing).
- The system works as if they were two unrelated resources in parallel from the production scheduling point of view.
- Raw material presents uncertainty in terms of its weight, so it is not possible to know precisely how many broccoli/cauliflower must be used to obtain a final product of a certain quality.
- The product that does not adjust to the desired weight will remove the circuit and will be treated as non-compliant.
- The non-conforming product can be reused for another type of product in this same work centre (same family) or to make another product such as the mix of cauliflower and broccoli trays.

12.2.2 Broccoli and Cauliflower Mix Trays Work Centre

- Product is elaborated whose raw material is the broccoli and the cauliflower in florets.
- Both materials are packaged separately in the same plastic tray.
- Formed by a conveyor belt in which the broccoli and the cauliflower are placed individually in florets.
- Several operations are carried out (cleaning, packaging, etc.).
- The system works as if it was a single resource from the point of view of production scheduling.

As a consequence of the operations of both work centres, it can be said that the flow of materials is as follows. In the packaging work centre (Process 1.1 or Process

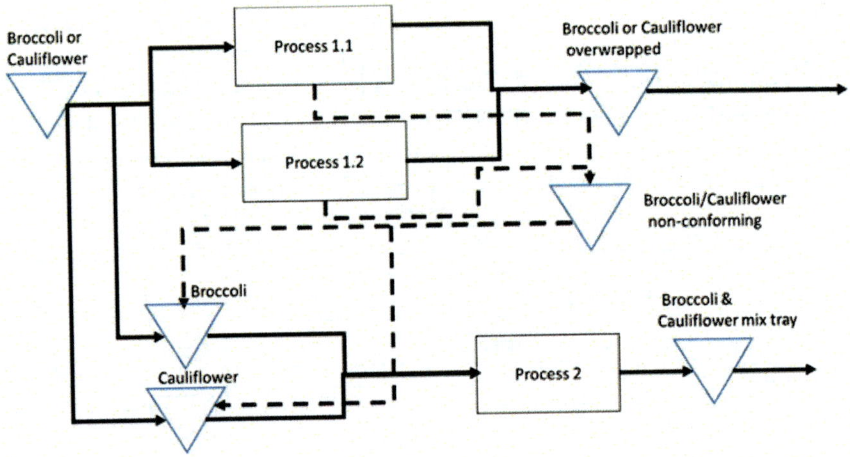

Fig. 12.1 Flow of broccoli and cauliflower in the packaging centre

1.2), the raw material, broccoli or cauliflower, is overwrapped if it meets the quality criteria. If the raw material does not meet quality criteria, it departs from the flow and is reconducted for Process 2. In the mix tray packaging work centre (Process 2), the raw material is prepared in florets. This raw material can be directly obtained from the field or from the product reconducted from Process 1.1 or 1.2 in the overwrapped centre. The representation of this flow can be seen in the following Fig. 12.1.

12.3 Scheduling Problem Modelling

For the sake of clarity, two models (mixed integer linear programme—MILP) are developed in order to decide the sequencing of jobs and the amount of raw material used in each product demanded to maximize the benefits. First model, consider all the issues except the lack of homogeneity on broccoli and cauliflower. In the second model, LHRM is included. As a result of the resolution of the models, the dates of completion of each job are also determined, and consequently, the start dates could be calculated in a simple way.

12.3.1 Model Hypothesis

The hypotheses for the two considered models are:

- It is assumed that there is an adjusted planning of the harvests and that both the cauliflower and the broccoli are available in a certain amount at the beginning of each day.
- A single period will be considered, e.g. an 8-h workday every day.
- Products could have different shelf-life determined by the farmer's delivery date and the time waiting in the warehouse.
- Each day that the product is company loses value which is reflected in an economic penalty.
- A milk-run distribution is considered. Due dates are not considered.

12.3.2 Notation

The information will be presented using the following indexes:

- i, l Index of the orders set N $\{1..n\}$
- j Index of the resources set M $\{1..m\}$
- t Index of the broccoli quality set Q $\{1..q\}$
- tt Index of the cauliflower quality set QQ {1..qq}

The parameters of the model are:

- Qb_t (integer) kg. of broccoli available at the beginning of the day for quality t.
- Qc_{tt} (integer) kg. of cauliflower available at the beginning of the day for quality tt.
- D_i (integer) kg. required for the product i.
- I_i (integer) expected income for the product i.
- L_i (integer) shelf-life available for the product i at the time of harvest (units of time).
- W_i (integer) time elapsed from the moment of harvest until the processing of the product i begins.
- PY_i (integer) cost of shelf-life economic penalty per time unit for the product i.
- BC_t (real) cost of raw broccoli of the quality t.
- CC_{tt} (real) cost of raw cauliflower of the quality tt.
- $BR_{t,i}$ (Boolean) 1 if quality of broccoli t is conform for the product i.
- $CR_{tt,i}$ (Boolean) 1 if quality t of cauliflower is conform for the product i.
- $WC_{j,i}$ (Boolean) 1 if work centre j can process product i.
- $P_{j,i}$ (integer) processing time of the product i in the work centre j.
- M (integer) large number.

The MILP models determine the value of the following variables:

- $xb_{t,i}$ (integer) broccoli (kg) of quality t assigned to the product i.
- $xc_{t,i}$ (integer) cauliflower (kg) of quality t assigned to the product i.
- $wc_{j,i}$ (Boolean) 1 if product i is processed in the work centre j, otherwise 0.
- $y_{i,l}$ (Boolean) 1 if product I is processed before product l.
- c_i (integer) completion time for the product i.
- u_i (integer) shelf-life available at completion time for the product *i*.
- q_i (integer) quantity of the product *i* produced.

12.3.3 MILP Model I

The objective is to maximize the benefit:

$$F.O.\max z = \left(\sum_{i=1}^{n}(q_i * I_i - (D_i - k_i) * \mathrm{PTY}_i - u_i * \mathrm{PY}_i - \left(\sum_{t=1}^{q}(\mathrm{xb}_{t,i} * \mathrm{BC}_t)\right) + \sum_{tt=1}^{qq}(\mathrm{xc}_{tt,i} * \mathrm{CC}_{tt})\right)\right\}$$

$$F.O.\max z = \left\{\sum_{i=1}^{n}\left(q_i * I_i - \left(\sum_{t=1}^{q}(\mathrm{xb}_{t,i} * \mathrm{BC}_t)\right) + \sum_{tt=1}^{qq}(\mathrm{xc}_{tt,i} * \mathrm{CC}_{tt})\right) - u_i * \mathrm{PY}_i\right\} \tag{12.1}$$

The constraints of the model are presented below in two sets (12.2–12.5 and 12.6–12.13), each representing a type of system restriction. The model is subject to:

$$\mathrm{Qb}_t \geq \sum_{i=1}^{n} \mathrm{xb}_{t,i} \quad \forall t \tag{12.2}$$

$$\mathrm{Qc}_{tt} \geq \sum_{i=1}^{n} \mathrm{xc}_{tt,i} \quad \forall tt \tag{12.3}$$

$$D_i \geq q_i \quad \forall i \tag{12.4}$$

$$q_i = \sum_{t=1}^{q}(\mathrm{xb}_{t,i} * \mathrm{BR}_{t,i}) + \sum_{tt=1}^{qq}(\mathrm{xc}_{tt,i} * \mathrm{CR}_{tt,i}) \quad \forall i \tag{12.5}$$

$$\sum_{j=1}^{m} \mathrm{wc}_{j,i} = 1 \quad \forall i \tag{12.6}$$

$$\sum_{j=1}^{m}\left(\mathrm{wc}_{j,i} * \mathrm{WC}_{j,i}\right) = 1 \quad \forall i \tag{12.7}$$

$$\mathrm{wc}_{j,i} + \sum_{r=1/r\neq j}^{m} \mathrm{wc}_{r,l} + y_{i,l} \leq 2 \quad \forall i \forall l \forall j \tag{12.8}$$

$$c_l - c_i + M * \left(3 - y_{i,l} - \mathrm{wc}_{j,i} - \mathrm{wc}_{j,l}\right) \geq P_{j,l} {*} \mathrm{q}_l \quad \forall i, \forall l, \forall j \tag{12.9}$$

$$c_i - c_l + M * \left(2 + y_{i,l} - \mathrm{wc}_{j,i} - \mathrm{wc}_{j,l}\right) \geq P_{j,i} * q_i \quad \forall i, \forall l, \forall j \tag{12.10}$$

$$c_i + M * \left(1 - \mathrm{wc}_{j,i}\right) \geq P_{j,i} * q_i \quad \forall i \forall j \tag{12.11}$$

$$(L_i - W_i - c_i) \leq u_i \tag{12.12}$$

$$u_i \geq 0 \tag{12.13}$$

Constraints 2 and 3 prevent more raw material from being consumed than is available. Constrain 4 does not allow more final product to be prepared than demanded, although it allows not to meet a demand. Constrain 5 relates the quantity of final product with the raw material used. This relationship prevents counting quantities of the final product that do not use the corresponding predetermined qualities.

From now on, the term job will be used instead of product. In any case, it must be considered that a product is associated with a manufacturing order, and the effects of this work are the same entity. Constraint 6 requires that all orders be assigned to a work centre. And, 7 restricts to 6, forcing it to be one of the work centres where the product can be processed. Restriction 8 forces the variable $y_{i,1}$ can only be 1 when jobs i and l are assigned to the same work centre. Constraints 9 and 10 force the completion times of the jobs to correspond to the order of the sequence. Constraint 9 is aimed at forcing the completion times when two jobs i and l are assigned to the same centre and job i is sequenced before l. With Constraint 9, the solver could assign value 0 to any element in the set, since it would satisfy the equation. To avoid this, Restriction 10 forces that when a zero is assigned to the variable $y_{i,1}$, it is satisfied that the completion time of l is at least that of i plus the processing time of l. Constraint 11, the completion times of the first jobs of each work centre are at least equal to the time of their processing time. Constraints 12 and 13 allow associating the variable with the shelf-life hours of the product once the waiting and process times have been subtracted. This value will never be negative.

Table 12.1 **a** Q set and **b** Triangular matrix QBM

(a)

Q	1 (A)	2 (B)	3 (C)

(B)

QBM	1 (A) (%)	2 (B) (%)	3 (C) (%)
1 (A)	95	4	1
2 (B)	0	97	3
3 (C)	0	0	100

12.3.4 A MILP Model II: LHRW

The lack of homogeneity in raw material (LHRW) is introduced. This is understood as the fact that the number of kg of broccoli or cauliflower that are in accordance with the type of product requested from the farmer is not known precisely. This second model, uncertainty will be introduced in a very simple way, as a first step. The way chosen to represent uncertainty is considering the average yield of the raw material. For this, some specific variables are used, such as:

- With sets Q and QQ, the types of possible qualities of broccoli and cauliflower are modelled.
- The quality of the raw material delivered from in a crop harvested as quality t can be composed by several percentages of different qualities.
- Triangular matrix QBM includes the percentage of expected raw material of t quality and all the lower ones for broccoli (QCM for cauliflower). The yield of raw material for a final product.
- A product conforms to its quality if it uses raw material of the same or higher quality, which was already reflected in the BR and CR matrices.

An example of Q set and of Triangular matrix QBM are shown in these tables (Tables 12.1a, b), and an example of BR matrix is in Table 12.2.

In this new model, Constraint 5 should be modified so that the calculation of qi reflects the performance according to the quality of the raw material used and the desired final product:

Table 12.2 BR matrix

BR	Order 1 (QA)	Order 1 (QB)	Order 1 (QC)	Order 1 (TX)
1 (A)	1	1	1	1
2 (B)	0	1	1	1
3 (C)	0	0	1	1

$$q_i = \sum_{t=1}^{q}\sum_{t'}^{q}\left(xb_{t,i}{}^*\mathrm{QBM}_{t,t'}{}^*\mathrm{BR}_{t',i}\right) + \sum_{tt=1}^{qq}\sum_{tt'}^{q}\left(xc_{tt,i}{}^*\mathrm{QCM}_{tt,tt'}{}^*\mathrm{CR}_{tt',i}\right) \quad \forall i \tag{12.5'}$$

The objective function should include the cost of storage of the raw material that is not used during the current period but could be used in the next. $\mathrm{CT_i}$ is the storage cost of the raw material needed for the product i.

$$F.O.\max z = \left\{\sum_{i=1}^{n}\left(q_i * I_i - \left(\sum_{t=1}^{q}(\mathrm{xb}_{t,i} * \mathrm{BC}_t)\right) + \sum_{\mathrm{tt}=1}^{\mathrm{qq}}(\mathrm{xc}_{\mathrm{tt},i} * \mathrm{CC}_{\mathrm{tt}})\right) - u_i * \mathrm{PY}_i - \left(\sum_{t=1}^{q}\mathrm{xb}_{t,i} + \sum_{\mathrm{tt}=1}^{\mathrm{qq}}\mathrm{xc}_{\mathrm{tt},i} - k_i\right) * \mathrm{CT}_i)\right\} \tag{12.1'}$$

12.4 Conclusions and Future Research

As underlined by [3] "production planning is considered one of the most important processes to efficiently balance supply and demand in terms of quantity and due dates". This work presents an illustration of this balance in the agri-food sector. It concerns scheduling of the packing machine of broccoli and cauliflower considering both efficiency and sustainability. This line of research will be pursued by the authors by focusing on proposing alternatives to solve this model efficiently in order to apply them to the factories involved.

Acknowledgements Authors of this publication acknowledge the contribution of the Project 691249, RUC-APS "Enhancing and implementing Knowledge based ICT solutions within high Risk and Uncertain Conditions for Agriculture Production Systems" (www.ruc-aps.eu), funded by the European Union under their funding scheme H2020-MSCA-RISE-2015.

References

1. Garetti M, Taisch M (2012) Sustainable manufacturing: trends and research challenges. Prod Plan Control 23(2–3):83–104. https://doi.org/10.1080/09537287.2011.591619
2. Grillo H, Alemany MME, Ortiz A, Fuertes-Miquel VS (2017) Mathematical modelling of the order-promising process for fruit supply chains considering the perishability and subtypes of products. Appl Math Model 49:255–278. https://doi.org/10.1016/j.apm.2017.04.037
3. Mundi I, Alemany Díaz MDM, Poler R, Fuertes-Miquel VS (2019) Review of mathematical models for production planning under uncertainty due to lack of homogeneity: proposal of a conceptual model. Int J Prod Res 57(15–16):5239–5283. https://doi.org/10.1080/00207543.2019.1566665

4. Ortiz A, Alarcón F, Pérez D, Alemany MME (2019) Identifying the main uncertainties in the agri-food supply chain. In: New global perspectives on industrial engineering and management. Springer, pp 221–229
5. Shah NH, Chaudhari U, Cárdenas-Barrón LE (2018) Integrating credit and replenishment policies for deteriorating items under quadratic demand in a three echelon supply chain. Int J Syst Sci: Oper Logist 2674. https://doi.org/10.1080/23302674.2018.1487606
6. Vidal-Carreras PI, Garcia-Sabater JP, Coronado-Hernandez JR (2012) Economic lot scheduling with deliberated and controlled coproduction. Eur J Oper Res 219(2):396–404. https://doi.org/10.1016/j.ejor.2011.12.020

Chapter 13
Purchasing Process Consequences After In-house Additive Manufacturing Adoption

J. Morcillo-Bellido, J. Martínez-Fernández, and J. Morcillo-García

Abstract Additive manufacturing (AM) is nowadays a major and very relevant manufacturing key tool in many manufacturing companies that seek to improve business competitiveness by adjusting processes to highly demanding customer requirements. This particular study seeks to deepen the understanding of additive manufacturing adoption impact into aerospace companies purchasing process. In the study, it had been analyzed the potential consequences that would be expected on the aerospace companies purchasing processes, after the adoption of AM considering the scenario in which AM is implemented in-house the aerospace companies. Expected process changes identified in the study are both an important organizational opportunity and also a serious threat for quick AM adoption at aerospace companies.

Keywords Additive manufacturing · Aerospace industry · Purchasing process · 3D printing

13.1 Introduction

One of the most significant drivers in manufacturing new edge is the emergence of new manufacturing technologies that could enable companies to manage cost efficiency and small-scale production, such as personalization [21, 36, 37] and additive manufacturing [18, 46]. Additive manufacturing (AM) started few years

J. Morcillo-Bellido (✉)
Área de Ingeniería de Organización, Universidad Carlos III de Madrid, Avenida de La Universidad nº 30, 28911 Leganés (Madrid), Spain
e-mail: morcillo@ing.uc3m.es

J. Martínez-Fernández
Aerospace Platform Systems and Equipment, Paseo John Lennon, s/n, 28906 Getafe, Madrid, Spain

J. Morcillo-García
Facultad de Ciencias Económicas Y Empresariales, Área de Organización de Empresas, Programa de Doctorado Economía Y Empresa UNED, Universidad Nacional de Educación a Distancia (UNED), Paseo Senda del Rey, 11, 28040 Madrid, Spain

C. Avilés-Palacios and M. Gutierrez (eds.), *Ensuring Sustainability*,
Lecture Notes in Management and Industrial Engineering,
https://doi.org/10.1007/978-3-030-95967-8_13

ago to become a manufacturing methodology that it was successfully applied in different industries and companies, and quickly, it got a relevant displacement of other traditional manufacturing methodologies [3].

However, for many companies, AM is not more than a different way of manufacturing goods in small scale, rather than a revolutionary way to handle specific manufacturing operations which can get some additional advantage from this technology benefits, to improve competitiveness and customer service level [11].

Early seventies, Ciraud launches the first initial technology that could be considered as an indicator of the current AM technology [6]. Few years later, Kodama and Herbert launched the first devices that worked with a technology called "3D printing", and it was known later in a more comprehensive way like "additive manufacturing" [34]. In 1986, Hull developed the "stereo-lithography machine" [23], and that equipment could be considered as the first one really capable to manufacture pieces in three dimensions.

Additive manufacturing most relevant current application is "rapid prototyping", usually considered as the more extended application. To manage prototyping manufacturing, process includes a 3D model-based design to run with a computer-aided design software and combined with a manufacturing 3D printing technology unit [39]. Prototypes are ease to operate and feasible to make in many trials at an affordable cost. Since one piece manufacturing is not a hard issue, "small batches production" is applicable at situations in which an isolated unit or a rather limited number of units should be manufactured.

When 3D manufacturing is applied, production could be efficient even in case of production decentralization. This means that with AM [22, 42] would be possible a simplest and shorter supply chain, so a number of geographically distributed printers could meet local demands reducing significantly transport costs and order—to—delivery lead times. Industries like the fashion sector [8, 17] guest that AM could be applicable in a near future by certain companies as a breakthrough alternative to improve their customer service via lead times reduction.

Although it is possible to identify some disadvantages versus traditional subtracting manufacturing processes, as the limited volume of products to be executed since AM is not prepared till now to make many units of the same piece, also, AM is limited to certain raw materials and colors [38].

Adoption of AM appears to herald a future in which supply chains could be shorter, smaller, more localized, more collaborative and offer certain sustainability advantages over traditional manufacturing process [20]. In the frame of AM future applications, it is possible that additive manufacturing application could be extended to numerous industries [26]. Nowadays, main applications are: (i) consumer electronics products, including fashion, toys and jewelry, (ii) automotive industry, mainly for prototyping, (iii) medical and dental solutions, more than half of the hearing aids and orthotics are currently produced using 3D printing and (iv) aerospace industry, which could reduce the buy-to-fly ratio due to the possibility of replacing heavy components by elements made of titanium and nickel [24, 35].

In addition to these examples, other applications are being developed in areas as food-related applications [16, 28, 43] and some new research on healthcare applications as the study which is searching on biological structures that could be created in digital form somewhere and delivered worldwide to be finished and personalized anywhere the right equipment for additive manufacturing would be available [40, 41].

Also, it could be identified some sustainability benefits [18] as: (i) extended product life: based on technical approaches such as repair, remanufacture and refurbishment, (ii) improved resource efficiency: Improvements can be done easily at any moment of product cycle life and (iii) reconfigured supply chains: Shorter and simpler supply chains are feasible via new AM and distribution models.

Regarding aerospace industry, the use of AM technology is increasingly widespread in the aerospace industry [32, 35]. Aerospace industry has been one of the pioneer sectors on AM adoption. There are several examples as Boeing that introduced by the end of 2015 about 20,000 original parts based on AM technology [13, 30]. Airbus [1] produces several metal parts made on titanium using AM for its most modern aircraft as A350 XWB. It is an objective of this study to go deep into the consequences of AM adoption on aerospace industry with the particular aim of getting a higher knowledge of particular impact on purchasing organization process.

13.2 Objective and Study Methodology

This study is trying to deep into the understanding of to what extent additive manufacturing development and application is affecting to the aerospace industry supply chain, in particular in the purchasing processes. Study could be particularly important for the aerospace industry since a major part of their components is outsourced to third parties. Global research plan has different phases and includes several different scenarios, combining scenarios quite different from in-house manufacturing in which focal company manages all production processes to those in which outsourced manufacturing is performed by third-party suppliers.

In this case, the research is focused on the consequences for the aerospace manufacturer purchasing process when company decides to implement AM within its own in-house manufacturing versus a previous situation in which the components were fully manufacture by a third-party supplier. The analysis will be focused on the issues and potential benefits for the aerospace company after taken this in-sourcing AM decision.

Authors try to identify potential threads and benefits for the aerospace manufacturer when decided to make in-house components previously outsourced to third parties, trying to get benefit of AM technology implementation. Academic studies have tried to identify some benefits of AM in supply chain management [5, 19, 29, 49]. Some studies identified certain benefits of AM technology within the aerospace supply chain industry as: (i) inventory reduction, (ii) higher reaction speed and (iii) lead times significant reduction [27, 33]. AM technology applications increasingly widespread in the aerospace industry [32].

This study tried to go further and aims to understand the current and future applications of AM technology as a factor that could strongly transform the aerospace industry supply chain by influencing its purchasing process. Authors based the research analysis on a deep and detailed review of relevant available published data regarding aerospace sector companies practice and authors professional own field experience in AM projects and purchasing process reengineering during many years.

13.3 Additive Manufacturing Influence in Aerospace Supply Chain

To analyze the supply chain [44], it would be necessary to consider those processes identified by the supply chain operations reference model or SCOR model [7]. Main supply chain processes were defined as: make, source, delivery and return [15, 31, 45]. Following figure (Fig. 13.1) shows a simplification of the typical global process in the aerospace supply chain. Figure 13.1 shows the most relevant stages, from the design to the recycling at the end of its useful life.

An aircraft is built in its final assembly line based on major structural building blocks delivered by Tier 1 suppliers. Those building blocks include hundreds of parts supplied by Tier 2 suppliers. Supply chain management and purchasing process have become key success factors for the aerospace industry [12]. As example of its relevance, an A380 aircraft incorporates more than 2.5 million different parts, and 70% are sourced from 1500 suppliers [49].

The main functions involved in an aerospace purchasing process are described at the following Table 13.1.

These six functional levels are not the only ones in the whole purchasing process, but they are the most relevant ones. Those functions could be found on every purchasing process of any aerospace company. In this paper, the analysis is focused on first four levels (A to D). Levels E and F are transversal support function to A to D and having much lower exposure to any manufacturing technology change.

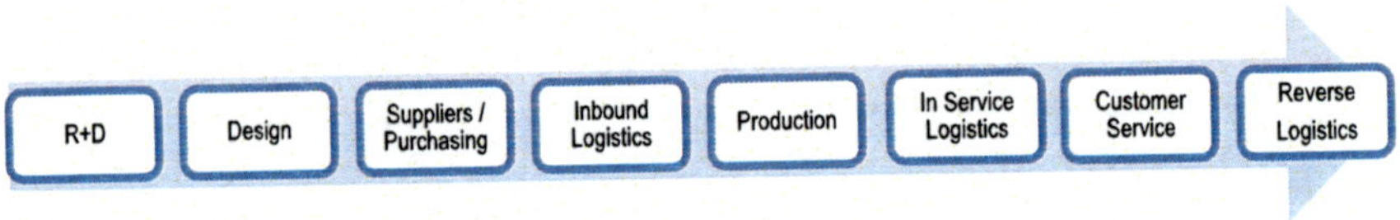

Fig. 13.1 Aerospace industry supply chain. *Source* authors own elaboration

Table 13.1 Main purchasing functions and responsibilities

LEVEL A—Strategic Purchasing. Responsible of leading the commercial negotiations, defining the purchasing strategy and maintaining the B2B high-level relationships
LEVEL B—Operational Purchasing. Accountable for supplier performance, deliveries follow-up, order release, inventory decisions and recovery plans when necessary
LEVEL C—Logistics Control. Responsible of logistics and warehousing, forwarding follow-up, warehouses management the incoming inspection
LEVEL D—Purchasing Quality. In charge of supplier quality performance, supplier's process control, audits and industrial capacity assessments
LEVEL E—Purchasing Coordination. Focus on the coordination between all purchasing functions. Level E implements processes and manages the prioritization of tasks
LEVEL F—Supplier Development. This function is transversal to the other purchasing functions and dedicated to develop strategic suppliers

Source authors elaboration

13.4 Discussion When AM is Applied In-House Aerospace Plant

If the aerospace manufacturer decides to produce a specific aircraft component in-house, they need specialized suppliers [2, 9, 10], which are companies with very specific technical capabilities. After the adoption of the AM by the aerospace manufacturer [4, 14], supplier manufacturing relation ends. Function of collecting base material and productive machinery will continue, but now managed within the aircraft manufacturer organization. Manufacturer´s purchasing process will take over all the tasks previously performed by the supplier, and this means a quite relevant organization change [25, 47, 48] to adapt from traditional final components purchasing to the purchasing of 3D printers and raw materials sourcing. The impact on each purchasing function of the aerospace manufacturer is analyzed in Table 13.2.

Level A purchasing function, before AM adoption by the manufacturer, it was responsible for tender launching to suppliers, negotiations and to improve constantly the value for money. After change over to AM, its job will be focused on negotiating 3D printers purchasing conditions, and the process not only undergoes a change in technical expertise, but also could cause a drastic reduction of its daily workload.

Level B controls the work in process and manages the inventory. After switching to AM, new vendors would no longer be as specialized as previous suppliers before AM. New vendors will be powdered metal or consumable yarn suppliers, depending on the type of printer. The number of references purchased would be significantly reduced, and inventory management would be simplified.

Level C, who deals with logistics control, reception, warranty confirmation and storage until the moment of internal consumption, would now manage fewer references, mainly consumable raw materials for printers. And finally, the Level D that is involved in a traditional manufacturing process as quality assurance control function when purchasing parts responsibility, it will move its tasks after the adoption of the

Table 13.2 Impact on aerospace company purchasing functions

Level A	Before AM	After AM	Level A_AM impact
- Negotiations strategy - Contracts man	- OEM continuous improvement	- 3D printers purchasing	- Tasks reduction - Cost reduction
Level B	Before AM	After AM	Level B_AM impact
- Procurement - Follow-up - Inventory man	- WIP control - Inventory and technology management	- Purchasing comp - Inventory management	- Lower WIP and less industrial expertise required
Level C	Before AM	After AM	Level C_AM impact
- Logistics - Inspections	- Logistic control - Warehousing	- Less logistics/space management	- Cost savings - Easer logistics management
Level D	Before AM	After AM	Level C_AM impact
- Process control (audits)	- OEM process control - Audits suppliers	- AM raw materials process control	- Cost control reduction - Standardization

Source authors own elaboration

in-house AM and will take responsibility only for the quality of the purchased raw materials. The process could be ever simplified by the reduction of references to be managed.

13.5 Conclusions

After the study that was performed, with the aim of going deep into the aerospace supply chain and particularly focused on its purchasing processes, it could be inferred that when aerospace manufacturers would implement in-house AM (from precious outsourced traditional manufacturing), the purchasing processes will be drastically modified and simplified. One of the clearest effects can be recognized as a quite shorter supply chain after in-house AM implementation, and this will lead to a relevant lead times reduction.

After analyzing the different functions included in the aerospace purchasing process of a typical company, several relevant changes are identified as it was described at previous analysis. Previous changes, by purchasing function, will mean some corresponding benefits such as lower inventories (consequence both from simpler inventory management and lower replenishment lead times due to the new technology). Another direct effect could be less space occupied at company warehouses.

Additional consequence that will be very relevant in term of cost involved will be linked to the amount of people needed to manage the "new purchasing role", a lower amount of human resources would be required. Initial estimations show that this would have a major impact in term both of cost reduction and new skills required on the future when aerospace company will be under an in-house AM environment. On top of that, an increase on the relevant suppliers' specialization (AM process design) based on their know-how, and a tendency to decouple between design and production activities, in such a competitive industry could also be foreseen.

According to the potential benefits, AM will probably be adopted in the aerospace industry gradually since the effect in term of people roll and skills is quite significant. It could be expected that initially main application will be focused on low mover parts, but in the long run, it will likely be most probably that this manufacturing technique even would be feasible and profitable in the mass production parts.

References

1. Airbus SAS (2014) Printing the future: Airbus expands its applications of the revolutionary additive layer manufacturing process. Retrieved from: http://www.airbus.com/presscentre/pressreleases/press-release-detail/detail/printing-the-future-airbus-expands-its-applications-of-the-revolutionary-additive-layer-manufacturig
2. Airbus (2020) Additive Manufacturing. De-carbonization meets digitalization. Retrieved from https://www.airbus.com/public-affairs/brussels/our-topics/innovation/3d-printing.html
3. Ashour Pour M, Zanoni S, Bacchetti A et al (2019) Additive manufacturing impacts on a two-level supply chain. Int J Syst Sci: Oper Logist 6(1):1–14
4. ASTM (2012). Committee F42 on Additive Manufacturing Technologies. Retrieved from https://www.astm.org/COMMITTEE/F42.htm
5. Attaran SA, Hassan A, Wahit MU (2017) Materials for food packaging applications based on bio-based polymer nanocomposites: a review. J Thermoplast Compos Mater 30(2):143–173
6. Beaman JJ, Leu MC, Rosen DA, Bourrell D & Borrel DL (2009). A brief History of additive manufacturing and the 2009 roadmap for additive manufacturing: looking back and looking ahead. RapidTech 2009: US_turkey workshop on rapid technologies
7. Beamon BM (1999) Measuring supply chain performance. Int J Oper Prod Manag 19(3):275–292. https://doi.org/10.1108/01443579910249714
8. Bingham GA, Hague RJ, Tuck CJ, Long AC, Crookston JJ, Sherburn MN (2007) Rapid manufactured textiles. Int J Comput Integr Manuf 20(1):96–105
9. Boeing (2017) 3D printing done right. Innovation Quarterly 17Q4. Retrieved from https://www.boeing.com/features/innovation-quarterly/nov2017/feature-thought-leadership-3d-printing.page
10. Boeing (2019) Innovation in manufacturing. Innovation quarterly 19Q4. Retrieved from https://www.boeing.com/features/innovation-quarterly/2019_q4/btj-additive-manufacturing.page
11. Bourell DL, Rosen DW, Leu MC (2014) The roadmap for additive manufacturing and its impact. 3D Print Additive Manuf 1(1):6–9
12. Braziotis C, Tannock JD, Bourlakis M (2017) Strategic and operational considerations for the extended enterprise: insights from the aerospace industry. Prod Plan Control 28(4):267–280
13. Catalano F (2015) Boeing files patent for 3D-printed aircraft parts—and yes. It's already using them, GEEKWIRE (Mar 6, 2015, 11:23 AM)
14. Council of Supply Chain Management (2020) Supply chain management definitions and glossary. https://cscmp.org/CSCMP/Educate/SCM_Definitions_and_Glossary_of_Terms.aspx
15. Delipinar GE, Kocaoglu B (2016) Using SCOR model to gain competitive advantage: a literature. Procedia Soc Behav Sci 229:398–406

16. Derossi A, Husain A, Caporizzi R, Severini C (2020) Manufacturing personalized food for people uniqueness: an overview from traditional to emerging technologies. Critic Rev Food Sci Nutr 60(7):1141–1159
17. Flynn A, Vencat EF (2012) Custom nation: why customization is the future of business and how to profit from it. BenBella Books
18. Ford S, Despeisse M (2016) Additive manufacturing and sustainability: an exploratory study of the advantages and challenges. J Clean Prod 137:1573–1587
19. Froes FH, Boyer R (eds) (2019) Additive manufacturing for the aerospace industry. Elsevier
20. Gebler M, Uiterkamp AJS, Visser C (2014) A global sustainability perspective on 3D printing technologies. Energy Policy 74:158–167
21. Guo N, Leu MC (2013) Additive manufacturing: technology, applications and research needs. Front Mech Eng 8(3):215–243
22. Handal R (2017) An implementation framework for additive manufacturing in supply chains. J Oper Supply Chain Manage 10(2):18–31
23. Hatch M (2013) The maker movement manifesto: rules for innovation in the new world of crafters, hackers, and tinkerers. McGraw Hill Professional
24. Holmström J, Partanen J, Tuomi J, Walter M (2010) Rapid manufacturing in the spare parts supply chain. J Manuf Technol Manage
25. Jackson B (2019) Airbus ramps up 3D printing. 3D Printig Industry. Retrieved from https://3dprintingindustry.com/news/airbus-ramps-up-3d-printing-with-liebherr-and-ultimaker-148901/
26. Janssen GR, Blankers IJ, Moolenburgh EA, Posthumus AL (2014) TNO: the impact of 3-D printing on supply chain management. The Hague, Netherlands, TNO
27. Khajavi SH, Partanen J, Holmström J (2014) Additive manufacturing in the spare parts supply chain. Comput Ind 65(1):50–63
28. Karantininis K, Sauer J, Furtan WH (2010) Innovation and integration in the agri-food industry. Food Policy 35(2):112–120
29. Kidd A, Sciacchitano J (2015) How 3-D printing could disrupt your supply chain
30. Kumar LJ, Nair CK (2017) Current trends of additive manufacturing in the aerospace industry. In: Advances in 3D printing and additive manufacturing technologies. Springer, Singapore, pp 39–54
31. Liu P, Huang SH, Mokasdar A, Zhou H, Hou L (2014) The impact of additive manufacturing in the aircraft spare parts supply chain: supply chain operation reference (scor) model based analysis. Prod Plan Control 25(13–14):1169–1181
32. Marchese G, Bassini E, Calandri M, Ambrosio EP, Calignano F, Lorusso M, Fino P et al (2016) Microstructural investigation of as-fabricated and heat-treated Inconel 625 and Inconel 718 fabricated by direct metal laser sintering: contribution of politecnico di torino and istituto italiano di tecnologia (iit) di torino. Met Powder Rep 71(4):273–278
33. Mashhadi AR, Esmaeilian B, Behdad S (2015) Impact of additive manufacturing adoption on future of supply chains. In: ASME 2015 international manufacturing science and engineering conference
34. Matias E, Rao B (2015) 3D printing: on its historical evolution and the implications for business. In: 2015 proceedings of PICMET'15: management of the technology age
35. Morcillo Bellido J, Martínez Fernández J, Morcillo García J (2019) Impact of additive manufacturing in aerospace industry purchasing process. In: 13th international conference on industrial engineering and industrial management/XXIII Congreso de Ingeniería de Organización. Julio 2019, Gijón (España). Retrieved from http://hdl.handle.net/10651/51506
36. Neely A (2008) Exploring the financial consequences of the servitization of manufacturing. Oper Manag Res 1(2):103–118
37. Novais LR, Maqueira JM, Bruque S (2019) Supply chain flexibility and mass personalization: a systematic literature review. J Bus Industrial Market
38. Oettmeier K, Hofmann E (2016) Impact of additive manufacturing technology adoption on supply chain management processes and components. J Manuf Technol Manage
39. Pham DT, Gault RS (1998) A comparison of rapid prototyping technologies. Int J Mach Tools Manuf 38(10):1257–1287

40. Pavlovich MJ, Hunsberger J, Atala A (2016) Biofabrication: a secret weapon to advance manufacturing, economies, and healthcare. Trends Biotechnol 34(9):679–680
41. Ramola M, Yadav V, Jain R (2019) On the adoption of additive manufacturing in healthcare: a literature review. J Manuf Technol Manage
42. Srinivasan V, Bassan J (2012) 3D printing and the future of manufacturing. In: CSC leading edge forum
43. Sun S, Wang Y, Liu J, Cai H, Wu P, Geng Q, Xu L (2016) Sustainability assessment of regional water resources under the DPSIR framework. J Hydrol 532:140–148
44. Uhlmann E, Kersting R, Klein TB, Cruz MF, Borille AV (2015) Additive manufacturing of titanium alloy for aircraft components. Procedia Cirp 35:55–60
45. Van Weele AJ (2009) Purchasing and supply chain management: analysis, strategy, planning and practice. Cengage Learning EMEA
46. Velázquez DRT, Simon AT, Helleno AL, Mastrapa LH (2020) Implications of additive manufacturing on supply chain and logistics. Independent J Manage Prod 11(4):1279–1302
47. Vialva I (2020) 3D printing industry review 2019. 3D Printing Industry. Retrieved from https://3dprintingindustry.com/news/3d-printing-industry-review-of-the-year-october-2019-166755/
48. Wright A (2017) The airbus roadmap to additive manufacturing.Engineering.com. Retrieved from https://www.engineering.com/AdvancedManufacturing/ArticleID/15886/The-Airbus-Roadmap-to-Additive-Manufacturing.aspx
49. Wallace J (2007) A380 ride and factoids. Blog Seattle Pi. http://blog.seattlepi.com/aerospace/2007/03/29/a380-ride-and-factoids/

Chapter 14
Artificial Intelligence for Solving Flowshop and Jobshop Scheduling Problems: A Literature Review

P. Gomez-Gasquet, A. Boza, A. Navarro, and D. Pérez-Perales

Abstract With recent advances in artificial intelligence (AI), it is time to take a review of learning process as an approach for production scheduling. Neural networks, reinforcement learning, multi-agent systems, etc., have been successfully applied to solve a variety of complex problems. However, although combinatorial problems are also complex, it is not evident that the application of AI techniques can help to solve them in a satisfactory way and specifically in the field of production scheduling. At this time, it is interesting to know if researchers propose AI applications to solve scheduling problems in a global way and these are more efficient than those used up to now, or on the contrary, the dominant research lines focus on some partial aspect of the resolution. This paper makes a review of the different contributions that the AI field has made in recent years on the problem of the flowshop and jobshop scheduling. The work aims to see which are the AI methods that have been used, which have greater presence and what possibilities they offer in future.

Keywords AI · Machine learning · Jobshop · Flowshop · Scheduling

---

P. Gomez-Gasquet (✉) · A. Boza
Centro de Investigación de Gestión e Ingeniería de Producción (CIGIP), Universitat Politècnica de València. Cno/ de Vera S/N, 46022 Valencia, Spain
e-mail: pgomez@cigip.upv.es

A. Boza
e-mail: aboza@omp.upv.es

A. Navarro · D. Pérez-Perales
Departamento de Organización de Empresas, Universitat Politècnica de València. Cno/ de Vera S/N, 46022 Valencia, Spain
e-mail: alnaal@etsii.upv.es

D. Pérez-Perales
e-mail: dapepe@omp.upv.es

C. Avilés-Palacios and M. Gutierrez (eds.), *Ensuring Sustainability*,
Lecture Notes in Management and Industrial Engineering,
https://doi.org/10.1007/978-3-030-95967-8_14

14.1 Introduction

The main objective of this article is to review the methods and algorithms applied to the resolution of a set of production scheduling problems whose approach has modified its traditional approach to incorporate contributions from the discipline that studies AI.

AI is a discipline that could be divided into a wide number of fields, which in many cases are interconnected and share a wide spectrum of knowledge. One of these fields is machine learning, which provides methods to improve efficiency by automating learning processes. Other fields, linked to the previous one that have come to have their own space, are those related to multi-agent systems, rule-based systems or case-based systems. And also, we can mention lines of research within AI based on the improvement and expansion of the applications of certain transversal methods such as neural networks, fuzzy logic or Bayesian networks, to give a few examples. Within these lines, it is necessary to highlight the one that has to do with the development of algorithms based on artificial life or evolutionary computation (genetic, ant, bee algorithms, etc.) since it provides an important set of techniques that converge with others from classic operation research. Throughout this paper, it will be revealed how the field of machine learning is the one that arouses the most interest among researchers focused on production scheduling. A classic classification of machine learning establishes three main areas of action: supervised or unsupervised learning and reinforcement learning [17].

Scheduling problems have been widely studied in the past decade, and solutions based on bioinspired algorithms which maintains a strong relationship with artificial life of the AI have been very successful. However, in this paper this area will not be addressed, as it is well known, and the contribution of machine learning techniques and methods will be sought. In this sense, the use of transversal AI techniques that are not used from the ML approach is out of focus. For example, Sabuncuoglu [8] provides a set of applications to the flowshop and jobshop workshops with neuronal network but without the indicated approach.

The set of scheduling problems that has been considered is wide since it includes any variation that addresses a configuration in jobshop or flowshop. What we try is to analyse how AI has modified the structure of the resolution method, either by providing a way to perform pre- or post-processing of the solution of any of the algorithms commonly used in this type of problems, or either by changing the approach to any of its key functions, or by simply using a typical AI technique instead of those algorithms. In the following section, the methodology that has been applied is commented; finally, the analysis of the results is addressed, in Sect. 14.3 and conclusions in Sect. 14.4.

14.2 Methodology

Once the scope of study is established, the methodology proposed is as follows: (1) select the sources of information; (2) search contributions; (3) debug the results; (4) analyse the results.

According to the methodology, the first step has been to choose Scopus as a source of information, since within the scientific field it has a better balance between the quality of the contributions and the breadth of search. There are other alternatives, such as Google Scholar, with a greater number of references but including the ones of Scopus and the differential is usually achieved with marginal contributions.

The second step is to properly choose the keywords for the search. The words "artificial intelligence", "machine learning" and "reinforcement learning" were used, on the one hand, in combination with "jobshop" and "flowshop", and on the other hand, in combination with "production scheduling". As a consequence of the debugging process, in step three of the methodology, a set of filtering layer has been used. First, we filtered depending on the year because we discovered that most papers were published in the last ten years as it is shown in Fig. 14.1. Second, we got rid of papers that were not related to research.

Third, we took into account the title and the abstract of the paper to decide whether such paper talks about the topic we were concerned with. Fourth, we found out that, because of the huge paper database that we were managing, there were some papers which were duplicated; consequently, we removed them. Finally, a set of 21 papers which were directly related to the scope were found.

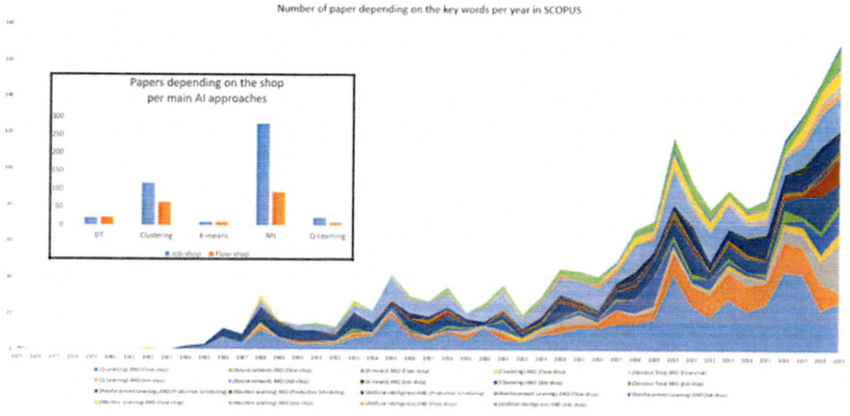

Fig. 14.1 Quantitative overview of papers found according to keywords

14.3 Results and Analysis

As a result of searching keywords in Scopus, we came up with an important number of papers which leads us to Fig. 14.1.

In Fig. 14.1, we can see two charts. The big one shows the evolution of the publication of papers in Scopus in a timeline taking into account the keywords that we looked for first. We clearly see that the publication is increasing and that in the last ten years it has dramatically gone up. Besides, the small chart represents the number of papers published in Scopus per shop within the main artificial intelligence approaches. We can see that the jobshop papers are greater in number to the flowshop ones and that neural network approaches are the most common ones in order to deal with both shops. The picture shows the tendency of the publications of papers towards this topic which is interesting to perceive that it presents a growing tendency.

Now, we were able to filter and keep those papers which are related to our field of study; but, also thanks to the findings we were also able to figure out what is currently the tendency within this topic (Table 14.1).

The set of selected paper have been classified as contributions in the context of supervised, unsupervised [14] or reinforcement learning [17]. The first two have been widely validated and used in other fields, and the reinforcement learning is a fashionable issue to address complex problems in recent years.

14.3.1 AI Contribution to the Flowshop Scheduling Problem

A flowshop scheduling problem consists in a set of jobs $J = \{1, 2, \ldots i, \ldots n\}$, and a set of operations $P = \{1, 2,..o,..P\}$ have to be carried out with the same route for each job. In this way, a set of stages $S = \{1, 2,..s,..S\}$ is considered such that Po is carried out in So and $P = S$. A sequence for each stage must be found in order to minimize any objective function. Any type of additional restriction (set-ups, buffer restrictions, multiple machines in each stage, etc.) will be welcome in this work since no additional hypotheses are established.

The few proposals that have been found in the field of supervised learning are aimed at predicting the behaviour of some element of the problem. The proposal

Table 14.1 Results of selected bibliography after debugging

	Supervised	Unsupervised	Reinforcement learning		
	No specific	Clustering	Q-Learning		
Flowshop	[13], [5], [7] (Bartosz Sadel 2016)	[22]	[2, 3, 6]		
		NN	Q-Learning	NN	DQN
Jobshop		[21]	[4, 12, 18, 19, 23],	[1, 9, 11, 16, 20]	[24][10]

of Pavelski et al. [13] tries to identify the metaheuristic (Hill Climbing, Simulated Annealing, Tabu Search, ILS) that lower makespan or flowtime get given an instance in a flowshop problem, which may be permutation, no-wait, no- idle. Although the proposed approach has presented low accuracy in some cases, the resulting models showed interesting relations between the problem features and metaheuristics characteristics. In the case addressed by Hao et al. [7], they are oriented to use the learning capacity to identify the probabilities that help to locate a job in a position of the sequence of one of the phases of the estimation distribution algorithm (EDA) in a dynamic permutation flowshop problem with the objective of minimizing the total tardiness. For Sadel and Sniezynski [15], the focus is on identifying the most suitable machine to execute an operation in a hybrid flowshop that considers a job arrival following a Poisson distribution with the aim of minimizing the idle time of the machines. In the contribution of [13], a learning methodology is proposed whose central element is the identification of statistically significant differences using the Kruskal–Wallis test. However, Sadel and Sniezynski [15] perform a classification process through a software agent architecture, where in addition to supervised learning, reinforced learning is used. The previous proposals focus their efforts on improving sequencing; however, [5] presents a contribution whose objective is not to improve the sequence but to predict the lead time of the jobs. In a mass customization environment, they use the data from a MES to predict lead time using several predictive learning algorithms (linear regression, tree models, support vector regression).

In the field of unsupervised learning, Wang and Tang [22] propose to address the problem of the classic permutation flowshop with the variant of being multi-objective using AI in the exploitation phase of the local search process. The contribution of learning is focused on a clustering method that efficiently groups the non-dominated solutions, improving the local search phase: "There are two main features in the proposed ML-MOMA. First, each solution is assigned with an individual archive to store the non-dominated solutions found by it and based on these individual archives a new population update method is presented. Second, an adaptive multi-objective local search is developed, in which the analysis of historical data accumulated during the search process is used to adaptively determine which non-dominated solutions should be selected for local search and how the local search should be applied" [22].

In the reinforcement learning field, all the papers found apply the Q-learning algorithm with the aim of minimizing makespan. In a first approximation, Fonseca-Reyna et al. [3] addressed the problem of permutation flowshop to later [2] apply what has been learned to the hybrid flowshop with unrelated machines, setup times depending on the sequence and machines not eligible for some jobs. In both cases, the algorithm is executed at each stage of the workshop, which is represented by an agent. The state is the representation of the sequence that has been constructed in each agent until a given instant. The action consists in deciding which job in the queue of the corresponding stage is chosen to be operated. The reward is the inverse value of the makespan of the sequence generated by the agent. In a non-deterministic hybrid workshop [6], redefine the problem in terms of a Markov process where the states are defined by the tuple (stage, job), the decisions are the machines where

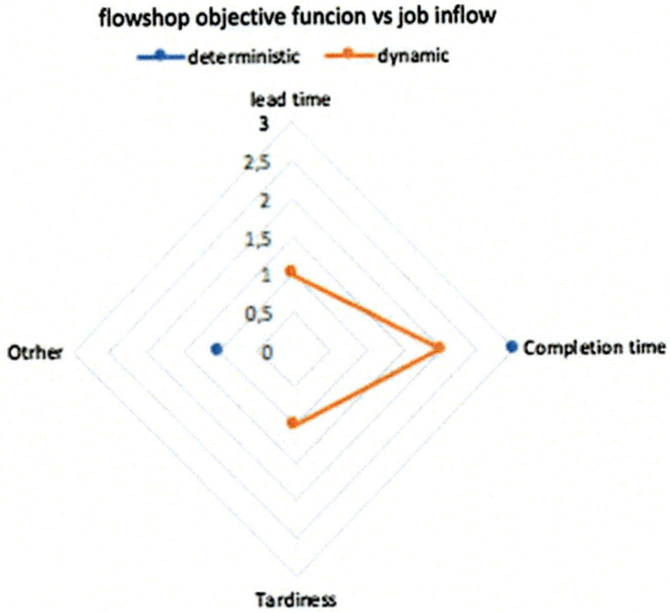

Fig. 14.2 Review of the objective function vs job inflow found in jobshop papers

it operates in the next stage, and the reward is a function inversely proportional to the waiting time of job in the next stage. The algorithm applies a greedy variable criterion that facilitates exploration at the beginning and reduces it as more has been learned. Two methods are used to adjust the greedy coefficient, one of them based on the concept of Boltzmann temperature (simulated annealing).

In addition, we have also analysed the objective function used and how the job arrival is considered in each paper that is shown in Fig. 14.2. As can be seen, the number of papers that consider deterministic and dynamic flowshop is balanced; however, the predominant objective function is the makespan.

14.3.2 AI Contribution to the Jobshop Scheduling Problem

A jobshop scheduling problem consists in a set of jobs $J = \{1, 2, \ldots i, \ldots n\}$, and a set of operations $P = \{1, 2,..o,..P\}$ have not to be carried out with the same route for each job. In this way, a set of stages $S = \{1, 2,..s,..S\}$ is considered such that Po is carried out in So and $P = S$. A sequence for each stage must be found in order to minimize any objective function.

In the jobshop context, no references of interest that use supervised learning have been found. The only proposal found in the field of unsupervised learning [21] is aimed at predicting makespan without the need to calculate it. To be able to specify the date, the system has the data generated by the real-time RFID system that feeds a deep belief network (DBN).

The rest of the contributions are concentrated in the field of reinforcement learning, and all of them apply a Q-learning method, a neural network or a combination of both. The authors who apply Q-learning mainly focus on static workshops, except [18, 19]. Thus, [12, 23] propose a multi-agent system that implements a Q-learning. In a more complete proposal, [23] defines that the states are determined by 4 characteristics of the scheduling and the workshop and are previously reduced to a K cluster in a deterministic shop for minimizing earliness and tardiness penalty. An adaptive scheduling is proposed, which consists in the strategy being modified according to the conditions of each moment, for which they propose a multi-agent system (agent work, agent machine, agent buffer, agent state). On this system, an adaptation of the Q-learning algorithm is proposed in which the status is determined by 4 characteristics of the programme and the workshop (average of penalties for early/late, average of the delivery factor, load and utilization ratio of the machines) and is previously reduced to a cluster K. The actions are the heuristic rules that can be used (EDD, SPT, FIFO, etc.). Reward is the negative value of the sum of the early and tardy or 1 otherwise. The contribution of [4] is very similar to that commented in the case of the flowshop. In a dynamic context, [18] decides so on the configuration of the batches and their sequence. In the first place, the problem focuses on when a batch of jobs is closed to be operated on a machine, considering that it is uncertain when each job arrives, and therefore, it is not known when the desirable size will be reached. Second, in what order the jobs are run. To do this, they propose to use a neural network based on the Q-learning algorithm in which the state is defined by 4 variables in each machine (the number of jobs being travelling to, being waiting in front of, being processed on machine and the estimated remaining time that machine m becomes idle), and the actions are tied to the chosen batch on a machine, including the option to not process any. And the reward is made up of two parts, one related to the chosen action associated with the waiting time and batch size, and the other to the cost of maintaining a job in the system. Meanwhile, [19] address a system with 2 agents dealing with a dynamic hybrid jobshop problem with the goal of minimizing makespan. A system with 2 agents is proposed. The first is responsible for creating sequences and acts when an event occurs that involves the location of a job. The solution proposal is based on a Q-learning method. In the proposal, the state is comprised of a vector with 3 parts: existence of an idle resource, existence of a job in queue and ratio between current and desired performance. Actions consist of assigning a job to a resource pool. The reward is a high and positive number if the end is reached with a throughput greater than the threshold and negative if this value is not reached; for intermediate states, it will be a much lower positive value and proportional to the throughput. The second agent uses a learning method in the form of a classifier that allows identifying the machines that are causing bottlenecks.

In the context of the authors who apply the neuronal network, a balance between dynamic and static problems is perceived. The contributions of [1, 11] address dynamic problems with minimization of makespan, being a flexible jobshop in the last case. In the proposal of Mao et al. [11] is addressed a computing system that resembles a dynamic jobshop whose objective is to minimize the average makespan. In this area, it is important to have systems that adapt to the variability of the load. For this, they propose a scalable reinforced learning system and various training methods. The use of a Neuronal Network whose the most novel aspect in the consideration of a set of job characteristics by means of an embedded graph that allows more or less attributes to be included without modifying the design of the network. On the other hand, [1] consider dependence on variable job and time. It also has the characteristics that jobs tend to be repetitive and have a similar duration. In the static context, [20] address a bi-objective jobshop and propose a reinforced neural network for each objective with a single layer of NxM neurons. The results are adjusted using a couple of complementary algorithms to avoid not complying with the precedence restrictions. The results are analysed by an experimental procedure. Setiawan [16] orient the problem as a critical path problem with the objective of minimizing makespan. The proposed solution consists of a reinforced learning based on temporary difference learning (TDL) based on a gradient descent method that is generated through a neural network. And [9] propose a reinforced learning method, which is implied in an agent. Due to the large size of the matrix (state, action), it is replaced by an estimator formed by fully connected neural network that uses policy gradient to train its learnable parameters.

Finally, there is a set of contributions that apply Q-learning using a neural network to calculate the matrix Q, known as DQN. In [24], a dynamic hybrid jobshop is addressed in order to reduce delays using deep reinforcement learning (DRL) in two phases. In phase A, it is applied individually in each work centre, and in phase B, it is done with the objective of balancing the workshop as a whole. In its training phase, the system compares its decisions with those of an expert system, supervised learning, and receives rewards based on the degree of affinity of the response. In proposal of Lin et al. [10] is desired to minimize the makespan. The authors propose the resolution of a jobshop with the aim of minimizing the makespan. The authors follow a strategy that assigns a dispatch rule (SPT, LPT, FIFO, etc.) to each station, and they may be different. There is centralized information in the cloud that allows information to be received and distributed from the entire plant. The authors use a multiclass DQL (MDQL) to assign a rule to each station. The MDQL is formed by a forward propagation neural network with 3 layers. In the input layer, characteristics of the orders and the system are considered. When the NN provides the rules, these are used to generate a programme that is then adjusted, the result is used to calculate the values of the matrix Q, and later the characteristics give input to the NN. Finally, an experimental analysis is carried out with standard data.

In addition, we have also analysed the objective function used and how the job arrival is considered in each paper that is shown in Fig. 14.3.

In the case of jobshop, there is a greater tendency to study deterministic cases than dynamic ones, and the predominance of makespan as an objective function is greater than in the case of flowshop.

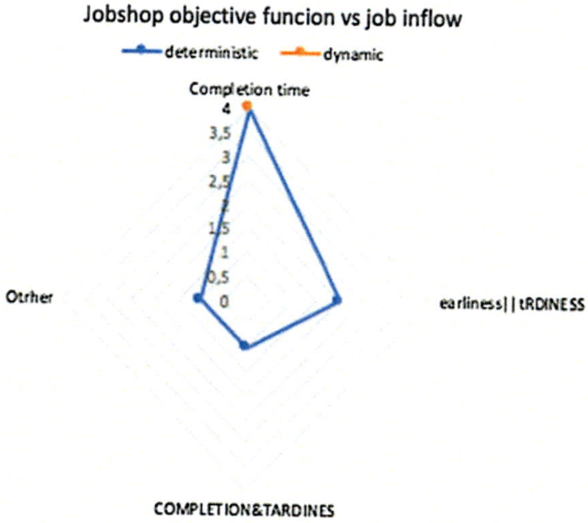

Fig. 14.3 Review of the objective function versus job inflow found in jobshop papers

14.4 Conclusions

This paper has made a review through the best contributions that have been found on a set of more than 1000 references. Most of them were published in the last 2–3 years. In general, it has been observed how the techniques that have been classified as reinforcement learning used for the generation of scheduling or selection of heuristic rules are the ones that present the most applications for the reviewed workshops, and as the rest of the techniques are in many cases used as an aid or complements to scheduling. In addition, it should be noted that the applications tend to focus on the most complex problems that are usually encountered in this field.

Because of all this analysis process, we can conclude that most of the research towards this topic is currently growing in the field of RL. It looks like these algorithms are able to fit the scheduling problem better than others, and consequently, research on this field is supposed to go up in the following years.

Finally, this paper is part of a research project within the application of AI techniques to production scheduling and it will lead us to develop our own RL algorithm in order to solve a flowshop scheduling problem. We have currently developed a Q-learning and SARSA algorithm without having very good results; but in future, we will work in a deep Q network algorithm to improve the accuracy of the scheduler.

Acknowledgements This research is part of the I+D+i project 'Integración de la Toma de Decisiones de los Niveles Táctico-Operativo para la Mejora de la Eficiencia del Sistema de Productivo en Entornos Industria 4.0 (NIOTOME)' (Ref. RTI2018-102020-B-I00) funded by MCIN/AEI/10.13039/501100011033/ ERDF A way of making Europe.

References

1. Cheong M, Lee H, Yeom L, Woo H (2019) SCARL: attentive reinforcement learning-based scheduling in a multi-resource heterogeneous cluster. IEEE Access 7:153432–153444. https://doi.org/10.1109/ACCESS.2019.2948150
2. Fonseca-Reyna YC, Martinez Y, Dervecia-Cabrera A, Rodriguez EA (2019) Optimization of heavily constrained hybrid-flexible flowshop problems using a multi-agent reinforcement learning approach. Investigacion Oper 40(1):100–111
3. Fonseca-Reyna YC, Martínez-Jiménez Y, Nowé A (2017) Q-learning algorithm performance for m-machine, n-jobs flow shop scheduling problems to minimize makespan. Investigacion Oper 38(3):281–290
4. Fonseca Reyna YC, Martinez Y, Bermudez JM, Mendez B (2015) A reinforcement learning approach for scheduling problems. Investigacion Oper 36(3):225–231
5. Gyulai D, Pfeiffer A, Nick G, Gallina V, Sihn W, Monostori L (2018) Lead time prediction in a flowshop environment with analytical and machine learning approaches. IFAC-PapersOnLine 51(11):1029–1034. https://doi.org/10.1016/j.ifacol.2018.08.472
6. Han W, Guo F (2019) A reinforcement learning method for a hybrid flow-shop scheduling problem. Algorithms 12(11):222. https://doi.org/10.3390/a12110222
7. Hao L, Tieke L, Bailin W, Zhiwei L (2017) Estimation of distribution algorithm with machine learning for permutation flowshop scheduling with total tardiness criterion. Proc 10th Int Symposium Comput Intell Des ISCID 2017(2):283–286. https://doi.org/10.1109/ISCID.2017.193
8. Ihsan, Sabuncuoglu (1998) Scheduling with neural networks: A review of the literature and new research directtions. Production Planning & Control 9(1) 2–12 10 (2010). https://doi.org/10.1080/095372898234460
9. Li F, Hu B (2019) DeepJS: Job scheduling based on deep reinforcement learning in cloud data center. In: ACM international conference proceeding series, pp 48–53. https://doi.org/10.1145/3335484.3335513
10. Lin C, Deng D, Chilh Y, Chiu H (2019) Smart Manufacturing scheduling with edge computing using multiclass deep q network. IEEE Trans Indus Inf 15(7):4276–4284. https://doi.org/10.1109/TII.2019.2908210
11. Mao H, Schwarzkopf M, Venkatakrishnan B, Meng Z, Alizadeh M (2019) Learning scheduling algorithms for data processing clusters. SIGCOMM. 10(1145/3341302):3342080
12. Palacio JC, Jiménez YM, Nowé A (2019) Multi-agent reinforcement learning tool for job shop scheduling problems. In: CEUR workshop proceedings, vol 2491, pp 1–3
13. Pavelski L, Delgado M, Kessaci ME (2018) Meta-learning for optimization : a case study on the flowshop problem using decision trees. In: IEEE international conference auto science Engineering. https://doi.org/10.1109/CEC.2018.8477664
14. Russell SJ, Norvig P (2010) Artificial intelligence—a modern approach. Third international edition. Pearson Education
15. Sadel B, Snieezynski B (2016) Online supervised learning approach for machine scheduling. Theoretical Found Mach Learn Kraków 25(December):165–176. https://doi.org/10.4467/20838476SI.16.013.6194
16. Setiawan K (2012) Progress in Business Innovation and technology management reinforcement learning combined with radial basis function neural network to solve jobshop scheduling problem. APBITM Society 002:31–38
17. Sutton RS, Barto AG (1998) Reinforcement learning: an introduction. MIT Press.
18. Tao OZ, Shufang X, Rose O (2018) Real-time batching in jobshops based on simulation and reinforcement learning. In: 2018 winter simulation conference (WSC). IEEE, pp 3331–3339
19. Thomas TE, Koo J, Chaterji S, Bagchi S (2018) MINERVA: a reinforcement learning-based technique for optimal scheduling and bottleneck detection in distributed factory operations1. In: IEEE international conference on automation science and engineering, pp 129–136. https://doi.org/10.1109/COMSNETS.2018.8328189

20. Tselios D, Savvas I, Kechadi MT (2013) RNN modelling for bi-objective MPM Job shop scheduling problem. IEEE Comput Intell Magaz 13–18. https://doi.org/10.1109/CICSYN.2013.38
21. Wang C, Jiang P (2019) Deep neural networks based order completion time prediction by using real-time job shop RFID data. J Intell Manuf Springer, US 30(3):1303–1318. https://doi.org/10.1007/s10845-017-1325-3
22. Wang X (2015) Tang L (2017) A machine-learning based memetic algorithm for the multi-objective permutation flowshop scheduling problem. Comput Oper Res Elsevier 79:60–77. https://doi.org/10.1016/j.cor.2016.10.003
23. Wang Y (2018) Adaptive job shop scheduling strategy based on weighted Q-learning algorithm. J Intell Manuf Springer US (1997). https://doi.org/10.1007/s10845-018-1454
24. Waschneck B, Reichstaller A, Belzner L, Altermüller T, Bauernhansl T, Knapp A, Kyek A (2018) Optimization of of global production scheduling deep reinforcement reinforcement learning. Elsevier. https://doi.org/10.1016/j.procir.2018.03.212

Part IV
Operations Research

Chapter 15
Optimal Telecommunications Network Expansion Using Mixed Linear Integer Programming

E. Fernández-Bravo, Á. García-Sánchez, M. Ortega-Mier, and T. Borreguero

Abstract Telecommunications networks have to be upgraded and redesigned to meet changing demand as new technologies are developed. This paper presents an optimization model (mixed integer linear programming) for deciding how to expand an existing network with the minimum cost, where relevant costs are both capital and operational expenditures. The efficiency of the model has been evaluated solving randomly generated instances for different sizes.

Keywords Location · Mixed integer programming · Telecommunications

15.1 Introduction

Deploying a new communication technology, such as 5G, requires installing the corresponding capacity to be able to meet the demand as this increases. Actually, demand evolves as new devices are ready for that technology.

This paper presents a mixed integer linear programming model for addressing this problem.

E. Fernández-Bravo · Á. García-Sánchez (✉)
Baobab Soluciones, SL. Calle José Abascal 55, Entreplanta Izquierda, Madrid, Spain
e-mail: alvaro.garcia@upm.es; alvaro.garcia@baobabsoluciones.es

E. Fernández-Bravo
e-mail: esther.fernandez@baobabsoluciones.es

Á. García-Sánchez · M. Ortega-Mier · T. Borreguero
Departamento de Ingeniería de Organización, Administración de Empresas y Estadística, Universidad Politécnica de Madrid, Madrid, Spain
e-mail: miguel.ortega.mier@upm.es

T. Borreguero
e-mail: tamara.borreguero@upm.es

C. Avilés-Palacios and M. Gutierrez (eds.), *Ensuring Sustainability*,
Lecture Notes in Management and Industrial Engineering,
https://doi.org/10.1007/978-3-030-95967-8_15

More specifically, the problem considers a given a target area (a city, a neighborhood, etc.) where there are users who demand the use of the network (for the traffic of their corresponding mobile devices) and there is infrastructure that meets that demand.

There is some capacity installed for existing technologies (such as 4G) or none (if 5G is to be deployed). The problem consists in installing enough capacity to meet all the demand with the minimum cost, both taking into account the capital expenditure (investment) for installing new capacity and the operational expenditure (operational cost for providing service to customers using the installed capacity).

This paper offers a first approach as to efficiency of a MILP approach to address it and what are the limits of this technique. Other approaches will be required to address problems beyond the size of those which are tractable by means of a using an off-the-shelf solver.

The rest of the paper is organized as follows. Section 15.2 presents the most important literature related to this problem. Section 15.3 contains a description of the problem. The formulation to model that problem is contained in Sect. 15.4. The experimentation is described in Sect. 15.5. Finally, conclusions and further research are outlined in Sect. 15.6.

15.2 Related Work

Network planning is a problem previously addressed in the literature; as in Zhao et al. [7], the authors studied the minimum cost cell planning in a heterogeneous network, where macro-base stations and relay nodes were involved and proposed an approximation algorithm in order to solve the problem. They solve problems of up to 5×5 km, with 50 sites and 400 demand nodes.

Wang and Chuang [5] developed a heterogeneous base station deployment strategy in order to address the eNB deployment with the minimum cost problem. They devised a for an area of size 16×16 km and the large group scenario considered a part of the downtown area in Kaohsiung city in Taiwan.

Authors in Arthi [3] analyzed the deployment of macro- and micro-evolved nodes and relay stations in a heterogeneous network and proposed a fuzzy logic scheme. They analyzed a geographic area of 10×10 km and a range of 100–400 user equipment. No budget constraints or considerations were included.

More recently, Mathew [2] proposed a strategy using K-means algorithm in order to minimize the placement cost of the sites for heterogeneous base stations while satisfying the requirements of user equipment. Solved problems were 10×10 km, and the number of user equipment was in the range of 200 to 300.

Recently, Abedi and Wichman [1] presented a Voronoi framework for finding the optimal network topology compatible with non-uniform demands distribution and meeting both capacity and coverage constraints. They used particle swarm optimization and tested it on an area of 14×18 km^2.

Finally, Wang et al. [6] developed a location optimization model for optimizing the urban service coverage of 5G BSs. They combined geographic information system for spatial modeling with visualization and an artificial immune system. The model was applied to Wuhan city in China.

What we lack in the literature is the following. First, the joint treatment of capital and operational expenditures. Second, the situation where a system already exists and the problem consists in upgrading it or adapting it to the introduction to a new technology.

15.3 Problem Description

A telecommunication network has to be deployed in order to meet the traffic demand it is designed for. The problem can be decomposed into a set of different relatively dense areas (metropolitan areas) with high demand. We will focus on problems that occur in each of these metropolitan areas (target area), which is divided into a set of small enough grid lots.

Technology evolves in such a manner that different technologies can coexist at a given time. We will refer to these technologies as nodes. For each grid lot and each node, there is an average traffic demand so that the installed capacity of the system should be able to meet the whole demand.

To meet the demand, antennas have to be installed. Antennas are installed in sites (such as rooftops). Each site can have the equipment installed for a given node and, thus, can meet the demand corresponding to the technology of that node. If a site has a node, it still needs cells that oriented in some direction. Depending on the technology, the power of the cell, the obstacles that might exist, a give cell can reach a given lot or not. If so, we say that the cell can light the corresponding lot. Therefore, for a given cell of a node in a given site, we can have the set of grid lots that it lights. This cell can provide traffic to meet the demand of those lots.

Figure 15.1 illustrates a specific case for two nodes located in two different grid lots of the target area. Site 1 corresponds to the antennas installed on a given rooftop. In particular, this node can satisfy the demand both of 4G and 5G of nearby customers. Each node has three cells. Site 2 only has one node (4G) with three cells. Cells are the actual antennas that have some orientation and given the orientation. Shaded lots in the grid correspond to those grid lots whose 4G demand can be met by means of Cell 1 of Node 1 (4G) of site.

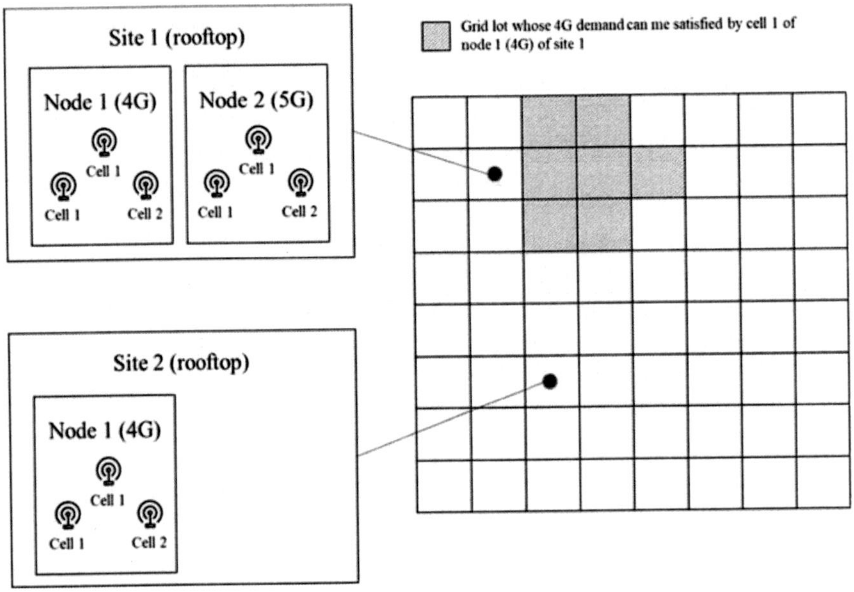

Fig. 15.1 Example of two sites, where Site 1 has two nodes and Site 2 has only one node

In a typical context, companies have already built sites, with nodes and cells. The problem consists in deciding if building new sites and if upgrading existing sites.

There are eligible locations for building new sites, and if new sites are built, nodes and cells can be built.

Existing sites can be upgraded in two ways. First, new nodes can be added to existing sites (if those sites do not have those nodes), and second, existing cells can be upgraded so that more demand can be satisfied with those cells.

New cells or upgraded cells can have a given maximum capacity. The traffic that a cell can provide for lots it can cover cannot exceed the final capacity, whereas all demand corresponding to every node has to be satisfied.

The objective is to minimize the total cost as the summation of capital expenditure (CAPEX) and operational expenditure (OPEX). Regarding the CAPEX, there are costs for installing a new site, for installing a new node and for upgrading an existing cell and for building a new cell. Every site has an operation expenditure, regardless of the cells or nodes it has.

15.4 Model Formulation

This section contains the MILP formulation for the problem presented in the previous sections. The following subsections contain, respectively, the sets, parameters and variables and, finally, the MILP model itself.

Sets $\mathcal{S}$: sites $\mathcal{N}$: sites SS: sites $\mathcal{L}$: sites Parameters A_sSS: 1 if site $s \in \mathcal{S}$ exists, 0 otherwise (potential site) B_{sn}SS: 1 if node $n \in \mathcal{N}$ at site $s \in \mathcal{S}$ exists, 0 otherwise N_{sn}SS: maximum number of cells of node $n \in \mathcal{N}$ at site $s \in \mathcal{S}$ C^U SS: cost for upgrading a cell C^CSS: cost for building a new cell	X_sSS: operational cost for site $s \in \mathcal{S}$ G_{snc}: 1 if cell $c \in \mathcal{C}$ of node $n \in \mathcal{N}$ at site $s \in \mathcal{S}$ exists, 0 otherwise CAP^O_{snc}SS: existing capacity for cell $c \in \mathcal{C}$ of node $n \in \mathcal{N}$ at site $s \in \mathcal{S}$ CAP^{MAX}_{snc}SS: maximum capacity for cell $c \in \mathcal{C}$ of node $n \in \mathcal{N}$ at site $s \in \mathcal{S}$ D_{ln}SS: traffic demand for node $n \in \mathcal{N}$ of lot $l \in \mathcal{L}$ Ω_{scl}SS: if site cell $c \in \mathcal{C}$ of site $s \in \mathcal{N}$ can reach lot $l \in \mathcal{L}$ C^S SS: cost for installing a new site C^NSS: cost for installing a new node
Binary variables α_sSS: 1 if new site $s \in \mathcal{S}$ is built, 0 otherwise β_{sn}SS: 1 if new node $n \in \mathcal{N}$ is built at site $s \in \mathcal{S}$, 0 otherwise γ_{snc}SS: if cell $c \in \mathcal{C}$ at node $n \in \mathcal{N}$ is built at site $s \in \mathcal{S}$, 0 otherwise δ_{snc}SS: if cell $c \in \mathcal{C}$ at node $n \in \mathcal{N}$ is upgraded at site $s \in \mathcal{S}$, 0 otherwise	**Continuous variables** c_{snc}SS: eventual capacity of cell $c \in \mathcal{C}$ of node $n \in \mathcal{N}$ at site $s \in \mathcal{S}$ u_{sncl}SS: traffic demand of lot $l \in \mathcal{L}$ met by cell $c \in \mathcal{C}$ of node $n \in \mathcal{N}$ at site $s \in \mathcal{S}$

The following MILP model represents the problem described in the previous section.

$$\min.\ C^S \sum_s \alpha_s + C^N \sum_s \sum_n \beta_{sn} + C^U \sum_s \sum_n \sum_c \delta_{nsc} + C^C \sum_s \sum_n \sum_c r_{snc} + X_s \sum_s \alpha_s \tag{15.1}$$

Subject to:

$$\alpha_s \leq 1 - A_s s \in \mathcal{S} \tag{15.2}$$

$$\beta_{ns} \leq 1 - B_{sn} n \in \mathcal{N}, s \in \mathcal{S} \tag{15.3}$$

$$\gamma_{nsc} \leq 1 - G_{snc} n \in \mathcal{N}, s \in \mathcal{S}, c \in \mathcal{C} \tag{15.4}$$

$$\delta_{nsc} \leq G_{snc} n \in \mathcal{N}, s \in \mathcal{S}, c \in \mathcal{C} \tag{15.5}$$

$$\beta_{ns} \leq A_s + \alpha_s s \in \mathcal{S}, n \in \mathcal{N} \tag{15.6}$$

$$\gamma_{nsc} \leq B_{ns} + \beta_{ns} c \in \mathcal{C}, s \in \mathcal{S}, n \in \mathcal{N} \tag{15.7}$$

$$\sum_{c} \gamma_{nsc} \leq N_{sn} - \sum_{c} G_{nsc} s \in \mathcal{S}, n \in \mathcal{N} \quad (15.8)$$

$$c_{snc} \geq CAP^{O}_{snc} s \in \mathcal{S}, n \in \mathcal{N}, c \in \mathcal{C} \quad (15.9)$$

$$c_{snc} \leq CAP^{O}_{snc}$$

$$+(CAP^{\text{MAX}}_{snc} - CAP^{O}_{snc}) G_{scn} \delta_{scn}$$

$$+(CAP^{\text{MAX}}_{snc} - CAP^{O}_{snc})(1 - G_{scn}) \gamma_{scn}$$

$$s \in \mathcal{S}, n \in \mathcal{N}, c \in \mathcal{C} \quad (15.10)$$

$$\sum_{l} \Omega_{scl} u_{sncl} \leq c_{snc} s \in \mathcal{S}\ n \in \mathcal{N}, c \in \mathcal{C} \quad (15.11)$$

$$\sum_{s} \sum_{c} \Omega_{scl} u_{sncl} \geq D_{nl} l \in \mathcal{L}, n \in \mathcal{N} \quad (15.12)$$

$$\alpha_s \in \{0, 1\} s \in \mathcal{S},\ \beta_{sn} \in \{0, 1\} s \in \mathcal{S},\ n \in \mathcal{N},$$
$$\gamma_{snc} \in \{0, 1\} s \in \mathcal{S},\ n \in \mathcal{N},\ c \in \mathcal{C},\ \delta_{snc} \in \{0, 1\}\ s \in \mathcal{S},$$
$$n \in \mathcal{N},\ c \in \mathcal{C},\ c_{snc} \geq 0\ s \in \mathcal{S},\ n \in \mathcal{N},\ c \in \mathcal{C},$$
$$u_{sncl} \geq 0\ s \in \mathcal{S},\ n \in \mathcal{N},\ c \in \mathcal{C},\ l \in \mathcal{L} \quad (15.13)$$

The objective function (Eq. 15.1) computes the total cost, corresponding to both capital expenditure (building new sites, new nodes, new cells and upgrading existing cells) and operational costs per site.

A set of constraints is referred to the system design and the corresponding installed capacity. Equations 15.2, 15.3 and 15.4 only allow to build new sites, nodes and cells, respectively, only if they do already exist. Equation 15.5 only allows upgrading existing cells. Equation 15.6 allows building a new node on an existing site or on a built site. Likewise, Eq. 15.7 allows building a new cell on an existing node or on a built node. Equation 15.8 upper bounds the total number of cells per node a site. Equations 15.9 and 15.10 bound the minimum and maximum capacities, respectively. The minimum capacity is the initial capacity of every cell. The maximum capacity is the initial capacity plus the capacity upgrade if the cell is upgraded or the capacity installed if the cell is newly built.

Two additional constraints refer to meeting traffic demand. Equation 15.11 limits the traffic every cell provides to the lots it lights according to its resulting capacity. Equation 15.12 ensures that the demand of every lot is fulfilled with the set of cells that light every lot. Finally, Eq. 15.13 is domain constraints according to the definition.

15.5 Experimentation

Instances of different sizes were randomly generated as rectangles ranging from 1 to 500 km^2, divided into 30 $\times$ 30 lots regardless of the target area. The cells of a site are supposed to light an area of 0.15 km^2, and each covers a sector of 120°. The number of existing sites was set as twice the total target area divided by 0.15 km^2, and the number of potential sites was half the existing sites. Two nodes were considered (4G and 5G). Both existing potential sites were scattered across the target area with some degree of randomness.

All instances were run on an Intel Core i5-7200 CPU 8 GB machine running Windows 10 using CPLEX 12.8.

As an illustration, Fig. 15.2 depicts an example of an area of Madrid. The top left image shows the demand (the darker, the larger the demand), and the top left represents the current 4G network.

The bottom map shows the optimal network plan that allows fulfilling all the demand.

Table 15.1 contains all relevant information for the generated instances. All of them were solved to optimality in less than 300 s. As expected, the solving time grows faster than linearly, whereas the cardinality of the sets and the number of variables and constraints do grow linearly with the area of the problem (Fig. 15.3).

According to the solving time growing pattern, areas such as those under study can be analyzed in reasonable times, corresponding to small to intermediate size cities. However, larger areas could not be addressed by means of this MILP model.

15.6 Conclusions

In this paper, we have formulated a model for deploying the infrastructure to meet the demand of a set of technologies with the minimum cost, and this approach is effective and efficient for addressing cities up to 500 km^2 in very short times.

There is room for improving the efficiency of the solving techniques so that larger instances can be solved in reasonable times. In particular, there are two ways worth exploring. One is a Benders decomposition where the master problem proposes a location scheme (what sites, nodes and cells are available), whereas the subproblem would assign capacity to lots. The second one is a heuristic-based approach. In particular, some location heuristic combined with a random biased optimization might provide near-optimal solutions in very short times.

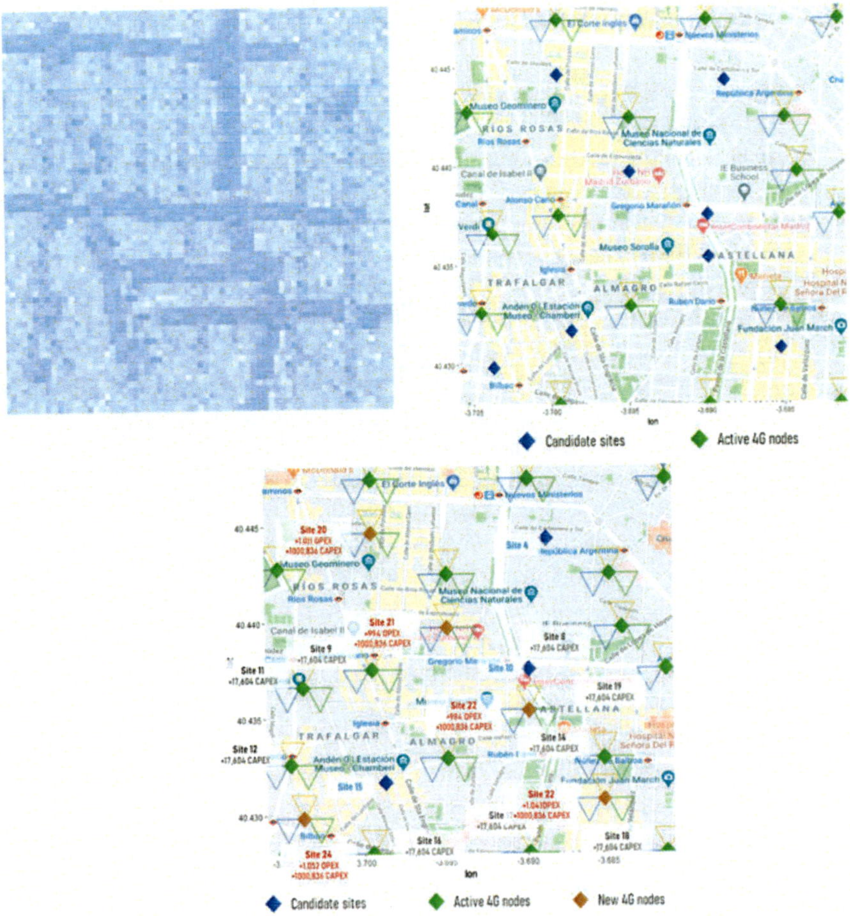

Fig. 15.2 Traffic demand (top left) and current network plan (top right) maps. Optimal network plan map (bottom)

Table 15.1 Computational results for different instances, with target areas ranging from 1 to 500 km^2

Area (km^2)	No. sites	No. lots	Total cost (k€)	Gap	Solving time	No. vars	No. bin. vars	No. cons
1	18	1089	343	0	0.33	12,917	258	5087
10	198	11,025	9161	0	2.48	98,087	2838	52,091
25	498	27,556	22,752	0	4.91	248,848	7138	130,315
50	999	55,225	47,921	0	9.11	537,631	14,319	261,198
75	1500	82,944	75,909	0	14.39	828,571	21,500	392,281
100	1998	110,889	103,727	0	17.56	1,116,414	28,638	524,147
150	3000	166,464	160,108	0	29.91	1,721,811	43,000	786,861
250	4998	277,729	266,210	0	53.68	2,944,317	71,638	1,312,507
400	7998	443,556	405,984	0	135.09	4,784,038	114,638	2,096,815
500	9999	555,025	510,691	0	213.23	6,010,408	143,319	2,623,398

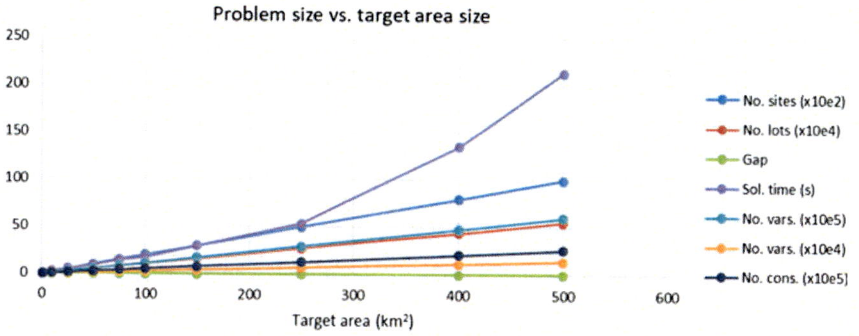

Fig. 15.3 Problem size for different target area values

References

1. Abedi M, Wichman R (2021) Cellular network planning via local regularity optimization. In: 2021 IEEE 93rd vehicular technology conference (VTC2021-Spring). IEEE, pp 1–7
2. Mathew AP, Arthi M, Vinoth Babu K (2017) An uniform clustering based coverage and cost effective placement of serving nodes for 5g. In: 2017 international conference on innovations in electrical, electronics, instrumentation and media technology (ICEEIMT). IEEE, pp 106–110
3. Murugadass A, Pachiyappan A (2016) A flexible and cost-effective heterogeneous network deployment scheme for beyond 4G. Arab J Sci Eng 41(12):5093–5109
4. Murugadass A, Pachiyappan A (2017) Fuzzy logic based coverage and cost effective placement of serving nodes for 4G and beyond cellular networks. Wirel Commun Mobile Comput 2017
5. Wang Y-C, Chuang C-A (2015) Efficient eNB deployment strategy for heterogeneous cells in 4g LTE systems. Comput Netw 79:297–312
6. Wang Q, Zhao X, Lv Z, Ma X, Zhang R, Lin Y (2020) Optimizing the ultra-dense 5g base stations in urban outdoor areas: Coupling GIS and heuristic optimization. Sustain Cities Soc 63:102445

7. Zhao W, Wang W (2013) Cell planning for heterogeneous cellular net-works. In: 2013 IEEE wireless communications and networking conference (WCNC). IEEE, pp 1032–1037
8. Zhao W, Wang S, Wang C, Wu X (2016) Approximation algorithms for cell planning in heterogeneous networks. IEEE Trans Vehic Technol 66(2):1561–1572

Chapter 16
Differences Between Static and Dynamic Home Care Routing and Scheduling Models. How to Design an Appropriate Model

À. Armadàs, A. Lusa-García, and A. García-Villoria

Abstract Routing and scheduling of health and socials workers to deliver care at patients' homes have been extensively studied. Nevertheless, this research generally deals with the idealistic static variant of the problem. Progressing from the static to the dynamic alternative is not a straightforward endeavour. Also, there has not been a clear-cut delimitation between static and dynamic models. This work proposes a clear differentiation between static and dynamic models and suggests a framework to design models appropriate for a broad range of home care routing and scheduling problems.

Keywords Home care · Routing · Scheduling · Dynamic · Planning

16.1 Introduction

Home care has experienced a rapid expansion among developed countries in recent years. Population ageing and social changes, such as the fragmentation of the traditional family group, are some of the causes [22]. A natural interest has ensued within the operations research community, which has been focused on static routing and scheduling, that is on devising methodologies to provide caregivers' routes and schedules targeting a given objective, provided demand was known and the number of caregivers fixed. Hence, this static problem can be regarded as a variant of the vehicle routing problem with time windows (VRPTW).

À. Armadàs · A. Lusa-García (✉) · A. García-Villoria
Dpto. de Organización de Empresas, Instituto de Organización y Control, Universitat Politècnica, de Catalunya. Av. Diagonal 647, p11, 08028 Barcelona, Spain
e-mail: amaia.lusa@upc.edu

À. Armadàs
e-mail: alexandre.armadas@upc.edu

A. García-Villoria
e-mail: alberto.garcia-villoria@upc.edu

C. Avilés-Palacios and M. Gutierrez (eds.), *Ensuring Sustainability*, Lecture Notes in Management and Industrial Engineering,
https://doi.org/10.1007/978-3-030-95967-8_16

However, home care organizations rarely operate within this idealized environment. Dynamic events are a core feature of real-world home care. New patients are added to the existing pool and caregivers resign or are hired. Past routing and scheduling decisions condition future planning. Also, short-term disruptions, such as sudden indisposition of caregivers, need to be handled appropriately.

As a result, the applicability of the research hitherto done remains very limited. This paper proposes a framework to design home care routing and scheduling models for real-world settings. To do so, the rest of the article is organized as follows: Sect. 16.2 contains a discussion on what differentiates dynamic from static models and a brief literature review. In Sect. 16.3, the framework is presented, and in Sect. 16.4, we provide conclusions and suggest future research directions.

16.2 Dynamic and Static Models. Brief Review of Existing Literature

16.2.1 What Is a Dynamic Model? Differences with Static Models

As we shall briefly see, a dynamic model must not be solely identified with other related concepts. We have found that, in the literature, *offline multiperiod models* are often mistaken as dynamic and *online models* are frequently considered the only possible dynamic models. We define these concepts as follows:

- **Offline multiperiod models** are those intended to be executed at regular time intervals. For instance most papers on home care routing and scheduling are designed to optimize routes and schedules regularly once a day.
- **Online models** consider real-time events, commonly referred to as disruptions, and adapt routes and schedules, while caregivers are on the move conducting their visits.

We discuss first offline multiperiod models. Are they dynamic? Not all of them. Consider this example: the planner of a home care organization designs, at the end of a working day, routes and schedules of caregivers for the following day. He/she does not take into consideration, in the decision-making process, past routes and schedules. In other words: routes and schedules are designed from scratch. The planner is just solving, every day, a static routing and scheduling problem. Hence, it follows that an offline multiperiod routing and scheduling problem is dynamic if past decisions on routes and schedules affect future routes and schedules.

There are two ways in which past decisions influence future planning. The first one is patient-related and has two typical non-exclusive possibilities:

1. Care consistency: in many settings, patients demand that services be performed always by the same caregiver, or a small set of caregivers, in order to build trust.

2. Schedule consistency: services' starting times are maintained. For instance a patient may receive the service on Mondays, always at 9:00 A.M.

The second way in which past decisions affect future ones are related to caregivers and has also two common non-exclusive versions:

1. Working hours: it is common in many countries that working hours be legally upper-bounded (for instance 40 weekly hours). In this case, an organization that was planning routing and scheduling, for instance on Wednesday, would need to consider Monday and Tuesday's hours already worked by caregivers.
2. Hires and dismissals of caregivers: when planning, the organization needs to achieve the highest possible rate of utilization of all caregivers on payroll. Thus, past hires and dismissals affect future routing and scheduling options.

Hence, to our understanding, what sets apart dynamic from static offline multiperiod models is not only the occurrence of dynamic events, but also the dependence on past decisions. Online models, according to what has been said, are always dynamic: routes designed before disruptions affect the way in which disruptions are handled. To sum up, home care routing and scheduling dynamic models are either (i) offline multiperiod models in which past decisions affect future planning; or (ii) online models.

16.2.2 Home Care Routing and Scheduling Brief Literature Review

This brief review is not an extensive survey. Its goal is to identify concepts relevant for the purpose of the paper and to provide appropriate context for the reader.

16.2.2.1 Static Models

Static models have planning horizons from one day, as in Akjiratikarl et al. [1], up to two weeks (see for instance [24]).

Objective functions can be: (1) single criteria, that is only one objective is tried to optimize ([19] is an example), (2) multicriteria, where several weighted terms compose the objective function, such as in Mankowska et al. [18], and (3) multiobjective functions aimed at obtaining Pareto frontiers (an approach adopted by Alves et al. [2]).

Articles enforce objectives by: (i) seeking optimization; (ii) imposing hard constraints; or (iii) introducing soft constraints (that can be violated at a penalty cost in the objective function). Objectives are aimed at satisfying one of the three following stakeholders: the home care organization, caregivers or patients. Organization objectives are related to cost minimization and efficiency; caregivers' objectives to working conditions, whereas patients' objectives are quality-related (care consistency, etc.). Organization's objectives are generally optimized in the objective function, such as total travelling time [12]; caregivers' objectives tend also to be optimized in the objective function and are frequently aimed at balancing workload among caregivers, as in Cappanera and Scutellà [6]; patients' objectives are more frequently imposed either via hard constraints, such as hard time windows to receive the service [3], or via soft constraints.

Routes and schedules of caregivers are the decision variables. Few papers discuss additional decisions, such as break time scheduling [16].

The size of problems tackled, measured by the number of caregivers, services or patients dealt with in numerical applications of the models is in general low. That is models are applied to relatively small instances of 30 or less caregivers, 100 or less patients or 200 or less services, as in Chaieb et al. [7]. Others, such as Mankowska et al. [18], cope with relatively greater instances, up to 300 caregivers, 500 patients and 1000 services. Studies going for even larger sizes are very few; an example is Mankowska et al. [17] (10,654 services and 1375 caregivers). Since the problem is well-known to be NP-hard, heuristics are the common solution approach.

16.2.2.2 Dynamic Models

Aspects already presented in the previous section are also applicable to dynamic models (objectives, constraints, etc.). Here, we present features specific to dynamic models or that need further comment.

Dynamic models tackle low- to medium-scale problems, from one caregiver problems [4] to 120 caregivers [20].

The way entries and exits of patients is modelled varies among papers. The pool of patients might be updated weekly [11], daily [13] or even at shorter periods, such as every two hours [25]; once the information is updated, (re)planning is conducted. Another updating mechanism is event-based, i.e. the arrival of new patients triggers updating and the consequent (re)planning. The simplest way is an event-by-event policy: every new request triggers its own service scheduling and routing [4]. Decision variables do not vary in essence from those of static models, except for Nasir and Dang [20] where hiring caregivers is a decision variable.

Models are applied to simulate relatively long periods of care delivery: from one week [21] to a year [8].

Algorithms (generally heuristics) that solve the problem may consider information on demand forecasts of eventual new patients when (re)planning. One typical objective under this future-sensible method is to preserve capacity, precisely for this forecasted demand [9]. However, the most common approach is future-blind.

As for past decisions that influence (re)planning, care consistency is imposed for instance, in Heching and Hooker [14] via hard constraints, whereas an example of schedule consistency is Demirbilek et al. [10].

16.3 Framework to Design a Suitable Model

Constructing a suitable model means making the right design choices for a given problem. In particular, here, we are going to answer:

- Should the model be static or dynamic?

  And, specifically for dynamic models:

- Which is the time interval or the appropriate event-based policy to update information and conduct rescheduling?
- Is it worth including demand forecasting in the model? Similarly, would forecasting on caregivers' resigning rate or absenteeism improve solutions?

Whether a model should be either static or dynamic comes straightforwardly at this point: online problems and offline past decisions-dependant problems necessitate a dynamic approach. In any other situation, the model must be static.

Updating policy in dynamic problems, to our understanding, must be based on two criteria: quickness of the response needed and degree of dynamism.

Some contexts necessitate a quick response. For instance in online settings, when there is a sudden indisposition of a caregiver, his/her services have to be quickly assigned to another caregiver. In this case, updating and (re)planning must be done fast, but it is also determined by the degree of dynamism (i.e. the rate at which dynamic events occur): if such events occur constantly, it is possible to accumulate some and conduct rescheduling by (probably small) batches or relatively short time intervals, so as to explore more routing options because of event grouping. If the event frequency is low, an event-by-event approach must be the choice. This quick response policy is depicted by the blue line of Fig. 16.1.

Other situations need a medium-paced response. For instance when the organization is legally obliged to schedule the first service of the patient at most two days after

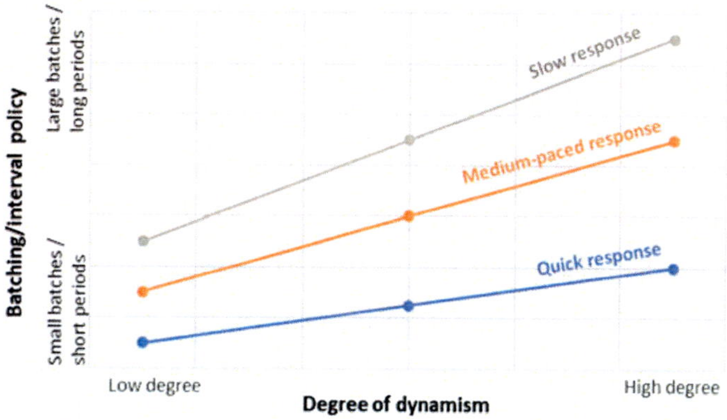

Fig. 16.1 Batching/interval policies

the request. This kind of clauses are typical on contracts when a Public Administration outsources the service to an organization. A logical approach would be to conduct (re)planning every two days, grouping requests into batches. Again, the degree of dynamism conditions the choice: if only two new patients request service in this 2-day interval, an event-by-event policy may be chosen: patients would promptly know the schedule, and the optimization possibilities lost are negligible. Medium-paced response is illustrated by the orange line of Fig. 16.1.

Finally, in some cases, the response can be relatively slow. If for instance new patients' requests can be scheduled for the next month, grouping all this one-month request for (re)planning would result in more possibilities for better routes and schedules. And, the more the degree of dynamism, the more the potential benefits of such a large grouping approach. This slow response policy is represented by the grey line in Fig. 16.1.

As for demand forecasting, Demirbilek et al. [9] showed that for a one-caregiver problem with a simple forecasting mechanism, considering possible future events when planning may yield better results as time goes on. Hence, it would be shocking if, for larger problems, contemplating future demand did not provide improved solutions. Nevertheless, research on home care demand forecasting has been scarce so far and only Restrepo et al. [23] integrate forecasting in a multicaregiver dynamic routing and scheduling problem.

As for forecasting on caregivers' resigning rate or absenteeism, there is, to the best of our knowledge, no research on the topic. Only some routing and scheduling papers deal limitedly with the issue; for instance Borsani et al. [5] apply their model to a dataset that contains information on unavailable caregivers due to illnesses and other causes. However, the phenomenon is treated as a known fact: these caregivers have no service assigned by default.

16.4 Conclusions and Future Research Directions

On the one hand, the more the degree of dynamism, the more appropriate a large batching policy and the more suitable long batching periods. On the other hand, when the response needs to be fast, small batching and short-period grouping policies are needed. However, the benefits of accumulating requests to conduct batch optimization might reach an upper bound: in a given setting, the gain of grouping one more day of new patients (for instance from 9 to 10 days accumulation) might be insignificant. Furthermore, even though larger grouping provides more optimization possibilities, complexity also increases. This added complexity may cause an algorithm to provide an inferior result caused by the increase in the search space, not because batching itself leads to worse outcomes. Hence, investigating these upper bounds and the trade-off between complexity and quality of results, in different settings and degrees of dynamism, would be useful.

Forecasting should be further investigated and models for dynamic settings tested to elucidate its benefits. A natural question arises in this regard: do the cost of devising and using forecasting models offset the benefits of the (allegedly) better planning it provides? To the best of our knowledge, only Lanzarone and Matta [15] investigated in this direction and found that the value of perfect information regarding demand, for routing and scheduling purposes, can be high. Finally, there is no research, and thus no assessment, on the utility of using forecasting models to predict caregivers' absenteeism or resignation for routing and scheduling aims.

References

1. Akjiratikarl C, Yenradee P, Drake PR (2007) PSO-based algorithm for home care worker scheduling in the UK. Comput Ind Eng 53(4):559–583
2. Alves F et al (2019) A multi-objective approach to the optimization of home care visits scheduling. In: ICORES 2019—Proceedings of the 8th international conference on operations research and enterprise systems. https://doi.org/10.5220/0007565704350442
3. Arabzadeh E, Mohammad-Taghi Fatemi-Ghomi S, Kariomi B (2016) A new mathematical model for multi period multi depot home health care routing scheduling problem. Nurs Pract Today 3(1):140–151
4. Bennett AR, Erera AL (2011) Dynamic periodic fixed appointment scheduling for home health. IIE Trans Healthcare Syst Eng 1(1):6–19
5. Borsani V et al (2006) A home care scheduling model for human resources. In: International conference on service systems and service management, pp 449–454
6. Cappanera P, Scutellà MG (2014) Joint assignment, scheduling, and routing models to home care optimization: a pattern-based approach. Transp Sci 49(4):830–852. https://doi.org/10.1287/trsc.2014.0548
7. Chaieb M, Jemai J, Mellouli K (2019) A decomposition—construction approach for solving the home health care scheduling problem. Health Care Manag Sci
8. Cire AA, Diamant A (2018) Dynamic scheduling of home health care patients to medical providers. SSRN Electron J. https://doi.org/10.2139/ssrn.3210339
9. Demirbilek M, Branke J, Strauss AK (2019) Dynamically accepting and scheduling patients for home healthcare. Health Care Manag Sci 22:140–155
10. Demirbilek M, Branke J, Strauss AK (2019) Home healthcare routing and scheduling of multiple nurses in a dynamic environment. Flex Serv Manuf J. https://doi.org/10.1007/s10696-019-09350-x
11. Gomes MI, Ramos TRP (2019) Modelling and (re-)planning periodic home social care services with loyalty and non-loyalty features. Eur J Oper Res 277(1):284–299. https://doi.org/10.1016/j.ejor.2019.01.061
12. Grenouilleau F et al (2019) A set partitioning heuristic for the home health care routing and scheduling problem. Eur J Oper Res 275(1):295–303
13. Güven-Koçak S, Heching A, Pinar Keskinocak AT (2019) Home health care routing and scheduling with consistency. Available at: https://www2.isye.gatech.edu/~atoriello3/hhcrs_v3.pdf
14. Heching A, Hooker JN (2016) Scheduling home hospice care with logic-based Benders decomposition. In: International conference on AI and OR techniques in constriant programming for combinatorial optimization problems, pp 187–197
15. Lanzarone E, Matta A (2009) Value of perfect information in home care human resource planning with continuity of care. In: 35th conference on operational research applied to health services (ORAHS 2009). Leuven
16. Liu R, Yuan B, Jiang Z (2017) Mathematical model and exact algorithm for the home care worker scheduling and routing problem with lunch break requirements. Int J Prod Res 55(2):558–575

17. Luna F, Cervantes A, Isasi P (2013) Large-scale home care crew scheduling with a parallel evolutionary algorithm. In: 2013 eighth international conference on P2P, parallel, grid, cloud and internet computing (3PGCIC), pp 588–593
18. Mankowska DS, Meisel F, Bierwirth C (2014) The home health care routing and scheduling problem with interdependent services. Health Care Manag Sci 17(1):15–30
19. Moussavi SE, Mahdjoub M, Grunder O (2019) A matheuristic approach to the integration of worker assignment and vehicle routing problems: application to home healthcare scheduling. Expert Syst Appl 125:317–332
20. Nasir JA, Dang C (2019) Quantitative thresholds based decision support approach for the home health care scheduling and routing problem. Health Care Manage Sci
21. Nickel S, Schröder M, Steeg J (2012) Mid-term and short-term planning support for home health care services. Eur J Oper Res 219(3):574–587
22. OECD (2008) Home care in Europe. The solid facts. Available at: http://www.euro.who.int/__data/assets/pdf_file/0005/96467/E91884.pdf?ua=1
23. Restrepo MI, Rousseau LM, Vallée J (2019) Home healthcare integrated staffing and scheduling. Omega (United Kingdom). https://doi.org/10.1016/j.omega.2019.03.015
24. Torres-Ramos A et al (2014) Mathematical model for the home health care routing and scheduling problem with multiple treatments and time windows. In: Proceedings of the 1st international conference on mathematical methods & computational techniques in science & engineering, pp 140–145
25. Xie Z, Wang C (2017) A periodic repair algorithm for dynamic scheduling in home health care using agent-based model. In: 2017 IEEE 21st international conference on computer supported cooperative work in design (CSCWD), pp 245–250

Chapter 17
Reformulation of the CVRP Based on the Equivalent Cyclic Polygon

Á. Herraiz, M. Gutierrez, and M. Ortega-Mier

Abstract We propose two new formulations for the objective function of the capacitated vehicle routing problem (CVRP) that are based on a geometric transformation of the feasible solutions. The transformation consists in conceptualizing a VRP solution as a polygon and transforming it into an inscribed polygon in which the corresponding edge lengths remain unaltered (equivalent cyclic polygon). Experimentation with a set of problems taken from the literature shows that for some cases, the new formulations lead to better results, which suggest the interest of further research on the proposed transformation.

Keywords Vehicle routing problem · VRP · Capacitated vehicle routing problem · CVRP · Equivalent cyclic polygon

17.1 Introduction

The vehicle routing problem (VRP) is a well-known and extensively studied combinatorial optimization problem. Generically, the problem consists in determining "the optimal set of routes to be performed by a fleet of vehicles to serve a given set of customers" [11]. There are multiple variants of the VRP. Among them, in this work, we focus on the capacitated VRP (CVRP).

Á. Herraiz (✉) · M. Gutierrez · M. Ortega-Mier
Universidad Politécnica de Madrid, C/ José Gutiérrez Abascal 2, 28006 Madrid, Spain
e-mail: alvaro.herraiz.solla@alumnos.upm.es

M. Gutierrez
e-mail: miguel.gutierrez@upm.es

M. Ortega-Mier
e-mail: miguel.ortega.mier@upm.es

C. Avilés-Palacios and M. Gutierrez (eds.), *Ensuring Sustainability*,
Lecture Notes in Management and Industrial Engineering,
https://doi.org/10.1007/978-3-030-95967-8_17

The CVRP aims at "determining a set of m vehicle routes satisfying the following conditions: (i) each route starts and ends at the depot, (ii) each customer belongs to exactly one route, (iii) the total demand of each route does not exceed the vehicle capacity Q, (iv) the total cost of all routes is minimized" [6]. The first formulation and optimization algorithm for the CVRP was proposed by Dantzig and Ramser [2]. Since then, there has been an extensive literature on the CVRP.

Laporte and Nobert [9] provide a thorough review of the methods to solve the CVRP problem as one of the relevant problems of the VRP family. Toth and Vigo [11] provide a thorough specific review of the available methods to solve the CVRP. Kumar and Panneerselvam [8] present a VRP survey, including the works dealing with the CVRP. Regarding the formulations directly related with our proposal, Kulkarni and Bhave [7] propose CVRP subtour elimination constraints following the Miller et al. [10] formulation for the traveling salesman problem (TSP). As detailed in the following section, Kara et al. [6] present some modifications to these constraints.

In this paper, we propose a geometric transformation of the VRP solutions that leads to a reformulation of the objective function of the CVRP. The remainder of the paper is organized as follows. In Sect. 17.2, we recall an exact CVRP formulation. In Sect. 17.3, we present the proposed VRP solution geometric transformation with the definition of an equivalent cyclic polygon that leads to the proposal in Sect. 17.4 of two modifications of the classic formulation of the objective function. In Sect. 17.5, we present the results of the short experimentation using the sets proposed by [1], including a brief discussion in Sect. 17.6. Finally, in Sect. 17.7, we summarize the conclusions drawn from the work.

17.2 CVRP Formulation

Let us follow the CVRP formulation as it is defined in [6], which adapts the [10] TSP subtour elimination constraints to the CVRP. Consider a graph $G = (V, K, E)$, where V is the set of depot and customers $V = \{0, 1, 2, \ldots, n\}$, K denotes the set of vehicles $K = \{1, 2, \ldots, k\}$, E is the set of links[1] $E = \{(i, j)|\ i, j \in N, i \neq j\}$, and c_{ij} is a distance parameter for each $(i, j) \in E$, as the distance between nodes i and j in any path p. We use a binary variable $X_{ij} = 1$ or 0 if the solution contains or not the link (i, j). We use an integer variable U_i associated to each customer i.

Indices and sets:

$i, j \in V$	depot and customers
$k \in K$	vehicles
$K = \{1, 2, \ldots, \lvert k\rvert\}$	set of vehicles
$V = \{0, 1, \ldots, n\}$	set of depot and customers
$V_c = \{1, \ldots, n\}$	set of customers.

[1] We use *nodes* and *links* to signify the CVRP graph, *vertices* and *edges* to signify the EIP, and the *arc* to signify to a circular arc.

Parameters:

c_{ij} distance between customers i and j
$|k|$ number of vehicles
Q capacity of vehicles
q_i weight of service for customer i.

Variables:

$X_{ij} \in \{0, 1\} \quad \forall (i, j) \in V$ 1 if use arc from i to j, 0 otherwise
$U_i \geq 0, \quad \forall i \in V_c$ weight of vehicles at node i.

Formulation:

$$\min z = \sum_{i,j} c_{ij} X_{ij} \tag{17.1}$$

such that

$$\sum_{j \in V_c} X_{0j} \leq |k| \tag{17.2}$$

$$\sum_{i \in V_c} X_{i0} \leq |k| \tag{17.3}$$

$$\sum_{j \in V} X_{ij} = 1, \quad \forall i \in V_c \tag{17.4}$$

$$\sum_{i \in V} X_{ji} = 1, \quad \forall j \in V_c \tag{17.5}$$

$$q_i \leq U_i \leq Q, \quad \forall i \in V_c \tag{17.6}$$

$$U_i - U_j + QX_{ij} + (Q - q_i - q_j)X_{ji} \leq Q - q_j, \quad \forall i, j \in V_c, i \neq j \tag{17.7}$$

Objective function (17.1) minimizes the total length of the fleet. Constraints (17.2) and (17.3) ensure the start and the ending of all vehicles, Constraints (17.4) and (17.5) ensure to visit each node no more than once, Constraints (17.6) and (17.7) eliminate subtours.

17.3 Equivalent Cyclic Polygon of a VRP Solution

We propose to extend the concept of equivalent cyclic polygon of a TSP tour [5], by its generalization to the VRP. Given a TSP Euclidean tour of nodes in a sequence and their corresponding links c_{ij}, it is possible to construct a polygon with vertices

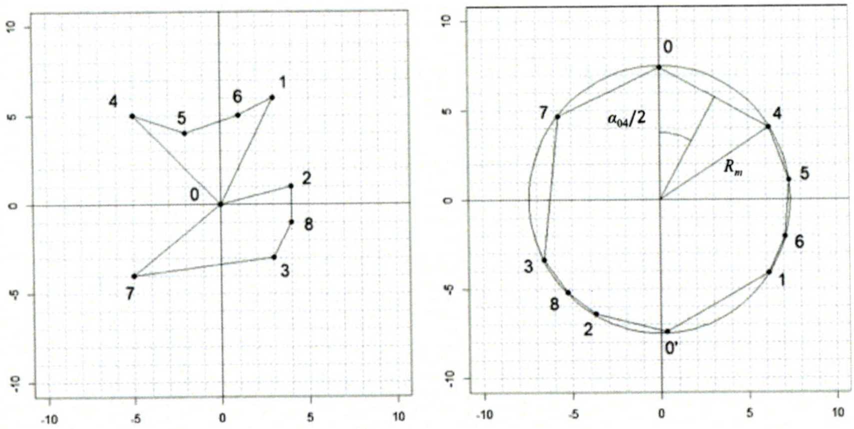

Fig. 17.1 Example of a tour or route (left) and its ECP (right)

inscribed in a circle in the same sequence order and edges of equal length, which is known as the equivalent cyclic polygon (ECP) [5]. This circle is the *mass circle*, defined by its *mass radius* R_m.

In the VRP, the geometric transformation needs a redefinition since typically the solution will include several closed routes. The generalization to the VRP is done as illustrated with the example of Fig. 17.1. As shown, the solution consists of two routes. One vehicle will follow a tour with the node sequence 0-4-5-6-1-0, while the second vehicle sequence is 0-2-8-3-7-0. In this case, ECP construction lies in the polygon that concatenates both sequences, denoting the depot visit between both routes with 0′; i.e., the VRP solution polygon will include the sequence 0-4-5-6-1-0′-2-8-3-7-0. Then, we follow the same procedure as in the definition of the TSP ECP, inscribing each link in a circumference such that all the lengths remain unaltered. Should only exist one route, then the VRP ECP thus defined yields the TSP ECP.

Then, let us consider the link from node i to node j belonging to a given route, where Eq. (17.8) is verified.

$$\frac{c_{i\,j}}{2} = R_m \sin\left(\frac{\alpha_{i\,j}}{2}\right),\ \alpha_{i\,j} = 2 \cdot \arcsin\left(\frac{c_{i\,j}}{2 \cdot R_m}\right),\quad \forall (i, j) \in E | X_{i\,j} = 1 \quad (17.8)$$

The sum of all edges in the ECP equals 2π (Eq. 17.9). Therefore, given a set of routes or tour, we can calculate R_m after solving this transcendental equation using mathematical software or an iterated procedure.

$$\sum_{i\,j} \arcsin\left(\frac{c_{i\,j}}{2R_m}\right) = \pi,\ R_m \geq \frac{c_{i\,j}}{2},\quad \forall (i, j) \in E | X_{i\,j} = 1 \qquad (17.9)$$

The optimal CVRP route allows us to construct what we will call an optimal ECP*, defined by an optimal $R_m{}^*$.

17.4 CVRP Reformulation

The geometric transformation of the former section allows us to reformulate the CVPR in a similar manner as for the TSP in [5]. The main idea is to define a fixed radius R_f and minimize the total arc measure that will yield each inscribed VRP polygon. If the mass radius of a VRP feasible solution is smaller than R_f, then the total arc measure of the VRP polygon will be less than 2π, whereas if it is greater than R_f, then the arc measure will exceed 2π. We can redefine the objective function (Eq. 17.1) of the VRP, minimizing the arc measure of each feasible solution for a generic fixed radius R_f. In order to assure the existence of the arcsin function, it is necessary to add constraints in Eq. (17.11) to complete the reformulation of the problem.

$$\min z = \sum_{i,j} X_{ij} \arcsin\left(\frac{c_{ij}}{2R_f}\right) \tag{17.10}$$

$$c_{ij}X_{ij} \le 2R_f, \quad \forall i, j \in V, i \ne j \tag{17.11}$$

Although the optimal $R_m{}^*$ is not known, it is possible to define values that are upper bounds or lower bounds for $R_m{}^*$. We call these new formulations the ECP_CVRP_summarcs and the ECP_CVRP_smR formulations whether we use an upper bound $R_m{}^{\mathrm{UB}}$ or a small radius $R_m{}^{\mathrm{S}}$. If we decide to use an $R_f = R_m{}^{\mathrm{UB}} \ge R_m{}^*$, optimality is guaranteed. If we use an $R_m{}^{\mathrm{S}}$, we may disregard links of the optimal solution, whenever $R_m{}^{\mathrm{S}} \le c_{ij}/2$ for any c_{ij} in E.

The CVRP includes weights of services for customers using several vehicles, and even the longest link may belong to the optimal solution. Due to this fact, link elimination might not be advisable with an ECP transformation in CVRP instances. If the mass radius $R_m{}^{\mathrm{S}}$ is equal or greater than the length of the greatest link divided by two $R_m{}^{\mathrm{S}} \ge \max(c_{ij})/2$, we avoid eliminating any link (i, j) in E. This is the most restrictive rule to avoid eliminating links. Under this condition, we can omit Eq. (17.11) in the ECP_CVRP_smR formulation because the range of the arcsine function $[-1, 1]$ allows an image for all links. We will use this rule expressed in Eq. (17.12).

$$\min z = \sum_{i,j} X_{ij} \arcsin\left(\frac{c_{ij}}{2R_m^S}\right), \; R_m^S \ge \frac{1}{2}\max(c_{ij}) \;\; \forall i, j \in V, i \ne j \tag{17.12}$$

Some other approaches may be feasible if we aggressively reduce the mass radius. It is possible to redefine the VRP objective function in order to use this reduced mass radius while also omitting constraints in Eq. (17.11). Let us call the radius parameter $R_m{}^{\mathrm{SS}}$ to any real number equal or smaller than the length of the minimum link in E divided by two, $R_m{}^{\mathrm{SS}} \le \min(c_{ij})/2$. Using this parameter in Eq. (17.10) implies all values are greater than or equal to one. Therefore, we take the inverse of the arcsin argument, that will be less than or equal to one while taking the inverse of the arcsin in order to minimize arc measure (Eq. 17.13). Let us call this formulation the ECP_CVRP_SmR_inv formulation.

$$\min z = \sum_{i,j} X_{i\,j}\left(\arcsin\left(\frac{1}{c_{ij}/2R_m^{SS}}\right)\right)^{-1}, \quad R_m^{SS} \le \frac{1}{2}\min\left(c_{i\,j}\right) \quad \forall i, j \in V, i \ne j \tag{17.13}$$

Figure 17.2 depicts the two proposed objective function transformations when taking the limit values $R_m{}^S = \max(c_{ij})/2$ in Eq. (17.12) and $R_m{}^{SS} = \min(c_{ij})/2$ in Eq. (17.13), respectively. The range of the transformation is $[c_{\min} = \min(c_{ij}), c_{\max} = \max(c_{ij})]$. In order to show the effect of the amplitude of this range, we represent two illustrative cases: (a) $c_{\max} = 5c_{\min}$ and (b) $c_{\max} = 2c_{\min}$. For the arcsin transformation (left of Fig. 17.2a, b), the x-axis represents the variable $x = c/c_{\max}$ for all possible values of c belonging to $[c_{\min}, c_{\max}]$; likewise, for the inverse arcsin transformation (right of Fig. 17.2a, b), the x-axis represents the variable $x = c/c_{\min}$.

The transformation ECP_CVRP_smR of Eq. (17.12) emphasizes the impact of higher costs c_{ij}. As we can see in both left graphs of Fig. 17.2a, b, the slope of the function is almost linear for values close to $c_{\min}$ and grows significantly as the costs approach $c_{\max}$. This will result in extra penalization for introducing high-cost links in the solution. The transformation ECP_CVRP_SmR_inv of Eq. (17.13) introduces a more than linear benefit for introducing links with smaller costs, while asymptotically approaching the linear increase for higher costs. Differences between the cases represented in Fig. 17.2a, b illustrate how, for both transformations, the effect is intensified when the proportion $c_{\max}/c_{\min}$ is slower. It is worth noting that even two problems with the same range $[c_{\min}, c_{\max}]$ will generally have different distribution of the costs in the range and thus might lead to different behaviors of the same trans-

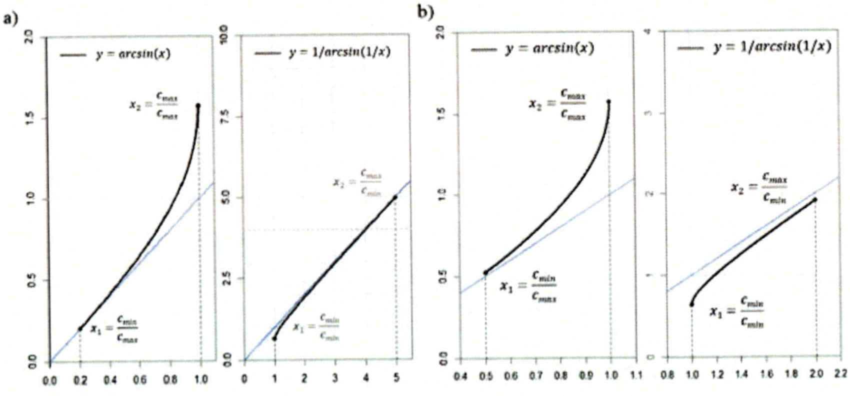

Fig. 17.2 Illustration of the proposed transformation of the objective function in Eqs. (17.12) and (17.13). Case **a** $c_{\max} = 5c_{\min}$; Case **b** $c_{\max} = 2c_{\min}$

formation. It will not be the same whether the costs distribute evenly throughout the range or whether, for instance, there is an accumulation of costs close to $c_{\min}$ and few costs spread in the rest of the range.

17.5 Experimentation

In order to assess the interest of the proposed transformations, we take the first twelve CVRP instances from [1], available at https://neo.lcc.uma.es/vrp/, and compare the results obtained with three different formulations:

- Model A: CVRP formulation of [6] as expressed in Sect. 17.2.
- Model B: ECP_CVRP_SmR formulation, using an R_m^{S} calculated as the longest link in E divided by two, $R_m^{\mathrm{S}} = \max(c_{ij})/2$ in (Eq. 17.12). All other constraints and formulation details remain as in Sect. 17.2.
- Model C: ECP_CVRP_SmR_inv formulation, using an R_m^{SS} calculated as the smallest link in E divided by two, $R_m^{\mathrm{SS}} = \min(c_{ij})/2$ (Eq. 17.13). All other constraints and formulation details remain as in Sect. 17.2.

The experimentation is carried out using a computer with Intel(R) Core (TM) i3-8100 CPU @ 3.60 GHz. Formulations are coded in Python 2.7, with Pyomo [4] and Gurobi 9.1 solver [3] to find the solutions. We establish a time limit of 10,000 s for all runs and an optimality gap limit in Gurobi (gap_G) of 0.01%. In Table 17.1, we show results for experimentation. Best-known solutions (BKSs) for the problem set are expressed in both the standard way (z^* EUC_2D) and their calculated float equivalents (z^* float), which are used to determine the solution gap in terms of distance $z(d)$. We note that BKSs refer to integer distances, while we use exact Euclidean float distances. In the proposed ECP formulation results, we also include the objective function values, respectively, in terms of radians z(angle) or their inverse $z(\text{angle}^{-1})$. The cases in which the optimality gap is achieved are distinguished in bold.

17.6 Discussion

The three formulations reduce the gap to the BKS to less than 3% in 10,000 s. Yet, there are some differences in terms of gap to BKS and solution time. Models (A) and (C) reach the optimal solution four times, while Model (B) does it three times. It is relevant to note that in three cases, Model (A) achieves better objective function value than using either Model (B) or Model (C), while in six cases, either Model (B) (2 cases) or Model (C) (4 cases) provides the best solution. These results point out the interest of further research in the proposed transformations.

Table 17.1 Results

Set	BKS		Model A		Model B		Model C	
	z^* EUC_2D	z^* float	$z(d)$ gap_d	gap_G	$z(d)$ gap_d	z(angle) gap_G	$z(d)$ gap_d	$z(\text{angle}^{-1})$ gap_G
A-n32-k5	784	787.08	**787.08** **0.0000**	14.02	788.65 0.1991	6.2454 11.87	789.17 0.2645	557.3101 33.81
A-n33-k5	661	662.11	**662.11** **0.0000**	14.10	662.26 0.0233	5.7408 13.99	**662.11** **0.0000**	660.9376 27.37
A-n33-k6	742	742.69	**742.69** **0.0000**	14.76	**742.69** **0.0000**	6.6473 15.69	**742.69** **0.0000**	741.8606 26.15
A-n34-k5	778	780.97	788.98 1.0148	18.88	**780.97** **0.0000**	7.1591 19.35	788.98 1.0148	788.3176 37.31
A-n36-k5	799	802.13	821.54 2.3625	23.69	819.02 2.0627	6.6946 24.97	822.06 2.4238	410.4160 35.37
A-n37-k5	669	672.59	**672.46** **−0.0191**	11.19	**672.50** **−0.0137**	5.6270 9.11	**672.52** **−0.0113**	299.2733 15.83
A-n37-k6	949	952.21	958.79 0.6859	26.74	969.01 1.7334	8.1469 26.76	**950.85** **−0.1436**	671.2608 35.33
A-n38-k5	730	734.18	735.05 0.1179	19.51	741.03 0.9239	6.3676 22.44	742.41 1.1083	741.6853 31.6604
A-n39-k5	822	828.98	832.64 0.4396	21.56	853.53 2.8755	7.2882 23.28	837.97 1.0718	837.3114 28.30
A-n39-k6	831	833.20	839.73 0.7777	20.32	842.33 1.0843	6.8645 20.61	835.44 0.2679	834.9029 29.2950
A-n44-k7	937	939.33	953.76 1.5132	10.68	964.64 2.6239	7.6824 12.01	944.41 0.5376	943.4669 33.15
A-n45-k6	944	944.87	969.22 2.5118	21.21	969.88 2.5784	7.3175 21.46	955.99 1.1633	674.9174 35.40

When we analyze the evolution of the gap solution during the executions, we find significant differences in the computation time necessary to reach quality solutions. We focus on the first hour of the Gurobi solver logs to check the evolution of the best solutions found by proposed models. In Fig. 17.3, we show this evolution during the first running hour for the 12 sets. Then, in the x-axes, we measure the Gurobi running time up to 3600 s. In the y-axis, we measure the calculated gap to the float BKS of the objective function value of each solution found. The lines in Fig. 17.3 plot the improvement of the respective objective functions for the three models (black for (A), green for (B), and blue for (C)).

If we focus on the first 3600 s (Fig. 17.3), Model (B) would be the best option four times, (A) and (C) two times each, (A) or (C) three times close to tie, and (A), (B), or (C) once tied. Therefore, using either Models (B) or (C) is useful in many cases. Hence, it may be worthwhile to increase experimentation to explore reasons for the use of Models (B) and (C). In what follows, we briefly comment on Fig. 17.3 case by case:

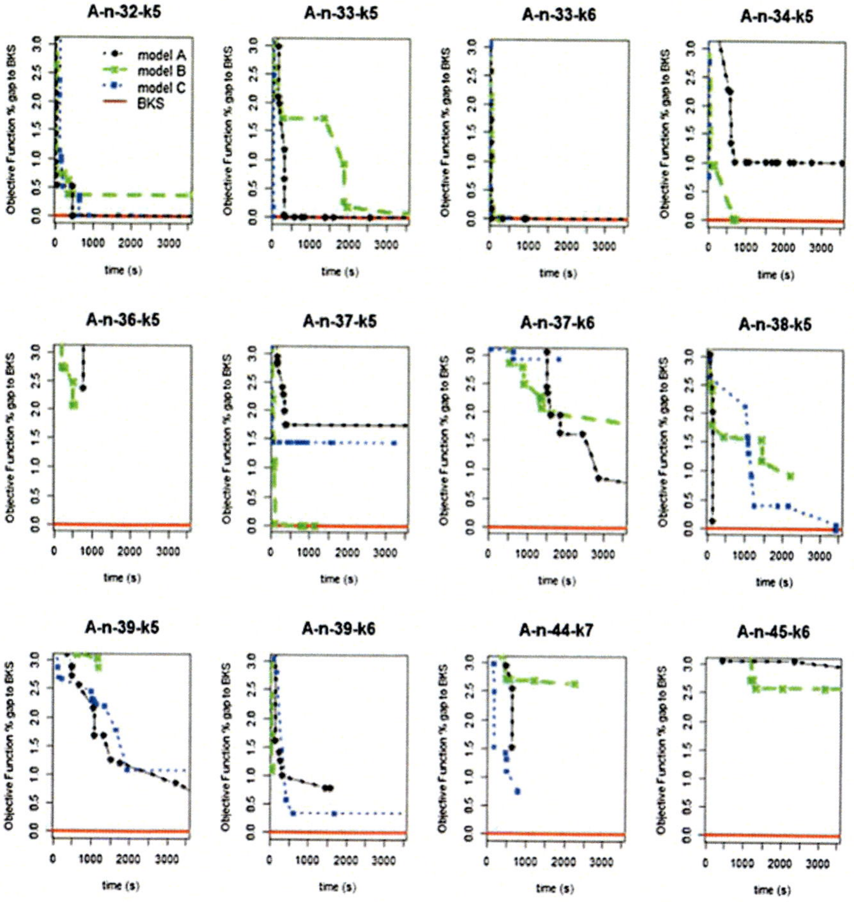

Fig. 17.3 Evolution of the gap improvement in Gurobi solver during the first running hour

- In A-n-32-k5, Models (A), (B), and (C) reduce the gap to BKS to less than 0.5% in less than 500 s, although Model (A) reaches the BKS, while (C) approximates it closely. Therefore, Model (A) would provide the best solution by a reduced margin.
- In A-n33-k5, Models (A) and (C) reach the BKS faster than Model (B).
- In A-n33-k6, Models (A), (B), and (C) perform very efficiently.
- In A-n34-k5, Model (B) finds the BKS in short time, while (C) decrease the gap to less than 1% very fast and (A) decrease the gap to 1% in less than 1,000 s, both consuming the remaining time without significant improvements.
- In A-n36-k5, Model (B) performs better in time and gap reduction (2%), while (C) requires more than 1 h to reduce to less than 3% gap.
- In A-n37-k5, Model (B) finds a very fast and close the BKS, showing better performance than (C) and (A).

- In A-n37-k6, Model (A) behaves better than (B) and (C), although it takes more than 1500 s to cross the 3% gap.
- In A-n38-k5, Model (C) behaves better than (A) and (B), reaching the BKS in less than 1 h, although (A) achieves a high-quality solution much faster.
- In A-n39-k5, Model (A) behaves better than (C) and much better than (B).
- In A-n39-k6, Model (C) behaves better than (A) and (B) both in time and gap reduction.
- In A-n44-k7, Model (C) behaves better than (A) and (B) both in time and gap reduction.
- In A-n45-k6, although Model (C) achieves the best results in 10,000 s, it requires more than 1 h to reduce the gap to less than 3%. If the time is reduced, (B) shows the best behavior followed by (A).

17.7 Conclusions

Through an equivalent cyclic polygon geometric transformation of CVRP possible solutions, we reformulate the CVRP objective function in terms of circumference angles and inversed angles. These transformations turn out to perform better than one classical formulation in some cases. Due to the great attention given to CVRP and the extensive practical applications, the fact that the transformations might lead to better results justify the interest for further research. The geometric transformation opens new paths for the development of efficient algorithms.

We detect the potential interest of both the arcsin and inverse arcsin formulations after the experimentation, with a better performance of the latter in the problem set subject to experimentation. Results suggest that if we take the distance parameters which have a natural linear impact in the classical objective function formulation and modify them by using the ECP geometric transformation, we may obtain some advantage from a nonlinear transformation.

Future research lines should focus on finding the key features that characterize the problems in which the transformations are likely to be beneficial for the solver. On the other hand, once the ECP polygon is defined for VRP solutions, it is immediate to extend the approach to other types of VRPs that might also benefit from the proposed transformations.

References

1. Augerat P, Belenguer JM, Benavent E, Corberán A, Naddef D, Rinaldi G (1995) Computational results with a branch and cut code for the capacitated vehicle routing problem. Research Report 949-M, Universite Joseph Fourier, Grenoble, France
2. Dantzig GB, Ramser JH (1959) The truck dispatching problem. Manage Sci 6:80–91. https://doi.org/10.1287/mnsc.6.1.80
3. Gurobi Op (2021) Gurobi optimizer reference Manual. https://www.gurobi.com. Accessed 30 July 2021
4. Hart WE, Carl DL, Watson JP, Woodruff DL, Hackebeil GA, Nicholson BL, Siirola JD (2017) Pyomo—optimization Modeling in Python, 2nd edn. Springer, p 67
5. Herraiz A, Gutierrez M, Ortega-Mier M (2021) Equivalent cyclic polygon of a euclidean travelling salesman problem tour and modified formulation. CEJOR. https://doi.org/10.1007/s10100-021-00784-z
6. Kara I, Laporte G, Bektas T (2004) A note on the lifted Miller–Tucker–Zemlin subtour elimination constraints for the capacitated vehicle routing problem. Eur J Oper Res 158(3):793–795. https://doi.org/10.1016/S0377-2217(03)00377-1
7. Kulkarni RV, Bhave PR (1985) Integer programming formulations of vehicle routing problems. Eur J Oper Res 20:58–67. https://doi.org/10.1016/0377-2217(85)90284-X
8. Kumar SN, Panneerselvam R (2012) A survey on the vehicle routing problem and its variants. Intell Inf Manag 4(3):66–74. https://doi.org/10.4236/iim.2012.43010
9. Laporte G, Nobert Y (1987) Exact algorithms for the vehicle routing problem. North-holl Math Stud 132:147–184. https://doi.org/10.1016/S0304-0208(08)73235-3
10. Miller CE, Tucker AW, Zemlin RA (1960) Integer programming formulations and traveling salesman problems. J Assoc Comput Mach 7:326–329. https://doi.org/10.1145/321043.321046
11. Toth P, Vigo D (2014) Vehicle routing: problems, methods, and applications, 2nd edn. SIAM, Philadelphia

Chapter 18
Scientific Trends in Artificial Neural Networks for Management Science

M. Jaca-Madariaga, E. Zarrabeitia, R. M. Rio-Belver, and I. Álvarez

Abstract The use of artificial neural network (ANN) is growing significantly, and their areas of application are varied. In this case, the main aim of the study is to present an overall view of trends and research carried out in ANNs specifically in management science. To this aim, the data of publications about ANN in the field of management through Scopus database have been analysed. Documents in the field of management science composed by: Business, Management and Accounting; Decision Sciences; Econometrics and Finance; and Social Sciences published from 2000 to 2019 have been obtained and downloaded. Then, text-mining and network analysis software have been applied to gather, clean, analyse and visualize article data. Thus, it has been found that the pioneer country in this research area is China, followed by the USA and India. The study allows to conclude that in the field of management science, ANNs are mostly used for: logistic regression, prediction, classification, forecasting, modelling, data mining and clustering, among others. In addition, it has also been found that the most used neural network is the convolutional neural network (CNN).

Keywords Artificial neural network · Deep learning · Scientific visualization · Network analysis · Management science

---

M. Jaca-Madariaga (✉) · E. Zarrabeitia · R. M. Rio-Belver · I. Álvarez
Industrial Organization and Management Engineering Department, Faculty of Engineering in Bilbao, University of the Basque Country (UPV/EHU), Leioa, Spain
e-mail: maite.jacamadariaga@ehu.eus

E. Zarrabeitia
e-mail: enara.zarrabeitia@ehu.eus

R. M. Rio-Belver
e-mail: rosamaria.rio@ehu.eus

I. Álvarez
e-mail: izaskun.alvarez@ehu.eus

C. Avilés-Palacios and M. Gutierrez (eds.), *Ensuring Sustainability*, Lecture Notes in Management and Industrial Engineering,
https://doi.org/10.1007/978-3-030-95967-8_18

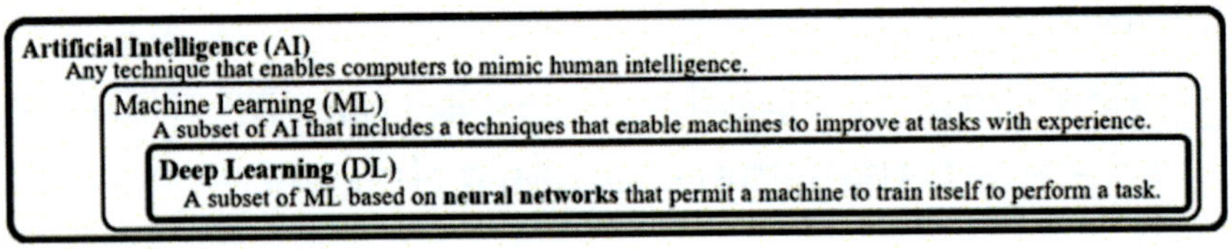

Fig. 18.1 Deep learning; machine learning; and artificial intelligence. *Source* Microsoft Azure [1]

18.1 Introduction

Artificial intelligence (AI) is considered as a fundamental way to process intelligence and listed at the first place in Gartner's Top 10 strategic trends in 2019 [3, 10]. Artificial neural networks (ANNs), the focus of this scientific study, are a very important field within AI. Inspired by the known behaviour of the human brain (mainly that related to neurons and their connections), they try to create artificial models that solve problems that are difficult to solve by conventional algorithmic techniques. Thus, ANNs are algorithms modelled as elementary units or neurons connected in such a way that form a network capable of solving complex nonlinear problems [2]. In Fig. 18.1, main concepts are summarized.

The history of ANNs can be divided into four periods of time considering the aspects of development, maturation and models that have been produced [6]. In the first period, from the 1940s to the 1950s, the first model of neural networks was developed by Walter Pitts. The second period (1960–1970) is when learning rules for networks with a single layer were developed. In the 1980s and the 1990s, after a reduced funding and interest in AI research, called AI winter, the interest in the field of neural networks was renewed developing learning rules for multi-layer networks and new applications. And, finally, the current period is the fourth one from 2000 until the present day, in which optimizations and improvements of the previous models are taking place [5].

Thus, ANNs areas of application are varied: astronomy, mathematics, physics, chemistry, earth and space sciences, life and medical sciences, social and behavioural sciences, economics, arts and humanities, and engineering [6]. In this case, the main aim of the study is to present an overall view of the trends and research carried out in ANNs, specifically in management science.

18.2 Methodology and Objectives

As previously mentioned, we are currently in the fourth historical period of the ANN. Nowadays, a lot of attention is being focused on the subject and large investments are being made to support more research, because of all that, it has been decided to retrieve the published documents of this last period and analyse them. Thus, documents in the field of management science composed by: Business, Management and Accounting; Decision Sciences; Econometrics and Finance; and Social Sciences published from 2000 to 2019 are studied with two main objectives: firstly, to visualize the science of ANNs in management science field: analysing the publication year (*when*); detecting the countries that are researching on them (*where*); authors that are working on them (*who*); and publications with the highest number of citations (*what*) and, secondly, to analyse the knowledge hotspots of the management science field (*what*). To do the latter, for each of the subfields, top 50 authors' keywords have been analysed to conclude the knowledge hubs in the management science field [7].

To prepare this work, the methodology shown in Fig. 18.2 has been followed.

The data for the analysis of ANNs in management science field were obtained from Scopus, a database owned by the company Elsevier, which contains 18,000 journals published by over 5000 international publishers [8]. Firstly, a query terminology was defined to obtain the target data set (Fig. 18.3). Thus, 16,101 records have been obtained and downloaded in their full record format.

The software used to gather, clean, analyse and visualize article data was VantagePoint and Gephi. The first one, VantagePoint v.12, "is a powerful text-mining tool for discovering knowledge in search results from patent and literature databases" [9]. Thus, it enables to analyse and visualize information to find patterns and relationships. And the second software used, called Gephi, is an open source "tool for data analysts and scientists keen to explore and understand graphs" [4].

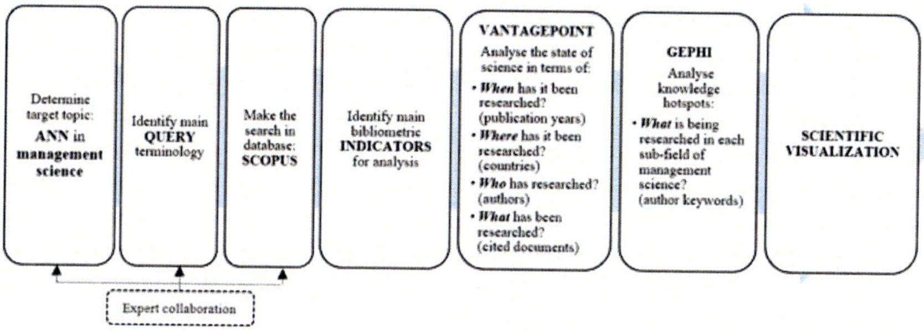

Fig. 18.2 Method of analysis. *Source* Own elaboration based on Zarrabeitia-Bilbao et al. [11]

Eq. (1) [General]	authkey ((neural and network*) or (neuronal and network*) or ("deep learning"))
Eq. (2) [Management]	and (subjarea(busi) or subjarea(deci) or subjarea(econ) or subjarea(soci))
Eq. (3) [2000-2019]	and (limit-to (pubyear > 1999 and pubyear < 2020))
Source	Scopus
Date of the search	March 5th, 2020
Results (∑ Equations)	16,101 articles

Fig. 18.3 Search query for ANN in management area at Scopus database

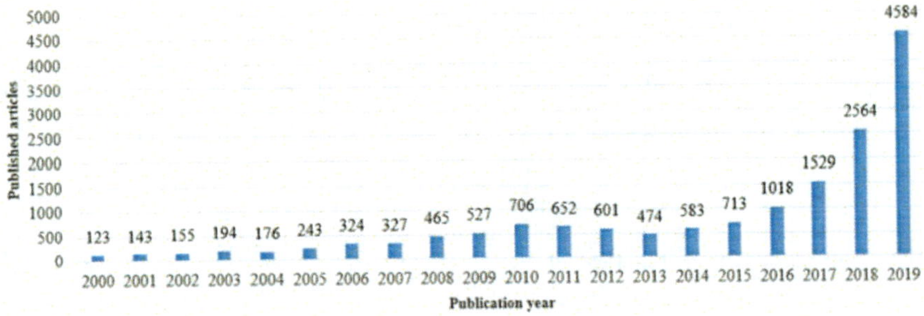

Fig. 18.4 2000–2019 annual number of publications in ANNs (management science)

18.3 Results and Discussion

In the following section, the results obtained are represented and explained.

18.3.1 Publication Year Analysis

The greatest activity has taken place in recent years, and it is evident how interest and intensity have increased in the publications of ANN in management research area (Fig. 18.4). From 2015 onwards, the growth in the topic has been increasing considerably year by year. Moreover, note especially the high growth in the last two years when the number of published documents has almost doubled.

18.3.2 Country Analysis

As it has been observed in the previous section, from the year 2015 on, there has been a great take-off in terms of publications. Due to this fact, the top 10 countries have been charted in Fig. 18.5 for that period.

The predominant country in terms of the highest number of publications by a long way from the rest is China, which more than doubles in every year the USA (second place), so the difference is meaningful. Even so, it is remarkable to say that India has been gathering more and more strength in the last few years as it has gone from 307 documents (year 2018) to 720 documents (2019), which is a significant growth. In addition, it can be seen how all the countries have been growing over the last five years.

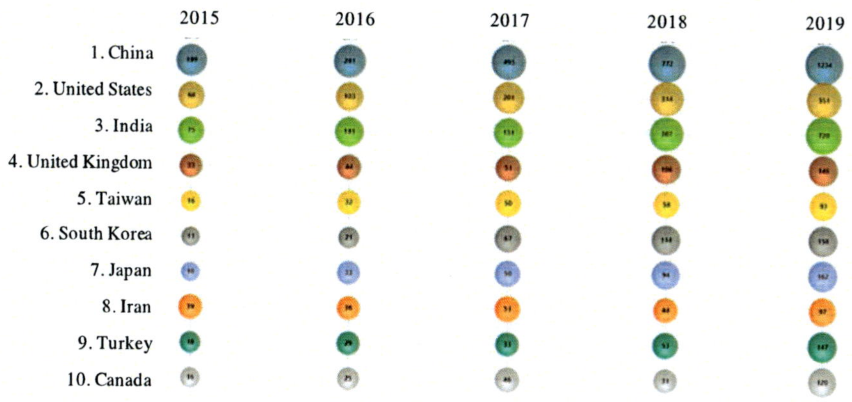

Fig. 18.5 10 most frequent countries from 2015 to 2019

18.3.3 Author Analysis

In Table 18.1, there are shown the 10 authors with the most occurrences. The leading author is Elmer P. Dadios from Philippines with 30 publications on the field. In the

Table 18.1 10 authors with the most occurrences

#	Author	h-index*	# of records	% of records	Shared publications	Country
1	Dadios, Elmer P	13	30	0.186	246/30	Philippines
2	Cao, Jinde	106	22	0.137	1198/22	China
3	Azadeh, Ali H	38	19	0.118	442/19	Iran
4	Pendharkar, Parag C	22	14	0.087	99/14	USA
5	Markovič, Vera V	11	14	0.087	126/14	Serbia
6	Marinkovic, Zlatica	11	13	0.081	90/13	Serbia
7	Melin, Patricia	55	13	0.081	685/13	Mexico
8	Buscema, Massimo	23	13	0.081	184/13	Italy
9	Castillo, Oscar	60	12	0.075	872/12	Mexico
10	Zhang, Peter G	23	12	0.075	43/12	USA

*: Data obtained in March 2020

second position is Jinde Cao from China, a country that, as it has been seen previously in the country analysis, plays a very important role. Note that in the third place there is an author from Iran, a country which, curiously enough, is doing quite a lot of research into this. And the fourth and tenth places are occupied by two different authors from USA, one of the leading countries in ANNs.

18.3.4 Keyword Analysis

Analysing the keywords of publications can help to identify which emerging topics within ANN are. Identifying the centres of knowledge in research and representing them in a co-occurrence network graph, it is possible to observe the relationships between them. In Figs. 18.6, 18.7, 18.8 and 18.9, the top 50 authors' keywords co-occurrence from 2000 to 2019 of each subfield of management science are represented (Business, Management and Accounting; Decisions Sciences; Econometrics

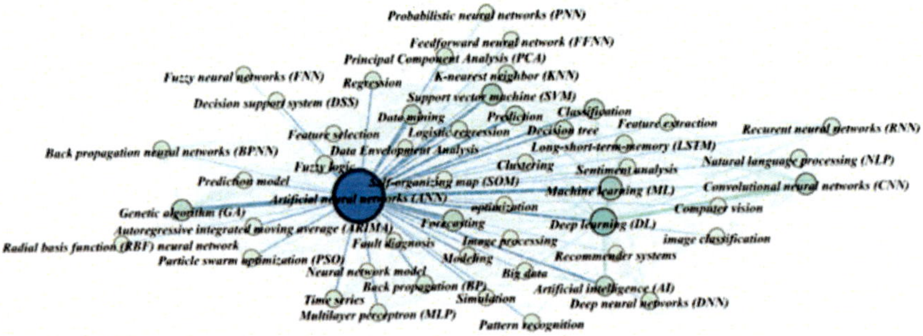

Fig. 18.6 Network of top 50 authors' keywords with most co-occurrence on business, management and accounting

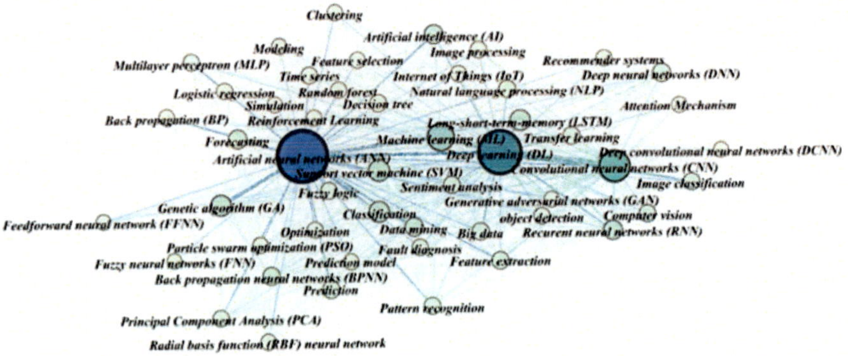

Fig. 18.7 Network of top 50 authors' keywords with most co-occurrence on decisions sciences

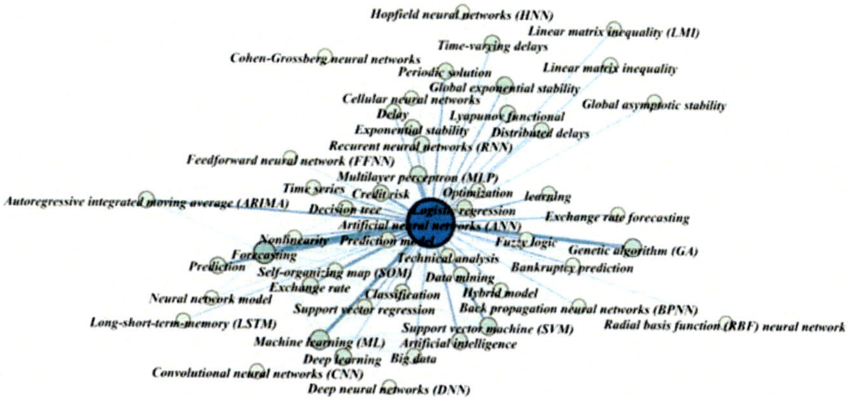

Fig. 18.8 Network of top 50 authors' keywords with most co-occurrence on econometrics and finance

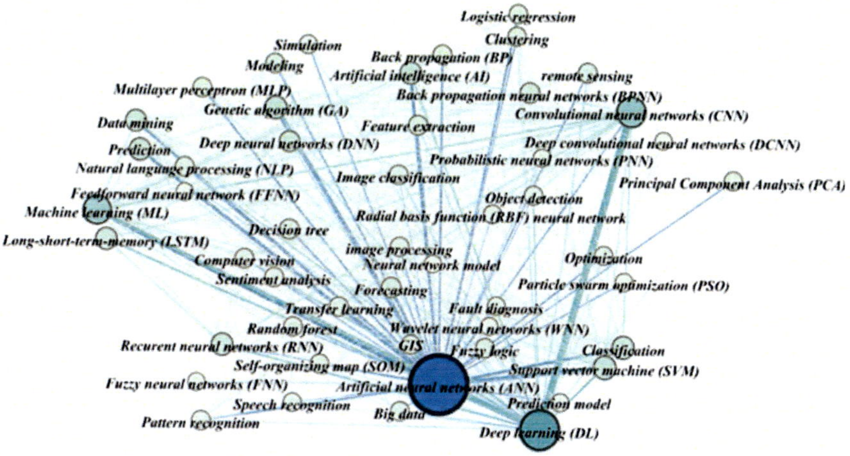

Fig. 18.9 Network of top 50 authors' keywords with most co-occurrence on social sciences

and Finance and Social Sciences, respectively). Having ANN as a nucleus, connections with the most related concepts can be seen. In all the subfields, with the exception of econometrics and finance, ANN is shown to be connected with the concept of deep learning and, in a minor degree but also appreciable, with machine learning.

In general, it is observed that the field of management science uses mostly ANN for logistic regression, prediction, classification, forecasting, modelling, data mining and clustering. For that, in business, management and accounting subfield (Fig. 18.6) it can be seen that different neural networks are used: probabilistic NNs, feedforward NNs, radial basis function NNs, deep NNs and recurrent NNs among others. But, the ones that have a lot of weight in this subfield are convolutional neural networks

(CNNs). In the case of decision sciences subfield (Fig. 18.7), it is remarkable the strength of CNN and the same occurs in social sciences (Fig. 18.9).

The subfield of econometrics and finance (Fig. 18.8) is the one that is the most different from the other. There is a big connection with genetic algorithm and even with forecasting and prediction (bankruptcy prediction is also present here). An important keyword that appears is the autoregressive integrated moving average (ARIMA), and the appearance of this keyword is reasonable as the goal of this model is to predict future securities or financial market moves. Other important concepts of ANN in this subfield are: credit risk, exchange rate and terms that are related in finding the stability as exponential stability, global asymptotic stability, linear matrix inequality, distributed delays and time-varying delays. In addition, in this subfield other types of ANNs appear: Cohen–Grossberg NNs and Hopfield NNs.

18.3.5 Sources of Knowledge

It was decided to analyse the 5 most cited publications as they can result as key resource in the knowledge of this scientific area (Table 18.2). As expected, four of these five documents are from the first part of the last period of ANN (2000–2009). And, the one that is on the second position has already been cited 710 times, even though it is a relatively new publication (2012).

All of these publications, except the first one, have the same fields of management in common (Decision Sciences and Business, Management and Accounting). In addition, as it can be seen in author keywords, three of them have used support vector machines (SVMs), a set of supervised learning algorithms usually used for classification and regression analysis of the data.

18.4 Conclusions and Future Work

Making use of text-mining software and combining it with visualization allows us to understand and draw conclusions of the evolution of a scientific field. Publications of ANN in the field of management sciences are growing year by year, and in the coming years, this growth is expected to continue. Nowadays, discussions and research in this field are occurring everywhere, and undoubtedly, today is a period of transition for ANN technologies as there is still a lot to research and invest in. In this last period, significant progresses have been made such as the development and improvement of the recurrent neural networks, convolutional neural networks and deep feedforward neural networks.

The pioneer country in this research area is clearly China, and it is followed by the USA and India. The size of China's population produces great advantages as it means a huge data traffic which favours the creation of algorithms for ANNs. In addition, from the results obtained by the analysis of countries, authors and sources

Table 18.2 5 most cited documents

TC	Title	Year	Authors	Affiliation	Author keywords
1287	Ensembling neural networks: Many could be better than all	2002	Zhou Z.H Wu J Tang W	Nanjing University, China	Neural network ensemble Machine learning Genetic algorithm Bagging Bias variance decomposition
710	A new Fruit Fly Optimization Algorithm: Taking the financial distress model as an example	2012	Pan W.-T	Oriental Institute of Technology, Taiwan	General regression neural network Data mining Financial distress Optimization problem Fruit fly optimization algorithm
700	Application of support vector machines in financial time series forecasting	2001	Cao L Tay F.E.H	Natl. Univ. Singapore, Singapore	BP neural network Support vector machines Structural risk minimization principle Generalization
565	Bankruptcy prediction in banks and firms via statistical and intelligent techniques -A review	2007	Ravi V Ravi Kumar P	Inst. For Development and Research in Banking Technology, India	Bankruptcy prediction Neural networks Intelligent techniques Fuzzy logic Support vector machine and soft computing
558	Credit rating analysis with support vector machines and neural networks: A market comparative study	2004	Huang Z Hsu C.-J Wu S Chen H Chen W.-H	CGU, Taiwan Univ. of Arizona, USA National Taiwan University, Taiwan	Backpropagation neural networks Support vector machines Bond rating prediction

(continued)

Table 18.2 (continued)

TC	Title	Year	Authors	Affiliation	Author keywords
					Data mining
					Cross-market analysis
					Credit rating analysis
					Input variable contribution analysis

of knowledge, it can be seen that Iran, Taiwan, UK, South Korea, Japan, Turkey and Canada are also actively researching on.

Besides, with the analysis of author's keywords, it has been possible to visualize the knowledge hotspots in each subfield of management. Therefore, the different uses of ANN can be seen: making classifications and regressions; making predictions and forecasting; taking decisions with the support of decision trees; discovering patterns in large data due to data mining and pattern recognition; modelling and simulating; making great optimizations; detecting which fault occurred; speech recognition; and image processing and classification, among others. The only management subfield that differs most from the others is econometrics and finance as most of the keywords are specially related to the financial area. Also with the support of network graphs, it is found that convolutional neural network (CNN) is the one that is most used by far; support vector machines (SVMs) have also a lot of strength mainly to solve problems of classification and regression. In addition, evidently, ANNs are so related to the fields of deep learning, machine learning and artificial intelligence. Besides that, terms like deep neural networks, deep belief neural networks and deep convolutional neural networks show that concepts are merging with each other and some new concepts are being born.

With the present work done, future work will include an analysis of the main journals and institutions that are working on with ANN. Moreover, the documents obtained will be studied in depth to make a case study with the most relevant contributions in the field of management science related to artificial neural networks. Thus, a summary that contains the objectives of these publications will be done; what type of network is used; to which problem it has been applied; a classification of these types of problems tackled by neural network applications: whether it is an operational application or a strategic application; and the findings obtained.

Finally, as a future extension of this study, it would be interesting to analyse the rest of the ANN fields apart from management. Probably, some technologies that are being used or developed in one field can be used in another one by collaborating and sharing knowledge with each other. And, a future possible work would also be to carry out an analysis of each model of ANNs and their applications.

References

1. Deep learning vs. machine learning—Azure Machine Learning | Microsoft Docs (n.d.) Retrieved 29 Oct 2020, from https://docs.microsoft.com/en-us/azure/machine-learning/concept-deep-learning-vs-machine-learning
2. Di Tollo G, Tanev S, Liotta G, De March D (2015) Using online textual data, principal component analysis and artificial neural networks to study business and innovation practices in technology-driven firms. Comput Ind 74:16–28. https://doi.org/10.1016/j.compind.2015.08.006
3. Gartner Top 10 Strategic Technology Trends for 2019—Smarter With Gartner (n.d.) Retrieved 28 July 2020, from https://www.gartner.com/smarterwithgartner/gartner-top-10-strategic-technology-trends-for-2019
4. Gephi—The Open Graph Viz Platform (n.d.) Retrieved 28 July 2020, from https://gephi.org/
5. Macukow B (2016) Neural networks-state of art, brief history, basic models and architecture. Lecture Notes in Computer Science (Including Subseries Lecture Notes in Artificial Intelligence and Lecture Notes in Bioinformatics), LNCS, vol 9842, pp 3–14. https://doi.org/10.1007/978-3-319-45378-1_1
6. Prieto A, Prieto B, Ortigosa EM, Ros E, Pelayo F, Ortega J, Rojas I (2016) Neural networks: an overview of early research, current frameworks and new challenges. Neurocomputing 214:242–268. https://doi.org/10.1016/j.neucom.2016.06.014
7. Río-Belver RM, Garechana G, Bildosola I, Zarrabeitia E (2018) Evolution and scientific visualization of Machine learning field, pp 115–123. https://doi.org/10.4995/carma2018.2018.8329
8. SCOPUS | FECYT (2020) Retrieved 16 Feb 2020, from https://www.fecyt.es/en/content/scopus
9. VantagePoint—The VantagePoint (n.d.) Retrieved from https://www.thevantagepoint.com/
10. Wang J, Ma Y, Zhang L, Gao RX, Wu D (2018) Deep learning for smart manufacturing: methods and applications. J Manuf Syst 48:144–156. https://doi.org/10.1016/j.jmsy.2018.01.003
11. Zarrabeitia-Bilbao E, Álvarez-Meaza I, Río-Belver R-M, Garechana-Anacabe G (2019) Additive manufacturing technologies for biomedical engineering applications: research trends and scientific impact, pp 1699–2407. https://doi.org/10.3145/epi.2019.mar.20

Chapter 19
Establishing a Fulfillment Costs Model for the *Subscription Box*

M. Rodríguez-García, I. González-Romero, C. Hernández-García, and J. C. Prado-Prado

Abstract Most of the companies that have decided on a subscription box model are startups that have gone bankrupt when they were still in their early days. One of the main reasons for this has been the high fulfillment costs they incurred. As subscription models undergo rapid growth, fulfillment costs per customer stay the same or even increase in the mid-term due to companies continually forced to increase the variety of items included in the subscription boxes, increase the types of boxes they offer, or even improve the logistics offer, for example, by offering greater flexibility in the order delivery windows. To respond to this, we developed a cost model for the fulfillment operations of a subscription box model, based on a real case study. The model analyzes the evolution of fulfillment costs, as these businesses expand their market to new populations, increase their number of customers, and vary their offer of products with new subscription boxes.

Keywords Subscription box · Subscription business model · Cost model · Fulfillment costs · E-commerce logistics

19.1 Introduction

Subscription boxes have become one of the most innovative and fastest growing e-commerce models in recent years: in the USA alone, sales increased from 57 million to 2.6 billion dollars between 2011 and 2016 [3]. The term *subscription box* is used for subscription models arising from digital service platforms that have managed to

M. Rodríguez-García (✉) · I. González-Romero · C. Hernández-García · J. C. Prado-Prado
Grupo de Ingeniería de Organización, Dpto. de Organización de Empresas y Marketing, Escuela de Ingeniería Industrial, Universidad de Vigo, c/Maxwell, s/n, 36310 Vigo, Spain
e-mail: miguelgarcia@uvigo.es

I. González-Romero
e-mail: iria.gonzalez.romero@uvigo.es

J. C. Prado-Prado
e-mail: jcprado@uvigo.es

C. Avilés-Palacios and M. Gutierrez (eds.), *Ensuring Sustainability*,
Lecture Notes in Management and Industrial Engineering,
https://doi.org/10.1007/978-3-030-95967-8_19

make the jump to physical product distribution. The industry covers a wide variety of what are usually niche products: food and drink, such as beer or fresh fruit and vegetables, pet food, beauty products, clothes, or even entertainment products such as video games and sports kit [6].

As subscription boxes are innovative, many of the firms that opt for the model are startups [9]. It appeals to them because it easily ensures recurring income and, logically, because demand for such services has grown in recent years. However, most of these companies disappear early on due to the major difficulties they have managing this type of business model, its supply chain, and the demands of the market [6].

There are two specific problems that lead to the closure of startups venturing into this business model: the high percentage of customers who cancel their subscription in the first months and difficulty of controlling fulfillment costs [6]. The *churn rate*, that is, the number of customers canceling their subscription is linked to the high expectations of these customers. Many customers choose subscription box products because of the originality or surprise value of receiving different items each month, which means that they require constant novelty in their boxes [2].

The second problem has to do with the high fulfillment costs involved in managing a subscription box model. Most firms opting for this business model consider marketing to be the main cost for their activity, which is very common among e-commerce models. Furthermore, as the companies are undergoing rapid growth, it is generally thought that fulfillment costs will tend to drop thanks to the economies of scale that will be reached in the future [1].

However, what actually happens is that the firms, in a bid to meet customer demands and control the *churn rate*, are continually forced to increase the variety of items included in the subscription boxes, increase the types of boxes they offer, and even improve the supply service, for example, by offering greater flexibility in the order delivery windows. All these actions mean that, in the mid-term, fulfillment costs per customer stay the same or even increase [6].

Despite these problems, there is practically no literature on fulfillment operations for subscription boxes. It is still a little-known industry, and most articles focus on marketing and describing the features of this new type of customer (see, e.g., [1–3]). Therefore, in order to fill this gap in the literature, our study answers the following research question:

RQ.1 How can fulfillment costs in a subscription box be modeled?

To respond to this question, our research focuses on modeling the most important fulfillment costs in the subscription box industry, in such a way that the model will allow firms using this type of service to evaluate their costs month by month. To do so, we take a real subscription box firm as a case study, which will allow us to glean specific knowledge about this particular industry. Furthermore, we will look at the most relevant literature on fulfillment operations.

19.2 Methodology

The company used as an example for our model offers an online subscription for pet food. In this model, customers receive the exact amount of food their pets require every week. The recipes offered by the firm represent the distinct options for the subscription box (which we will simply call the "box"). Thus, each box contains different types of food and different weights, as they have been adapted to suit each pet. The customer pays a monthly subscription although deliveries are weekly and outsourced to a third-party logistics company (3PL). The 3PL delivers the boxes from base to the main cities and then to each of the customers, with delivery tariffs based on the weight of each delivery.

To model the fulfillment activities and their associated costs, four interviews were initially held with the logistics manager at the firm, each lasting two hours. After that, two visits were made to the facilities of the firm. This combination of research based on observation and interviews is the most appropriate way of gaining the knowledge needed to model business processes in the retail industry (see, e.g., in [8]).

The fulfillment activities to be modeled had to include the complete flow: picking all the items for a box, preparation of the box itself, supply of the box to the customer, and any reverse logistics for the boxes that might occur [5]. Depending on the value of the boxes (not including the component items), many firms with subscription box models decide to set up reverse logistics flows to collect boxes from the homes of their customers for later reuse.

Regarding subscription timing, most are managed monthly, so the model evaluates the month-to-month fulfillment costs. Furthermore, considering the most important variables regarding subscription models, the customer has to be the center of our model [4]: This includes variables such as the number of customers renewing their subscription each month, new customers, what boxes are available, or the cities to which the boxes are sent.

Therefore, to obtain the cost, it is necessary to know how many customers have to be served each month [C_m]]. For this, we use the cohorts matrix created by [7], which combines the *churn rate* with the number of new subscribers joining. The columns in this matrix present the data for the months since the beginning of commercial operations (m), and the rows represent the group of customers that continue in a month (m'). For example, cell $C(m, m')$ represents the customers who joined in month (m) and who continue active in month (m'). Thus, to know how many customers have to be supplied in one month, the following formula can be used:

$$C_m = \sum_{m=1}^{m'} C\left(m, m'\right)$$

As mentioned, the type of box subscribed to and the population of customers are critical variables in this fulfillment costs model. Thus, we consider that information to be the input for adapting our cohorts matrix to each population k and each box

Fig. 19.1 Cohorts matrix from [7] adapted to each box *i* and population *k*

Cohorts Matrix for total number of boxes of type i ordered in population k

Acquisition Cohorts \ Time (months)	1	2	3	4	5	6
1	NB(1,1)	NB(1,2)	NB(1,3)	NB(1,4)	NB(1,5)	NB(1,6)
2		NB(2,2)	NB(2,3)	NB(2,4)	NB(2,5)	NB(2,6)
3			NB(3,3)	NB(3,4)	NB(3,5)	NB(3,6)
4				NB(4,4)	NB(4,5)	NB(4,6)
5					NB(5,5)	NB(5,6)
6						NB(6,6)

type *i*. In our adapted cohorts matrix, therefore, we represent the total number of boxes of type *i* ordered in population *k* in month *m* as: $\left[\mathrm{NB}_{i,k,m}\right]$ (Fig. 19.1).

19.3 Fulfillment Costs Model

Based on the above, our fulfillment costs model will consider the following cost items: purchase of components and supplies [COST G_m], preparation of components to go in the boxes [COST PREP CO_m], preparation of boxes [COST PREP BOX_m], delivery [COST D_m], and reverse logistics [COST RL_m]. The following sub-indices and variables have been used to formulate the model:

i: Type of box

j: Component

k: Population

m: Month

$\mathrm{NB}_{i,k,m}$ = Number of boxes of type i ordered in population k in month m

$\mathrm{NCO}_{i,j}$ = Number of components j for min g part of a box unit of type i

$\mathrm{NPrepCO}_{j,m}$ = Number of times component j is prepared in month m

$\mathrm{ND1}_{k,m}$ = Number of routes from base to population k in month m

WB_i = Weight of the type i box

$\mathrm{WCO}_{i,j}$ = Weight of component j for min g part of a box of type i

WE_i = Weight of packaging of each box of type i

$W_{k,m}$ = Total weight delivered from base to population k in month m

$WD1_{k,m}$ = Total weight delivered in each journey from base to population k in month m

$WInv1_{k,m}$ = Total weight delivered from population k to base in month m

Ce_i = Cost of packaging for each box of type i

Cco_j = Cost of component j

$CprepVCO_j$ = Variable cost of preparing component j

$CprepFCO_j$ = Fixed cost of preparing component j

$CD1_{k,m}$ = Cost of delivering boxes from base to population k in month m

$CD2_{k,m}$ = Cost of delivering boxes from population k to customers in month m

$CDInv_{k,m}$ = Cost of delivering returned containers from population k to base in month m

The monthly supply cost we call "cost of goods" [COST G_m], and it will be associated with the purchase of goods that will be sold in the month. To formulate it, we consider the number of boxes of each type that are sold in the month $\left[NB_{i,k,m}\right]$ and the cost of packaging for each box [Ce_i]. With a box reutilization model, some firms only have to buy boxes for new customers in the month (m). This is calculated by subtracting the number of boxes delivered to population k in the previous month $\left[NB_{i,k,m-1}\right]$ from number of boxes recorded for population k in the current month $\left[NB_{i,k,m}\right]$. To this must be added the components that each type of box i carries $\left[NCO_{i,j}\right]$, multiplied by the unit cost for each component j $\left[Cco_j\right]$ and by the total number of boxes of that type that were sold in month m in each population k $\left[NB_{i,k,m}\right]$.

$$CG_m = \left[\sum_{i=1}^{n}\sum_{k=1}^{K}\left(NB_{i,k,m} - NB_{i,k,m-1}\right) \times Ce_i\right] + \left[\sum_{j=1}^{l}\sum_{k=1}^{K}\sum_{i=1}^{n} NCO_{i,j} \times NB_{i,k,m} \times Cco_j\right]$$

In businesses of this type, we differentiate the purchase cost of the components from that for preparing them because the firm often bears the cost of adapting the component before boxing it. This is called "cost of preparing the components" [COST PREP CO_m], and it can be formulated by considering the total number of components that go in each box $\left[CO_{i,j}\right]$ and the number of boxes of each type to be prepared in the month $\left[NB_{i,k,m}\right]$. Thus, this cost is variable $\left[CprepVCO_j\right]$ because it depends on the number of components that are going to be put in each box a month. To this must be added the fixed cost of preparing each component $\left[CprepFCO_j\right]$, defined as the cost of devoting part of the preparation time to adapting the material.

Because of economies of scale, each month, there will be a specific number of times that this activity is accounted for $\left[\mathrm{NPrepCO}_{j,m}\right]$, and it will be defined so that it covers box supply needs.

$$\mathrm{COST\ PREP\ CO}_m = \left[\sum_{j=1}^{l}\sum_{k=1}^{K}\sum_{i=1}^{n}\mathrm{NCO}_{i,j} \times \mathrm{NB}_{i,k,m} \times \mathrm{CprepVCO}_j\right] + \left[\sum_{j=1}^{l}\mathrm{CO}_{j,m} \times \mathrm{CprepFCO}_j \times \mathrm{NPrepCO}_{j,m}\right]$$

In the same way that there is a monthly cost for component preparation, there is another linked to preparing the boxes that will be delivered called "cost of preparing boxes" [COST PREP BOX_m]. Only the variable part of this cost is used as the fixed cost of preparing each type of box is assumed to be negligible. It can, therefore, be said that the cost of preparing the boxes will mainly depend on the number of boxes to be prepared, which means that the total cost of box preparation in month m will be given by the number of boxes of type i $\left[\mathrm{NB}_{i,k,m}\right]$ sold in all the populations in month m, multiplied by the cost of preparing each box of type i $\left[\mathrm{CprepVB}_i\right]$.

$$\mathrm{COST\ PREP\ BOX}_m = \sum_{i=1}^{n}\sum_{k=1}^{K}\mathrm{NB}_{i,k,m} \times \mathrm{CprepVB}_i$$

To determine the monthly cost of deliveries, called "cost of delivery" [COST DDir_m], it must be considered that each month, the company transports a quantity of parcels from its base to each population k ($D1$) and from each population k to the various customers ($D2$). For that reason, the formula is divided into two parts:

$$\mathrm{COST}\ D_m = \mathrm{COST}\ D1_m + \mathrm{COST}\ D2_m$$

To calculate the monthly cost of deliveries from base to each population k or "cost of delivery 1" [COST $D1_m$], we must multiply the number of deliveries made from base to population k in month m $\left[\mathrm{ND1}_{k,m}\right]$, which is generally a standard set by the firm to optimize transport and cover the service, by the cost of delivering the parcels to that population $\left[\mathrm{CD1}_{k,m}\right]$.

$$\mathrm{COST}\ D1_m = \sum_{k=1}^{K}\mathrm{ND1}_{k,m} \times \mathrm{CD1}_{k,m}$$

The basis for calculating the monthly cost of each of the deliveries made to each of the populations $\left[\mathrm{CD1}_{k,m}\right]$ is the weight of deliveries, considering the weight-based tariffs most often used by 3PLs. Therefore, it is necessary to calculate the total weight in kg delivered to population k in month m $\left[W_{k,m}\right]$, which is obtained by multiplying the number of boxes of Type I that are sent to population k in month m $\left[\mathrm{NB}_{i,k,m}\right]$ by the weight of each box $[\mathrm{WB}_i]$. Thus, we obtain the number of kg delivered to each population k in a certain month m:

$$W_{k,m} = \sum_{i=1}^{n} \mathrm{NB}_{i,k,m} \times \mathrm{WB}_i$$

Furthermore, to obtain the weight of each box type $[\mathrm{WB}_i]$, we consider the total weight of the components $\left[\mathrm{WCO}_{i,j}\right]$ comprising it and the weight of the packaging $[\mathrm{WE}_i]$:

$$\mathrm{WB}_i = \sum_{j=1}^{l} \mathrm{WCO}_{i,j} + \mathrm{WE}_i$$

As mentioned, subscription box businesses enjoy relatively stable demand over them month, which means that a standard number of deliveries from base to each population can be defined $\left[\mathrm{ND1}_{k,m}\right]$. Thus, the weight of each particular delivery will be given dividing the number of kg sent to population k during month m $[W_{k,m}]$ by the number of deliveries sent to that population per month $\left[\mathrm{ND1}_{k,m}\right]$. In this way, we can define the weight of the trucks that go to each population in month m $\left[\mathrm{WD1}_{k,m}\right]$.

$$\mathrm{WD1}_{k,m} = \frac{W_{k,m}}{\mathrm{ND1}_{k,m}}$$

With this detail and the weight-cost tariffs per population, the costs associated with the deliveries to each population k in month m $\left[\mathrm{CD1}_{k,m}\right]$ can be determined.

To supply the boxes within a population, the cost for each box delivered must be considered, assuming that each customer orders one box a month. Thus, the formula for this monthly cost, which is called "cost of delivery 2" [COST $D2_m$], depends on the number of boxes of type i sent to each population in month m $\left[\mathrm{NB}_{i,k,m}\right]$ and the cost of each delivery of box i within a population k in month m $\left[\mathrm{CD2}_{i,k,m}\right]$. To calculate the monthly cost of each delivery of box i in population k in month m $\left[\mathrm{CD2}_{k,m}\right]$ and assuming a weight-based delivery cost within the city, we simply have to take into account the weight of each type of box $[\mathrm{WB}_i]$ calculated previously.

$$\mathrm{COST}\ D2_m = \sum_{i=1}^{n} \sum_{k=1}^{K} \mathrm{NB}_{i,k,m} \times \mathrm{CD2}_{i,k,m}$$

Finally, the cost of reverse logistics $[\mathrm{COST\ RL}_m]$ will have three constituent parts:

$$\mathrm{COST\ RL}_m = \mathrm{COST\ RL1}_m + \mathrm{COST\ RL2}_m + \mathrm{COST\ RL3}_m$$

The first, called [COST $RL1_m$], corresponds to the cost of collecting the box from the home of each customer. In line with the monthly subscription, this collection is made when the last order of the month is delivered, which means that [COST $RL1_m$] tends to zero. The second part, [COST $RL2_m$], is determined by the cost of storing the boxes in a transit warehouse in each population k before returning them to base. In most subscription models, it is reasonable for the customer to store the boxes in their home, which means that we will assume [COST $RL2_m$] as zero. Finally, [COST $RL3_m$] refers to the cost of sending the boxes back from population k to our company. To determine that cost, it is necessary to know the tariffs for each population. Those tariffs will link the weight of each delivery $\left[\mathrm{WDInv}_{k,m}\right]$ with its cost $\left[\mathrm{CDInv}_{k,m}\right]$. The formulas to calculate the cost associated with reverse logistics are as follows:

$$\mathrm{COST\ RL3}_m = \sum_{k=1}^{K} \mathrm{CDInv}_{k,m}$$

The weight of the delivery from each population k to base $\left[\mathrm{WDInv}_{k,m}\right]$ is calculated by multiplying the number of boxes i sent to k in month m by the weight of each box i:

$$\mathrm{WDInv}_{k,m} = \sum_{i=1}^{n} \mathrm{NB}_{i,k,m} \times \mathrm{WE}_i$$

So, by aggregating all the costs, we obtain the total monthly fulfillment cost for a subscription box model, whose processes can be seen in Fig. 19.2:

$$\begin{aligned}\mathrm{LOG\ COST}_m &= \mathrm{CG}_m + \mathrm{COST\ PREP\ CO}_m + \mathrm{COST\ PREP\ BOX}_m \\ &\quad + \mathrm{COST}\ D_m + \mathrm{COST\ RL}_m\end{aligned}$$

19.4 Conclusions

Our costs model is a major contribution to the field of subscription boxes fulfillment, as it is the first, based on a real case study, to focus on the particular logistics features of the industry.

The model includes the most important variables that this type of firm operates with when fulfilling orders, such as the populations they offer the subscription service to, the different types of subscription boxes they offer, and the number of subscribing customers from month to month. Furthermore, it includes other more specific features, such as the components that make up each box on offer or the reverse logistics for the boxes. This contribution is applicable to firms that are thinking of setting up a subscription model or even those that have already established them.

Fig. 19.2 Fulfillment processes comprising the cost model for a subscription box

With the proposed model, it is possible to estimate the monthly fulfillment costs based on different demand scenarios and the range of products and services that each firm defines.

Future research could incorporate other costs that have not been considered here, such as the costs of storing the components, prepared boxes and empty boxes, or delivery alternatives (e.g., different timetable windows) that would affect the last-mile cost. Furthermore, future versions of the cost model should consider synergies in deliveries to customers who are subscribed to more than one box.

Finally, an important aspect that could be included in the future studies is how the fulfillment costs of subscription box models evolve throughout the different evolutionary stages of startups.

Compliance with Ethical Standards The results of this work are supported by a set of four face-to-face interviews and two visits to the company's facilities. The participant of the research study voluntarily agreed and gave informed consent to participate in the interviews and to guide the researchers during the two visits. All data was anonymized before publication. In addition, the respondent was informed about the study objectives and the impact of their responds for modeling business processes. Given the nature of the methodology, the Research Ethics Committee of the University of Vigo states that it does not need further ethics evaluation.

References

1. Baxter K (2016) Subscription business models are great for some businesses and terrible for others. Harvard Bus Rev
2. Bischof SF, Boettger TM, Rudolph T (2019) Curated subscription commerce: a theoretical conceptualization. J Retail Consumer Serv, 101822.
3. Chen T, Fenyo K, Yang S, Zhang J (2018) Thinking inside the subscription box: new research on e-commerce consumers. McKinsey & Company
4. Cook RL, Garver MS (2002) Subscription supply chain. Mid Am J Bus 17(2):37–45
5. Kosior J, Strong D (2006) Supply/demand chain modeling utilizing logistical-based costing. J Enterp Inf Manag 19(3):346–360
6. McCarthy D, Fader P (2017) Subscription businesses are booming. Here's how to value them. Harvard Bus Rev
7. McCarthy DM, Fader PS, Hardie BG (2017) Valuing subscription-based businesses using publicly disclosed customer data. J Market 81(1):17–35
8. Pires M, Pratas J, Liz J, Amorim P (2017) A framework for designing backroom areas in grocery stores. Int J Retail Distrib Manage 45(3):230–252
9. Randall C, Lewis A, Davis A (2016) How subscriptions are creating winners and losers in retail. Harvard Bus Rev

Part V
Project and Process Management

Chapter 20
Why Do Traditional Project Management Methods Hinder the Competitiveness of the Construction Industry?

J. I. Ortiz-Gonzalez, A. Duran-Heras, and G. Castilla-Alcala

Abstract Traditional project management methods are no longer able to foster the competitiveness of construction companies. Variability and uncertainty within the construction projects put both efficiency and productivity into jeopardy. Additionally, the present scenario that construction firms need to confront in Spain is determined by systemic factors (globalization and the aftermath of the 2008 financial crisis); within this context, the production approach must shift from increasing revenues to enhancing competitiveness. To that end, the industry should embrace a new management paradigm. The new model should encourage cooperation among the different parties (owners, design team, contractors). Still, at the same time, it needs to take into account that activities interdependence is a critical factor when it comes to production planning and control.

Keywords Competitiveness · Management · Client · Value · Strategy

20.1 Introduction

The future of the construction sector goes through innovation and improving competitiveness [14]. Its possibilities for expansion will not be based again, at least for a long period, on sales increases above the growth of the economy with reduced and even falling levels of productivity.

A number of factors, along with those mentioned above, characterize the current competitive environment in which construction companies must develop:

J. I. Ortiz-Gonzalez (✉) · A. Duran-Heras · G. Castilla-Alcala
Dpto. de Ingeniería Mecánica, Universidad Carlos III de Madrid, Av. de La Universidad, 30, 28911 Leganés, Spain
e-mail: joortizg@ing.uc3m.es

A. Duran-Heras
e-mail: alfonso.duran@ing.uc3m.es

G. Castilla-Alcala
e-mail: castilla@ing.uc3m.es

C. Avilés-Palacios and M. Gutierrez (eds.), *Ensuring Sustainability*,
Lecture Notes in Management and Industrial Engineering,
https://doi.org/10.1007/978-3-030-95967-8_20

- The high growth rates of infrastructure investment of all kinds in emerging countries.
- The transformative effect that the financial crisis of 2008 had on public and private investment capacity in the West [6].

20.2 Objectives and Methodology

This document focuses on the first phase of a broader project. From the previous premises, the project as a whole has two objectives: (i) to identify the weaknesses derived from the organizational structure underlying the current methods of construction project management and (ii) to make proposals to overcome them and therefore improve the competitiveness of construction companies in the context described above. The present document describes the research conducted and the conclusion drawn regarding the first objective.

Thus, the first phase of the project, whose description is the goal of the present document, encompasses two parts: (1) An analysis of the intrinsic characteristics of the production under construction will be carried out first, regardless of the model of productive organization adopted; (2) the traditional approach to construction project management will be characterized, particularly for Spain's case.

The authors conducted an extensive literature review to characterize the intrinsic features of the production organization in the construction industry and the traditional approach to construction project management.

20.3 Intrinsic Characteristics of the Construction Production Process

Beyond the peculiarities that the different cultural factors of each country introduce in the system of execution of construction projects, their essential characteristics are universal and independent of their sociocultural environment. Indeed, since these characteristics derive from factors intrinsic to the very essence of the construction project, they have a structural nature.

The product of a construction project has intrinsic characteristics (long service life, uniqueness [1], it is complex and capital intense [4]). Besides, the production process also has specific inherent characteristics (long duration, reduced possibility of relocation [7] and complexity). These characteristics of the product and the production process under construction determine levels of uncertainty and variability that are superior to those of other production environments.

Undoubtedly, the above factors structurally introduce risk, risk inherent in the construction project concept itself. The proper conceptualization of risk is, therefore, vital to deal with any construction project with guarantees. Managing a construction project is, in short, equivalent to managing uncertainty and variability.

The method, tools, and philosophy of project management should reduce the uncertainty and variability inherent in the concept of construction without adding additional risks.

20.4 The Traditional Construction Project Management System in the Private Sector

In this section, the authors will analyze the traditional delivery models of construction projects commonly used.

The most common method is based on hiring different companies to design and construct the building or infrastructure. It is the model known as design–bid–build [9]. It is a well-known method in which the role of each member of the project team is evident (the owner defines the requirements, the design team designs, and the general contractor builds). This method gives owners control over the process since design and product characteristics are defined before the selection of the contractor.

The fundamental characteristics of the design–bid–build model are described below.

- Organizational structure. The project team consists of members of different organizations; at least three: owner, design team, and general contractor [15]. The owner defines the requirements and hires the design team, and once this team has completed the design, a competitive bid takes place to select the general contractor. The contracts between the parties are the formal and legal reflection of the aspects previously discussed. The contract is the formalization of a project delivery method and the organizational structure to manage it. The typical contract between the owner and the contractor (and between general contractor and subcontractors) is a transactional contract, whereby both parties formalize a series of agreements that reflect both the specific responsibilities assumed about the project and the consequences of breaching them. The contracts include agreements on price, term, quality, scope, and guarantees of compliance and usually establish penalties for the contractor in case of breach of the acquired commitments [11].
- Product design is the sole responsibility of the design team, and the contractor plays a secondary role in this regard.
- In all cases, the general contractor and its subcontractors design the production process once the design team completed the design of the product, since the choice of the general contractor is after the design [2]. It seems evident that this is so, however, in other industries, products and processes are usually designed at the same time since the interrelation between the two is undoubted. On the other hand, the standard type of subcontracting contract relies on the liability required rather than the search for efficient solutions, which constrains process design.
- Project time and cost management. The construction project must meet its objectives within the framework defined by the cost and time constraints defined by the requirements. There are numerous tools and methodologies to manage time and

cost of a project. However, one idea highlights above the others: Traditionally, professionals assume that the total time and cost reduction is a direct consequence of the reduction of the time and cost of execution of each of the parts of the project. This assumption is incorrect because it ignores something fundamental: the interdependence between the different activities of the project [16]. The consequence of a management model based on the previous premise is the increase in variation and uncertainty, and therefore usually the appearance of deviations in cost, time, and project quality.

- Risk management. It is usual for the owner to try to transfer most of the risks to the general contractor, sometimes without analyzing whether the organization that assumes the risk can address the negative consequences of its concretion and if it has tools to avoid or mitigate [11]. No doubt, this strategy is consistent with the traditional non-integrated vision of project management.
- Supply chain management. Production within traditional methods of construction project management leans on the extensive use of subcontracting, especially in the phase of project execution [5]. General contractors can provide management capacity, financial capacity, and the ability to assume the risks transferred by the owner; however, general contractors substantially transfer those risks to their subcontractors. Besides, each subcontractor usually plans its activities on an individual basis, without considering the interdependence between their activities and other subcontractor's [3].

However, on some occasions, design–bid–build model leads some owners to somewhat frustrating situations (deviations in costs, deadlines, and quality), which motivates the use of other models—also traditional: design–build, full delivery, and others, which nevertheless share most of the previous characteristics.

20.5 Weaknesses Derived from the Underlying Organizational Structure to the Traditional Models of Construction Project Management

Construction Management Association of America (CMAA) is an American organization with more than 6300 members: project managers, owners, contractors. CMAA's website states that its mission is to promote and improve leadership, professionalism, and excellence in the management of development and construction of projects and programs.

CMAA annually organizes a meeting of owners (Owners Forum). During the 2005 meeting, a survey was carried out among the owners, revealing that they demanded lower costs and delivery times from the projects, as well as a better fulfillment of the requirements, explicit and implicit, and less uncertainty about the achievement of the project objectives [13]. The market ultimately demands more value, so a general improvement in competitiveness is highly necessary. Unfortunately, things have changed little since then.

One of the areas that explain the previous deficiencies is the organizational structure model that underlies traditional management approaches; the weaknesses of this typical organizational structure around three fundamental aspects are analyzed below: the relationship between the members of the project team, subcontracting, and the contractual model.

20.5.1 The Relationship Between Project Team Members

The relationships between people from the different companies that make up the construction project teams under the traditional approach are determined on a few occasions by the undisguised existence of mutual distrust [11].

Concerning clients' distrust is reflected in the possibility of imposing painful penalties to the contractors in case of breach of their responsibilities. On the other hand, the risk posed by the character, sometimes excessive, of these penalties is covered by the contractor with the generation of contingency reserves, more or less explicit [10].

Mistrust can cause cost overruns and can lead to the discrepancy, tense relationships, and litigation, which in turn leads to a lack of active cooperation and, therefore, a lower probability of success or optimization of the initial objectives.

Mistrust is the answer to inefficient management of uncertainty and variability. The client appreciates uncertainty regarding the achievement of the objectives of the project. Why?

The traditional model obliges the contractor to assume responsibility for achieving the objectives of the project but does not assign the capacity to manage all the factors that influence the possibility of fulfilling it. Indeed, there are internal factors (requirements changes, design errors) or external factors (legal modifications, modification of binding requirements of external stakeholders), which can have a decisive influence on the success of the project.

In summary, the client appreciates uncertainty because it exists and because the current management model does not reduce it sufficiently. The solution would be to reduce this uncertainty, not to assume it. The solution would go through a relationship based on trust, an integrated relationship, an organizational structure based on a project team in which each party globally visualizes the project as a whole and therefore all parties collectively assume losses and rewards based to global performance [11, 12]. The solution may well be to adopt both an alternative type of organizational framework approach that incorporates the above features and a production planning method that takes into account interdependence between activities.

20.5.2 Subcontracting

General contractors routinely transfer the previous framework of the relationship between the clients and then to their relationship with subcontractors. Therefore, its negative impact on the project is conceptually similar.

It is common for contractors not to analyze the real capacity of subcontractors to meet those objectives, so, sometimes, optimization by subcontracting is only apparent. Thus, increasing the number of relationship scenarios based on the lack of trust might even lead to additional growth of contingency reserves and, therefore, a higher cost overrun for the project. The responsible involvement of subcontractors in product and process design, the establishment of relationships between contractors and subcontractors based on trust and the implementation of active supply chain management policies, would allow the extension of practical cooperation to a broader level, leading in turn to a scenario in which the chances of success and optimization of the project objectives would increase significantly [11].

20.5.3 Contractual Model

The transactional contract, customary in the traditional method of project delivery, is the formal and legally binding expression of the project's organizational framework [11].

Alternative organizational approaches will undoubtedly require different contractual models. Thus, it may be of great interest to explore the possibilities of the relational contract as a concretion of an integrated relationship based on trust, seeking sincere cooperation [8].

20.6 Conclusions

The organizational structure model of the construction project typical of traditional management methods cannot significantly increase the efficiency of the construction process. Taking into account that construction processes have uncertainty and variability as intrinsic attributes, traditional project management models lack the potential to drive the essential improvement of the competitiveness of construction companies.

Such a kind of organizational structure has several features that can underpin the deficiency in the competitive capacity of construction companies, namely: lack of integration of the project team, lack of collaboration, lack of active cooperation, lack of authentic teamwork, lack of trust among the members of the project team.

Overcoming this reality requires that all parties embrace an innovative mentality focused on the adoption of an organizational framework with a certain level of integration between the parties (owners, contractors, designers). This new organizational framework's primary goal is to create value efficiently; to that end, it should pursue these objectives: (1) Create a teamwork environment to overcome the lack of trust between the parties as well as foster cooperation; (2) share losses and benefits between the parties; (3) induce a global vision of the project; (4) properly manage activities interdependence.

In practice, the implementation of the following tools could help accomplish the above objectives:

- Integrated project delivery (IPD) focused on optimizing customer value [8].
- Last Planner System® [3]. It is a pull-type collaborative planning system among all the companies involved in the project. The aim is to achieve a smooth and predictable workflow, controlling through cooperative management the variation due to interdependence between activities.

Acknowledgement This work has been supported by the Madrid Government (Comunidad de Madrid-Spain) under the Multiannual Agreement with UC3M in the line of Excellence of University Professors (EPUC3M20), and in the context of the VPRICIT (Regional Programme of Research and Technological Innovation).

References

1. Andi A (2004) Appropriate allocation of contingency using risk analysis methodology. Civil Eng Dimension 6(1):40–48
2. Ballard G (1999) Work structuring. Lean Construction Institute White Paper, vol 5
3. Ballard HG (2000) The last planner system of production control, Doctoral dissertation, University of Birmingham
4. Ballard G (2005) Construction: one type of project production system. In: 13th international group for lean construction conference: proceedings. International Group on Lean Construction, p 29
5. Creedy GD 2006 Risk factors leading to cost overrun in the delivery of highway construction projects, Tesis doctoral. Queensland University of Technology
6. Economics O (2015) A global forecast for the construction industry to 2030
7. Howell GA (1999) What is lean construction-1999. In: Proceedings of the 7th annual conference of the international group for lean construction (IGLC 7), Berkeley, USA
8. NASFA; COAA, APPA; AGC; AIA (2010) Integrated project delivery for public and private owners
9. Pakkala P (2002) Innovative project delivery methods for infrastructure. Finish Road Enterprise
10. Russell MM, Hsiang SM, Liu M, Wambeke B (2012) Causes of time buffer in construction project task durations. In: Proceedings of the 20th annual conference of the international group for lean construction (IGLC 20), San Diego, USA
11. Slauson NP (2005) The effectiveness of the construction contract. AACE Int Trans; ABI/INFORM Complete pg. PM121
12. Smith JP, Ribkowski Z (2012) Literature review on trust and current construction industry trends. In: Proceedings of the 20th annual conference of the international group for lean construction (IGLC 20), San Diego, USA

13. Thomsen C et al (2010) Managing integrated project delivery. Informe. CMAA
14. Underwood J, Khosrowshahi F (2012) ICT expenditure and trends in the UK construction industry in facing the challenges of the global economic crisis. J Inf Technol Constr 17:26–42
15. Vieira A, Cachadinha N (2011) Lean Construction—how reducing waste can contribute to sustainability. World in Progress Ed.2. Soares da Costa
16. Wambeke BW, Hsiang SM, Liu M (2011) Causes of variation in construction project task starting times and duration. J Constr Eng Manag 137(9):663–677

Chapter 21
A Review of Tools and Techniques in Uncertainty Management

D. Curto, J. de Antón, D. Poza, and F. Acebes

Abstract Risk management has become a topic of interest in the management literature in general and particularly in the project management discipline. Notwithstanding, there is not widespread agreement on the concept of risk. An extensive literature review allows us to identify four types of uncertainty: stochastic, aleatoric, epistemic and ontological. The use of appropriate techniques and tools is basic for quality risk management. This paper aims to present a summary of the tools and techniques recommended to deal with each type of uncertainty.

Keywords Project management · Risk · Risk management · Uncertainty · Uncertainty management

21.1 Introduction

Many companies and organizations have found in specialization through projects the solution to their economic problems, especially aggravated by the last economic and financial crisis. Many surveys, such as World Economic Forum [47], conclude that risk management, in the scope of project management, is one of the most demanded hard skills for recruiters and hiring managers in the last decade.

Regarding the engineering corporations, the World Economic Forum [46] affirms that the requirements of efficient risk management must be directed by a "common riskmanagement strategy". To address the normally broad objectives of a project, this "strategy"? should be implemented as soon as possible in the project life cycle, as exposed in Meredith et al. [30].

The main goal of this paper is to present a proposal of the tools and techniques commonly used in risk management, based on the criterion of the different types of

D. Curto · J. de Antón · D. Poza · F. Acebes (✉)
GIR INSISOC - Dpto. de Organización de Empresas y CIM, Escuela de Ingenierías Industriales, Universidad de Valladolid, Pso. Del Cauce 59, 47011 Valladolid, Spain
e-mail: fernando.acebes@uva.es

J. de Antón
e-mail: juan.anton@uva.es

C. Avilés-Palacios and M. Gutierrez (eds.), *Ensuring Sustainability*, Lecture Notes in Management and Industrial Engineering,
https://doi.org/10.1007/978-3-030-95967-8_21

uncertainty. To achieve this objective, we have made a review of the concepts of risk and uncertainty, and the different classifications identified in the literature. Likewise, we did research on different tools and techniques used for the management of risks and uncertainties.

The remainder of the article is structured as follows. In the next section, we introduce our methodology. As a third section, we present a review of the historical meaning of the concept of risk in the scope of project management. In the fourth section, we describe the identified four categories of uncertainty: stochastic, aleatoric, epistemic and ontological. Therefore, an analysis of the most frequent tools, methods and techniques commonly used in risk management is presented in Sect. 21.5. Finally, we summarize the most relevant findings and conclusions.

21.2 Research Methods

In drawing up this report, we mainly searched for papers, articles, books and publications in general, related to the definitions of risk, uncertainty and different categories of uncertainty. Subsequently, we also searched for publications of methods and techniques frequently used in risk management. The terms used in the scientific journals databases of Scopus, Web of Science and Google Scholar were "risk", "uncertainty", "risk/uncertainty management" and "project management" as well as "'tools/techniques/methods' in 'risk/uncertainty management'". The methodological search process is depicted in Fig. 21.1.

In Phase 2, we selected those articles with a higher relevance, concentrating on the number of citations and paying particular attention to the newer ones. Henceforth, we focused on scientific journals of high impact, i.e., International Journal of Project Management, European Journal of Operational Research, International Journal of Production Research and Journal of Risk Research, inter alia.

As a third phase, we did research on backward and forward citations into the most relevant publications. Unlike the research on backward citations, which is focused

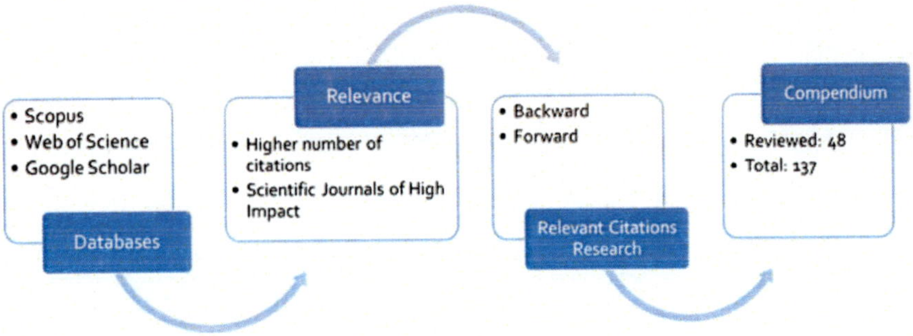

Fig. 21.1 Graph of the methodological search process

on the origins of the field of study, the research on forward citations aims to know the current state-of-the-art of the concepts reviewed.

To present this paper, 48 papers were reviewed out of the 137 documents reported by the research.

21.3 Risk Management

Although there is no broad consensus in the definition of neither risk nor uncertainty [40, 45], the Project Management Institute [42] defined individual risk as "*an uncertain event or condition that, if it occurs, has a positive or negative effect on one or more project objectives*". In this sense, Jaafari [28] perceived risk as the probability of losing or winning. Chapman and Ward [11] took the same view, understanding opportunities and threats as the two sides of a same coin. In Table 21.1, a summary of the views of "risk" by different authors can be seen.

On the other hand, Hillson [19] affirmed that risk was traditionally understood as an event which involves adverse outcomes, such as loss, danger or damage. This vision is present in Dowie [15], as long as risk is associated with negative consequences. By contrast, Chia [12] highlighted the common ground in some features of risk. In this regard, risk means a future event with a probability of occurrence and unexpected-unplanned repercussions.

The main conclusion to be drawn from this literature review is that risk is a living and changing concept. Considering the common principles of the most recent definitions of risk, it would be considered as an event that could lead to a threat or an opportunity [25].

Table 21.1 Summary of the views of risk by different authors

Author	Risk
Williams [45]	No consensus in the definition
Dowie [15]	Negative consequences
Jaafari [28]	Probability of losing or winning
Hillson [19]	Traditionally understood as adverse outcomes, such as loss, danger or damage
Chapman and Ward [11]	Opportunities and threats as the two sides of a same coin
Chia [12]	Common ground in some features of risk
Perminova et al. [40]	No consensus in the definition
Hillson and Simon [25]	Threat or opportunity
Project Management Institute [42]	Event or condition with positive or negative effects

21.4 Uncertainty

The definition of risk made by Hillson [21] is "*uncertainty that matters*". This idea is related to the definition of "general risk"', proposed by the Project Management Institute [42]: "*the effect of uncertainty on the project as a whole, arising from all sources of uncertainty including individual risks*". Based on this idea of connected concepts, Hillson [20] affirmed: "*risk is measurable uncertainty*". In this sense, any assessable uncertainty would be a source of risks.

Owing to uncertainty's fickle nature, it would be necessary to answer "*how uncertainties might 'matter'*" [22]. According to the idea of Alleman et al. [6], the core of any risk is the uncertainty which causes it. It is widely accepted that risk emerges from uncertainty [25] and risk might be a subset consequence of uncertainty [6, 23].

Hence, several problems may emerge; three issues were put in the spotlight: "Uncertainty is everywhere", "not all uncertainties matter" and "not all uncertainties that matter are bad" [24]. In addressing these questions, many classification criteria were reviewed [4, 17].

Hillson [20] drew attention to the relationship between knowledge and awareness in the identification of risks and uncertainties. The author relied on the "Johari Window" a tool from the cognitive psychology developed by Luft and Ingham [29] that describes four quadrants that will correspond with the four types of uncertainty studied. Traditionally, uncertainty was classified into two types: epistemic, related to the lack of knowledge, and aleatory, due to the randomness [16]. Hillson [22] proposed the next broadened classification of uncertainty—"non-event" that matters: stochastic, aleatoric, epistemic and ontological.

21.4.1 Stochastic Uncertainty

Stochastic uncertainty—also called "event risk"—is defined as "*future possible events*" [22]. An "event risk" could be understood as an "action" risk, which means that if the risk cannot materialize—"non-action"—it would not have any impact on the scope and objectives of the project.

The particular way of dealing with stochastic uncertainties necessitates an individual application for each event studied. As for the imprecise accuracy in determining the measures of the stochastic events, statistical techniques are recommended in stochastic treatment [37].

The stochastic nature of a project could be modeled with probability distributions of task durations and costs [2]. The main methods and techniques for dealing with stochastic uncertainty should be stochastic branches [25].

21.4.2 Aleatoric Uncertainty

The aleatoric uncertainty is described by variability, which means there is a wide range of possible results. Its nature is stochastic, in the sense of independence from the system and the environment. Thus, aleatoric uncertainty is irreducible.

This "variability risk" is related to planned events or activities [22]. The difference between the expected and the obtained results would correspond to this effect of variability. From the moment the event provoked by aleatoric uncertainty is known, Hillson proposed managing this kind of uncertainty through quantitative risk models.

The most common procedure to prevent and address aleatoric uncertainty is to create a margin in terms of costs, times and technologies [7]. This concept, as well as different methods of quantitative risk management, will be discussed below.

21.4.3 Epistemic Uncertainty

By contrast, epistemic uncertainty is related to lack of knowledge. This means it could be addressed by getting information about the system and the environment. As a result of living in a deterministic world, epistemic uncertainty relies on real-world limitations [14]. To that effect, expanding the boundaries of knowledge of the organization is just the settlement.

On account of the nature of epistemic uncertainty, the connection established between a system and its environment tends to introduce dependencies synergies. Therefore, the epistemic uncertainty is related to lack of knowledge.

Due to the lack of information, Hopkinson [26] points out the necessity of the formation of a risk maturity model (RMM) in the organization to prevent unexpected epistemic events. This kind of uncertainty would get reduced if information and communications technology (ICT) was further developed.

21.4.4 Ontological Uncertainty

On the other hand, nothing is knowable about ontological uncertainty. It transcends the physical limitations of a system and the environment. Ontological uncertainty is the unknown knowledge of what is impossible to know.

Commonly known as "Black Swans", since Taleb [44] first coined the term, the ontological uncertainty would correspond to those uncertainties which cannot be knowable [7]. The ontological uncertainty is located "*outside of our current mindset or cognisance*" [22]. The ontological uncertainty cannot be sought because it is impossible to know where to start. In general, the consequences of the occurrence of an ontological uncertainty event are usually negative.

21.5 Tools and Techniques

There are good reasons for categorizing tools, methods and techniques by the nature of the event treated. Zarikas and Kitsos [48] conclude that most experts are prone to misjudging threats, while they exaggerate gains. The crucial milestones of a project used to be in the presence of uncertain deadlines [9]. As they explain, becoming aware of this uncertainty results in estimations with a higher accuracy, which will increase the opportunities for the success of the project.

In the risk management process, only identified risks could be treated [23]. Consequently, the identification stage precedes the management phase. In this study, we attempted to show the relevance of determining the type of uncertainty to assign the proper tool, method or technique.

As for the occurrence of a risk, one form of certainty is to observe the deviation of the project control indicators. When a control parameter is outside the tolerance limits of the project design, we could think that risk is occurring [2].

To exemplify this, a huge number of methods were reviewed. In Table 21.2,

Table 21.2 Summary of the methods commonly used in detecting and classifying new risks

Tool, method and technique	Author(s)
Brainstorming	Chapman and Ward [11]
Cause and effect diagram	Project Management Institute [42]
Change analysis (ChA)	Mullai [34]
Checklist	Mullai [34], Project Management Institute [42]
Expert judgment	Project Management Institute [42]
Heuristic methods	Rojas-Delgado et al. [43]
Human reliability assessment (HRA)	Mullai [34]
Incident reporting	Cinotti [13]
Multi-criteria decision methods	Ishizaka and Nemery [27], Hillier and Lieberman [18]
Risk breakdown matrix (RBM); risk breakdown structure (RBS)	Hillson [19]
Risk probability and impact matrix (RPIM)	Project Management Institute [42]
Strengths, weakness, opportunities and threats (SWOT) analysis	Project Management Institute [42]
What-if analysis	Project Management Institute [42]
"5 Why's" technique	Mullai [34]

Table 21.3 Summary of the methods commonly used in managing risks and addressing uncertainty

Tool, method and technique	Author(s)
Decision tree analysis	Project Management Institute [42]
Delphi	Project Management Institute [42]
Event tree analysis (ETA)	Mullai [34]
Expected monetary value (EVM)	Pajares and López-Paredes [38]
Fault tree analysis (FTA)	Aven and Zio [8],Pramanik et al. [41]
Failure mode and effects analysis (FMEA)	Mullai [34]
Fuzzy logic	Aven and Zio [8], Pramanik et al. [41]
Linear programming	Hillier and Lieberman [18], Ishizaka and Nemery [27]
Margin	NASA [35], Alleman et al. [7]
Monte Carlo	Acebes et al. [1–4]
Multi-criteria decision	Ishizaka and Nemery [27], Hillier and Lieberman [18]
Pareto analysis (PA)	Mullai [34]
Sensitivity analysis	Acebes et al. [2]
Statistical experiments	Montgomery [31, 32]

techniques typically used in identifying and categorizing new risks are exposed. Besides, tools commonly used in managing risks and addressing uncertainty are described in Table 21.3. Furthermore, in Table 21.4, methods are classified by the type of uncertainty they could address, due to the nature of the variables treated in the methods and the nature of the type of uncertainty.

In the knowledge of its limitations, both quantitative and qualitative techniques are appropriate to anticipate and prevent the occurrence and recurrence of risks. It should be noted that the concept of quantitative techniques cannot be a hotchpotch of tools. By contrast, they are substantiated by different logical and mathematical basis. Conversely, the qualitative techniques focus on the analysis of qualitative variables: characteristics, categories, attributes and features [33].

Both qualitative and quantitative techniques [10, 39] and design of statistical experiments methods [31, 32] were reviewed.

Similarly, multi-criteria decision methods, such as analytic hierarchy process (AHP), analytic network process (ANP), ELimination Et Choice Translating Reality (ELECTRE), Preference Ranking Organization METHod for Enriched Evaluation (PROMETHEE) and Technique for Order of Preference by Similarity to Ideal Solution (TOPSIS), as well as linear programming [18, 27] were studied.

Table 21.4 Summary of the classification of methods and types of uncertainty

Tool, method and technique	Uncertainty[a]		
	Epistemic	Aleatoric	Stochastic
Decision tree analysis	X		
Delphi	X		X
Event tree analysis (ETA)	X	X	X
Expected monetary value (EVM)		X	X
Fault tree analysis (FTA)	X		
Failure mode and effects analysis (FMEA)	X	X	X
Fuzzy logic	X		
Linear programming		X	X
Margin		X	
Monte Carlo		X	
Multi-criteria decision	X		X
Pareto analysis (PA)	X	X	X
Sensitivity analysis	X	X	X
Statistical experiments		X	
Structured what-if technique (SWIFT) analysis	X		

[a] Ontological uncertainty is not included considering that only contingency and disaster plans could address it

Based on fuzzy and Boolean logic, fuzzy logic and event and fault tree analysis, proposed by Mullai [34], Aven and Zio [8], Pramanik et al. [41] were revised. In the scope of heuristic methods, genetic algorithms, taboo search, greedy randomized adaptive search procedure (GRASP), simulated annealing and artificial neural network [43] were also reviewed.

The main handling tool in aleatoric uncertainty management is margin, which is the tolerance given to the deviations of "*budget, projected schedules and technical performance parameters*" [35].

In addressing the variability, Monte Carlo method is recommended instead of less-accurate methods of critical path method (CPM) or Gantt Chart [1, 2, 36].

Monte Carlo is used in treating the subsequent risks generated by the aleatoric uncertainty. In fact, it is possible to obtain the necessary information to know the contribution that the measured aleatoric uncertainty makes to the whole project [3, 4].

On the other hand, the earned value management (EVM) technique monitors the deviation between the real and the estimated data from the execution of a project [38].

Likewise, other mathematical methods such as human reliability assessment (HRA) [34], Pareto analysis [34] and sensitivity analysis [2] were also reviewed.

In addressing stochastic uncertainty, the method of decision tree analysis [42] was revised. The methods of Delphi [42], risk breakdown matrix and structure [19]

and strengths, weakness, opportunities and threats—SWOT analysis [42] —focus on acquiring information.

The remaining methods reviewed are incident reporting (IR) [13], risk probability and impact matrix [42], what-if analysis and structured what-if (SWIFT) analysis, failure mode and effects analysis (FMEA) and "5 Why's" technique [34].

As explained by Ale et al. [5], ignorance in not treating ontological uncertainty—Black Swans—cannot be an excuse for failed management. They made a call for organizing "contingency and disaster plans" to avoid threats and promote opportunities.

21.6 Conclusions

In this paper, we present some of the most outstanding definitions and classifications of risk and uncertainty. There is no denying the fact that risk has traditionally been understood as a threat to the achievement of the objectives of a project. Delving into the most current definitions, it is certain to affirm that risk is considered to be bi-directional, in the sense of threats and opportunities.

Within the framework of controlling risk in projects, we suggested the main tools and techniques commonly used in risk management. The most common qualitative and quantitative tools have been reviewed for this paper. They include multi-criteria decision methods, heuristic, logical, matrix, mathematical or diagram-based techniques. Other methods commonly used to search for risks have been reviewed.

In consequence, we adopted an appropriate taxonomy of uncertainty that allowed us to assign the most suitable tools, methods and techniques to each type of uncertainty. In this study, we have endeavored to show the importance of correctly identifying the uncertainty of any risk to assign it a tool, technique or method that fits its nature. Managing uncertainty has, therefore, become a key factor for the success of any project.

Our findings show that, due to the particular nature of each type of uncertainty, there are specific tools and techniques for uncertainty management. We have attempted to show the importance of identifying the correct uncertainty to assign the proper tools, methods and techniques for the success of any project.

References

1. Acebes F et al (2014) A new approach for project control under uncertainty. Going back to the basics. Int J Project Manage 32(3):423–434
2. Acebes F et al (2014) Exploring the influence of seasonal uncertainty in project risk management. Procedia Soc Behav Sci 119(April):329–338
3. Acebes F et al (2015) Stochastic earned value analysis using Monte Carlo simulation and statistical learning techniques. Int J Project Manage 33(7):1597–1609

4. Acebes F, Pajares J, López-Paredes A (2015) Gestión de riesgos del proyecto. Desde la gestión de riesgos a la gestión de incertidumbre. In: XIX international congress on project management and engineering, July 2015, pp 15–17
5. Ale BJM, Hartford DND, Slater DH (2020) Dragons, black swans and decisions. Environ Res 183:109–127
6. Alleman GB, Coonce TJ, Price RA (2018) Increasing the probability of program success with continuous risk management. Coll Perform Manage Measur News 4:27–46
7. Alleman GB, Coonce TJ, Price RA (2018) What is risk? Coll Perform Manage Measur News 1:25–34
8. Aven T, Zio E (2018) Knowledge in risk assessment and management, 1st edn (eds: Aven T, Zio E). Wiley, Chichester, UK
9. Bordley RF, Keisler JM, Logan TM (2019) Managing projects with uncertain deadlines. Eur J Oper Res 274(1):291–302
10. Cagliano AC, Grimaldi S, Rafele C (2015) Choosing project risk management techniques. A theoretical framework. J Risk Res 18(2):232–248
11. Chapman C, Ward S (2003) Project risk management: processes, techniques and insights, 2nd edn. Wiley, Chichester, UK
12. Chia ES (2006) Assessment framework for project management. IEEE, pp 376–379
13. Cinotti R (ed) (2004) La Gestione del Rischio nelle Organizzazioni Sanitarie, 1st edn. Il Pensiero Scientifico Editore, Rome
14. Damnjanovic I, Reinschmidt K (2020) Data analytics for engineering and construction project risk management. Springer International Publishing (Risk, Systems and Decisions), Cham
15. Dowie J (1999) Against risk. Risk Decision and Policy 4:57–73
16. Elms DG (2004) Structural safety–Issues and progress. Prog Struct Mat Eng 6(2):116–126
17. Hazır Ö, Ulusoy G (2019) A classification and review of approaches and methods for modeling uncertainty in projects. Int J Prod Econ
18. Hillier FS, Lieberman GJ (2015) Introduction to operations research, 10th edn. McGraw-Hill Education
19. Hillson D (2002) Extending the risk process to manage opportunities. Int J Project Manage 20:235–240
20. Hillson D (2004) Effective opportunity management for projects—exploiting positive risk. Taylor & Francis
21. Hillson D (2009) Managing risk in projects, 1st edn. Gower Publishing Limited, Farnham, UK
22. Hillson D (2014) How to manage the risks you didn't know you were taking. In: PMI® Global Congress, pp 1–8
23. Hillson D (2016) The risk management handbook: a practical guide to managing the multiple dimensions of risk, 1st edn (ed: Hillson D). Kogan Page Publishers
24. Hillson D (2020) Capturing upside risk: finding and managing opportunities in projects, 1st edn. Taylor & Francis, Boca Raton
25. Hillson D, Simon P (2012) Practical project risk management. The ATOM methodology, 2nd edn. Management Concepts Press
26. Hopkinson M (2011) The project risk maturity model. Measuring and improving risk management capability, 1st edn. Gower Publishing limited, New York
27. Ishizaka A, Nemery P (2013) Multi-criteria decision analysis, 1st edn; Multi-criteria decision analysis: methods and software, 1st edn. Wiley, Chichester, UK
28. Jaafari A (2001) Management of risks, uncertainties and opportunities on projects: time for a fundamental shift. Int J Project Manage 19(2):89–101
29. Luft J, Ingham H (1961) The Johari window. Human Relations Training News 5(1):6–7
30. Meredith JR, Shafer SM, Mantel SJ (2017) Project management: a strategic managerial approach, 10th edn. Wiley
31. Montgomery DC (2017) Design and analysis of experiments, 9th edn (ed: Arizona State University). Wiley, Hoboken, NJ
32. Montgomery DC (2019) Supplement text material. In: Design and analysis of experiments. Wiley

33. Montgomery DC, Runger GC (2004) Probabilidad y Estadítica aplicadas a la Ingeniería, 2nd edn. McGraw-Hill Education, México City
34. Mullai A (2006) Risk management system—risk assessment frameworks and techniques, European Union (European Regional Development Fund) (ed: Turku). DAGOB Publication Series
35. NASA (2014) NASA space flight program and project management handbook, 1st edn. NASA, Washington, D.C.
36. Ortiz Pimiento NR, Diaz Serna FJ (2018) The project scheduling problem with non-deterministic activities duration: a literature review. J Ind Eng Manage 11(1):116
37. Padalkar M, Gopinath S (2016) Are complexity and uncertainty distinct concepts in project management? A taxonomical examination from literature. Int J Project Manage 34(4):688–700
38. Pajares J, López-Paredes A (2011) An extension of the EVM analysis for project monitoring: the cost control index and the schedule control index. Int Project Manage 29(5):615–621
39. Pellerin R, Perrier N (2019) A review of methods, techniques and tools for project planning and control. Int J Prod Res 57(7):1–19
40. Perminova O, Gustafsson M, Wikström K (2008) Defining uncertainty in projects—a new perspective. Int J Project Manage 26(1):73–79
41. Pramanik D, Mondal SC, Haldar A (2020) 'A framework for managing uncertainty in information system project selection: an intelligent fuzzy approach. Int J Manage Sci Eng Manage 15(1):70–78
42. Project Management Institute (2017) A guide to the project management body of knowledge (PMBOK® Guide), 6th edn. Project Management Institute
43. Rojas-Delgado J, Trujillo-Rasúa R, Bello R (2019) A continuation approach for training Artificial Neural Networks with meta-heuristics. Pattern Recogn Lett 125:373–380
44. Taleb NN (2007) The black swan: the impact of the highly improbable, 1st edn. Random House
45. Williams TM (1995) A classified bibliography of recent research relating to project risk management. Eur J Oper Res 85:18–38
46. World Economic Forum (2016) Shaping the future of construction. A breakthrough in mindset and technology. World Economic Forum (WEF)
47. World Economic Forum (2018) The future of jobs report. World Economic Forum (WEF)
48. Zarikas V, Kitsos CP (2015) Theory and practice of risk assessment (eds: Kitsos CP et al). Springer International Publishing

Chapter 22
Overview of Methods for Measuring Technological Maturity

R. Meza, G. Garechana, R. M. Rio-Belver, and E. Cilleruelo

Abstract Businesses with a technology management office can monitor and evaluate their technologies more efficiently. Companies measure technological maturity in their processes and products with the purpose of fulfilling properly their customers' needs. There is a wide range of methods for measuring technological maturity, where two major currents stand out. On one side, it is the Technology Readiness Level (TRL), which was developed by the National Aeronautics and Space Administration (NASA), and the methods made from its foundations. On the other side, there are the methods mainly made from foundations such as technology life cycle. This paper aims at giving a brief review of the main methodologies used for measuring technological maturity. Some of the features of these methods to be described will be a description of their performance, their inputs and where they have been implemented by their creators.

Keywords Technology maturity · Technology readiness level · Technological maturity evaluation · Technological characterization · Technological maturity assessment

R. Meza (✉) · G. Garechana · R. M. Rio-Belver · E. Cilleruelo
Dpto. de Organización de Empresas, Universidad del País Vasco, C/ Nieves Cano 12, 01006 Vitoria, Spain
e-mail: rameza002@ikasle.ehu.es

G. Garechana
e-mail: gaizka.garechana@ehu.es

R. M. Rio-Belver
e-mail: rosamaria.rio@ehu.eus

E. Cilleruelo
e-mail: ernesto.cilleruelo@ehu.eus

C. Avilés-Palacios and M. Gutierrez (eds.), *Ensuring Sustainability*,
Lecture Notes in Management and Industrial Engineering,
https://doi.org/10.1007/978-3-030-95967-8_22

22.1 Introduction

These days, in a rapidly changing market, many enterprises need to evaluate technologies validly and robustly. In this context of technology evaluation, the concept of technological maturity is widely used [8]. It has been defined as the description of the evolution of technology over time with respect to its application domains [2] and as a concept which refers to the availability of a technology or technology systems in the process of research and development [34].

Nieto et al. [20] said that the use of methods to determine the maturity of technology is useful for identifying the situation in which a certain technology is found at any given moment. Many times, companies measure technological maturity as a part of their technology management. Technology in the different phases of its maturity has pros and cons. For instance, emerging technologies have tremendous competitive potential. Nevertheless, technologies in this phase are very uncertain about how they will evolve in future.

There are two main ways to make a method for measuring technological maturity. The first one is based on aspects such as the technological life cycle. The technology life cycle is composed of different stages: emergence, growth, maturity and saturation. There are different types of growth curves to analyse the technology life cycle. The use of these curves assumes that technology evolution follows an S-shaped growth curve, which is referred to as a sigmoid function [37]. The second one is based on the method created by NASA, which is named the TRL. An accurate definition for this methodology is a measurement system that assesses systematically the maturity of a technology and allows to compare the maturity among different sorts of technologies [27].

Companies where technology is professionally managed have a competitive advantage. This competitive advantage takes the form of a decrease in costs and an increase in internal capabilities, among other aspects. All these competitive advantages allow companies to meet their customers' needs better than their competitors.

22.2 Methodology

For accomplishing the literature review on methods for measuring technological maturity, the most relevant articles on this topic were searched in databases, such as Scopus and Web of Science. The search was narrowed using various terms, such as "technology maturity", "technology readiness level" and "technological maturity assessment". The most quoted articles were selected for being part of this piece of research. The purpose of this paper is to provide readers with the most significant research on methods for measuring technological maturity.

22.3 Background and Significance of Measuring the Technological Maturity

First of all, it is appropriate to mention that technological maturity is composed of important parameters such as its factual versus theoretical capability, diffusion or usability among potential users. This maturity can be represented by the position of the technology within the technology life cycle. Technology life cycle models build patterns over time of certain technology parameters [1].

For accomplishing an accurate assessment of technological maturity is especially important to know when to introduce new technology into a system [17]. Among the wide range of methodologies used to measure technological maturity, TRL is the most common method to do it. Some scientists had defined this method as a tool useful for evaluating the maturity of technologies during their development through their operations [31]. Companies that use the TRL to measure the technological maturity of their technologies can take advantage of the benefits that this method offers such as: first, it appears that TRL "is useful to facilitate the communication of the status of technology", second, its "scale proved to be efficient at technology selection and procurement in the areas where there are guidelines developed", and third, the implementation of the TRL "scale in assessment guidelines explicitly promotes testing and verification" [12]. However, as mentioned above, the TRL has some gaps in its performance. That is the reason why other methods have been invented to assess technology maturity in specific cases.

Many of the alternative methods to measure the technology maturity use the maturity scale implemented for the TRL. This scale is shown in Table 22.1.

Table 22.1 NASA technology readiness level scale [22]

TRL	Definition
9	Actual system "flight-proven" through successful mission operations
8	Actual system completed and "flight qualified" through test and demonstration (ground or flight)
7	System prototype demonstration in a target/space environment
6	System/subsystem model or prototype demonstration in a relevant environment (ground or space)
5	Component and/or breadboard validation in a relevant environment
4	Component and/or breadboard validation in a laboratory environment
3	Analytical and experimental critical function and/or characteristic proof of concept
2	Technology concept and/or application formulated
1	Basic principle observed and reported

22.4 Methods for Measuring Technological Maturity

At present, there are two universally accepted methodological approaches for assessing technological maturity. The first one is future-oriented technology analysis (FTA) whose maturity assessment tools come mostly from technology forecasting. And the second one is named TRL that is mainly assessed by expert surveys [15].

22.4.1 *Future-Oriented Technology Analysis (FTA)*

FTA is a set of forecasting, foresight and assessment techniques that are technology oriented, though not in a predictive sense. This approach assumes that the future is not predetermined, but it can evolve in different directions. Its evolution depends on actions taken, and decisions made by different players. Many of the successful methods which are still used were developed during the 1950s and 1960s [5].

This group of techniques allows to investigate the interaction of technology with the environment and the identification of factors which affect the development of the technology, together with enabling to determine the maturity level of a certain technology [11].

These techniques include many approaches, even though the most widely used to assess technological maturity is technology life cycle analysis. In this methodology, S-curve represents the life cycle of the technology. In its representation, S-curves commonly cascade with a new one which begins where the last one leaves off. The beginning of a new curve represents that a new product replaces the old one or its equivalent when new technologies replace old technologies [19]. That is to say, the S-curve model says that technological change is cyclical. Each new S-curve starts behaving unstably, after this initial period, its performance has a rapid improvement, then its returns diminish, and finally, it is displaced by a new technological discontinuity [30]. One of the effects caused by the emergence of a new technological discontinuity might be the change in the order or the importance of the existing competitors of industry, creating new leaders and new losers.

Through the life cycle analysis (LCA) of a certain technology, it can be determined their level of maturity. LCA also has arisen with the purpose of managing TRL's constraints. LCA and TRL have proven to have a strong relationship. Since on the one side, the TRLs present a solid maturity index. On the other side, nevertheless, the way to determine them relies nearly exclusively on expert evaluation. In addition, on the one hand, the FTA methods enlarge the techniques for determining maturity. However, it is commonly based on LCA, whose stages and transitions are less established than TRL [25].

According to Ford and Ryan [7], the possibility of achieving competitive advantages strongly depends on the maturity of a certain technology. As shown by Reinhart and Schindler [24] there is an inverse relationship between "competitive potential" and "maturity". Maturity increases over time while competitive potential decreases.

Taking all this into consideration, it follows that TRLs 1–4 may be in the first stage, TRLs 5–8 might be in the second stage, and TRL 9 might be in the third stage. The last stage does not include TRL, as this stage goes beyond TRL intended assessment [25].

It is important to mention that FTA techniques are used with other methods to determine technological maturity [5]. The main pieces of research in this domain are as follows:

Reinhart and Schindler [24] created a method that measures the maturity of manufacturing technology from a strategic point of view. This methodology is based on the life cycle of technology and the creation of a profile of technological maturity. The technological maturity profile implies that each level of the TRL has both an index of maturity and a limit of that maturity. Starting at the first level, if the index of maturity is higher than its limit, then the same evaluation must be repeated at the next level. Technological maturity is reached at the level where the index of maturity is lower than its limit. The input of this method is expert opinion.

Lezama-Nicolás et al. [15] developed a bibliometric method to assess technological maturity. This method estimates the TRL as an approximation of technology life cycle stages, which in turn can be obtained through bibliometric records. This estimation was supported by the findings of a previous piece of research conducted by Watts and Porter [35] where each stage of technology life cycle was allocated an indicator. The inputs of this method were scientific papers stored in databases.

Gao et al. [9] devised a method, which is based on patent documents to analyse the life cycle of object technology and its development potential. This object technology is named test technology. The first step is the selection of some well-known training technologies. The next step is to choose 13 patent-based indexes that can be quantified to measure the technology life cycle stages of the object technology. Then, the nearest neighbour classifier is applied to process the 13 indexes by calculating the nearest distance between a test point and training points, to find out which of the indexes of the training technologies most closely resembles the test technology index. As a result, the training technology whose points have the closest distance to the points of the test technology determines the phase of maturity in which the test technology is.

Yoon et al. [37] proposed a method to assess the technological maturity and to identify the core and brokering technology classes within the overall technology network. In order to calculate the technological maturity, a growth curve was fit to yearly patent registration data. Then, with the purpose of identifying the core and brokering technology classes within the studied technology, it was combined a patent co-classification analysis and a social network analysis.

Albert et al. [1] proposed a method for measuring technological maturity, which was based on blog content. The wording of these blogs contains the expert's perception of the maturity of technologies. They operationalized a technology life cycle model with its three states of pacing, key and basic technologies defined. Then, some terms were defined. These terms were specifically used in the context of either pacing or basic technologies. Both concepts come from the selected technology life cycle model. Six technologies were tested in the method, and then the fuzzy logic approach was applied to generate an assessment of the technological maturity of each technology.

Greitemann et al. [10] developed a quantitative model for assessing and forecasting the maturity of a certain technology. The methodology of this method is made up of three steps. In the first step, it evaluated the maturity of the technology. The current maturity of the technology was assessed by using a questionnaire related to principles, activities, concepts or prototypes of a certain technology, which were answered by experts. In the second step, the total between the number of patents and publications is approximated by logistic functions. The procedure adopted in this second step is supported by the findings of the research made, first, by Whiston [36] that concluded that a logistic function describes the pattern of many technology life cycles, and second, by Linden and Fenn [16] that explained that the maturity of technology is highly correlated to the number of patents and the number of publications. And in the third step, it forecasted the development of the maturity of the studied technology by making a predictive model based on the repetition of steps 1 and 2 with a number of patents updated each time.

Below there is a table with the summary of the methods mentioned (Table 22.2).

22.4.2 The Technology Readiness Level (TRL)

The TRL is the most recognized method which provides a systematic and objective measure of the maturity of a certain technology [32]. It originated at NASA. This methodology started as a method for measuring how far was technology from being implemented in space [4].

In 1969, NASA was making plans for the post-Apollo era. With the purpose of checking if the technology was ready to start such a program, a technology readiness review was suggested [18].

It was not until the year 1989 when the concept of "readiness level" was introduced within a new technology development strategy proposal. That strategy was aimed at a safer development model for space technology [12].

Originally, the method had seven levels [26]. In 1995, levels 8 and 9 were included in the TRL, which expanded its range of maturity measurement [23, 27]. The scale starts with a technology in its most basic scientific form and progresses towards a technology proven in an operating environment [22].

Despite the great utility of this method, the TRL has some constraints. The metric of TRL is sufficient at its very basic level in evaluating technology readiness, but it is deficient in other aspects [3]. Although the TRL provides a high-level knowledge of technology maturity, it lacks accuracy and precision [6]. Some of the main limitations that the methodology TRL have are as follows:

1. It does not provide a complete representation of the integration of a certain technology or technological subsystems into an operational system.

Table 22.2 Main methods for measuring the technological maturity based on future-oriented technology analysis

Author	Short description	Method used	Type of method
Bildosola et al. [15]	Input: Scientific literature, patents, and news databases applied to: AM technologies	Evaluating mathematically a technology and assigning a TRL	Quantitative
Reinhart and Schindler [24]	Input: Expert opinion applied to: manufacturing technology	Evaluating the technological maturity level of manufacturing processes	Mixed
Albert et al. [1]	Input: Expert opinion applied to: six different technologies	Evaluating the level of maturity of technologies considering the information of prestigious blogs in the technological area	Mixed
Greitemann et al. [10]	Input: questionnaires answered by experts applied to: two technologies, the first one is "laser beam cutting of carbon fibre-reinforced-polymer (LBC), and the second one is "adiabatic cutting (AC)"	Evaluating the level of maturity of technologies and forecasting its maturity. This method combined the TRL with the technology life cycle model	Mixed
Gao et al. [9]	Input: Patent documents applied to: an object technology	A model for calculating the technology life cycle for an object technology based on multiple patent-related indicators	Mixed
Kim et al. [37]	Input: Patent documents applied to: Printed electronics	A growth model to identify the current technology maturity ratio, estimate the number of potential patents that will be granted in future and forecasting the expected remaining life	Quantitative

2. It includes no guidance related to the uncertainty that may be expected in moving through the maturation levels of TRL.
3. It assimilates no comparative analysis technique for alternative TRLs [23].

A well-known method which was created to overcome the first limitation mentioned below it was the System Readiness Level (SRL). This method, which was invented by Sauser et al. [28, 29], provides insight into the maturity of a system.

Each technology in a technological system is connected to, at least, one other technology [3]. This methodology performs, in the first stage, assessing the technological maturity of the individual technology components of a system. Then, in a second stage, this method assesses the integration between those components. The method Integration Readiness Level is used for measuring the maturity of the integration of the components of the technological system [28, 29]. Sauser et al. [27] created the Integration Readiness Level (IRL) metric to evaluate integration maturity. This method uses a scale that is like the one of TRL. This method was proposed with the objective of being used in conjunction with the TRL scale and another method for assessing technological maturity.

Other methods have been created to fill the gaps left by TRL and meet specific requirements to evaluate technological maturity. The main available methods in this field are going to be described below.

Nolte et al. [21] proposed a method to assess technological maturity. This methodology is based on the traditional TRL, but it avoids using expert opinion. Instead, this method uses the Microsoft Excel spreadsheet application which allows the users to answer a series of questions about a technology project. Once the questions have been answered, the calculator displays the TRL achieved.

Hicks et al. [13] created a method to assess the technological maturity of the liquid crystal display (LCD). This methodology based its scale on the scale of TRL. Therefore, in the first stage, the scale used as a reference was redefined according to the needs of this method. And on a second scale, the range of the scale of TRL was enlarged. As a result, the scale of this method considers the same range considered by the TRL scale, but also an extra part, which starts once the technology arrives on the market and embraces the potential that such technology possesses to develop new product families. To measure the potential that a certain technology possesses to develop new product families, the technological maturity scale of this method includes a new level which is subdivided into eight sublevels.

Vella et al. [33] developed a method to assess maturity in micro and nano-manufacturing processes. This methodology starts to identify a portfolio of research and development (R&D) projects in a technological field to have access to a rich and validated knowledge repository. The next step is to define a technology maturity scale, and the TRL scale was chosen in this study. The next step is to identify maturity indicators. In this study, a Delphi type study was conducted to identify key indicators for each maturity stage. The fourth step is to design a questionnaire to obtain information from R&D projects regarding their maturity indicators. The next step is the completion of an online questionnaire by researchers working on projects within the portfolio. The last step is to analyse the results to obtain technology profile information about the distribution of R&D efforts in the maturity scale.

Jæger et al. [14] elaborated a method to measure the maturity of the use of Internet of things (IoT) technologies. This method could assist manufacturers in giving directions for adopting new technologies. The proposed methodology comprises eight levels. To reach each level, the assessed company must meet some requirements. On level one, the companies have implemented little technology, and on level eight, they have implemented the technologies required to be part of the Industry 4.0. The

Table 22.3 Main methods for measuring the technological maturity based on the technology readiness level

Authors	Short description	Method used
Nolte et al. [21]	Input: questionnaires answered by experts applied to: hardware and software technologies	The TRL calculator is basically a Microsoft Excel spreadsheet application that based on the answers of a questionnaire display the TRL achieved
Hicks et al. [13]	Input: bibliometric data applied to: liquid crystal display	This method extended the existing TRL approach. First, redefined the TRL scale doing it more general. Then, the representation of the existing TRL scheme was enlarged. This extra part of this method includes the potential of technologies to develop new product families
Vella et al. [33]	Input: expert opinion applied to: micro and nano-manufacturing processes	This method based its scale on the TRL scale. Then the method Delphi was used to identify key indicators for each maturity phase. Then a questionnaire was made to obtain information about the maturity indicators of the projects under study. The questionnaires were answered by the managers of the projects. And finally, the answers were analysed to determine the maturity of the projects
Jæger et al. [14]	Input: expert opinion applied to: technological company Authors	This method comprises eight maturity levels. This scale is based on the TRL one. Level eight is reached when the company has an optimal development on the Internet of things

higher the level of maturity of the company, the higher the level of implementation of IoT technologies that the company has (Table 22.3).

22.5 Conclusions

Ever-increasing technological development results in the rapid obsolescence of many technologies. It is particularly important for companies to have enough tools to measure the maturity of all technologies that are of their interest. Having the appropriate method is especially important for companies since in this way they can efficiently manage their technologies. This literature provides an overview of the methods which are available in scientific journals and whose purpose is to measure technological maturity.

Despite the fact that TRL initially had a great acceptance, over the years certain limitations have been encountered. These constraints have opened up the possibility for other methods to be created. Another iconic method that tried to overcome the limitations of the TRL was the SRL. This method, unlike TRL, has the advantage that it can measure the technological maturity of a system. Despite this, the SRL, like TRL, has the disadvantage that the technological maturity calculated for a system cannot be compared with the technological maturity of other systems. As both methods have constraints, in some situations it will be necessary for other methods to be created.

Regarding the new methods that have been created, there are two large groups in which they could be classified. On one hand, there is a group composed of methods based on the TRL scale. These methods have in common that they start using the TRL scale, but they modified this scale to adjust it to their needs. After defining their scale, they continue to develop their method. On the other hand, there is a group composed of methods based on FTA, mainly in the analysis of the technology life cycle.

References

1. Albert T, Moehrle MG, Meyer S (2015) Technology maturity assessment based on blog analysis. Technol Forecast Soc Change 92:196–209. https://doi.org/10.1016/j.techfore.2014.08.011
2. Ardilio A, Warschat J, Spath D (2012) Customized technology readiness: introducing the application specific technology readiness model. In: 2012 Proceedings of Portland international center for management of engineering and technology: technology management for emerging technologies, PICMET'12. IEEE, pp 1260–1272
3. Azizian N, Sarkani S, Mazzuchi T (2009) A comprehensive review and analysis of maturity assessment approaches for improved decision support to achieve efficient defense acquisition. Lect Notes Eng Comput Sci 2179(1):1150–1157
4. Beims RF, Simonato CL, Wiggers VR (2019) Technology readiness level assessment of pyrolysis of trygliceride biomass to fuels and chemicals. Renew Sustain Energy Rev 112:521–529. https://doi.org/10.1016/j.rser.2019.06.017
5. Cagnin C et al (2008) Future-oriented technology analysis. Springer, Berlín, Heidelberg. https://doi.org/10.1007/978-3-540-68811-2
6. Cornford SL, Sarsfield L (2004) Quantitative methods for maturing and infusing advanced spacecraft technology. In: IEEE aerospace conference proceedings, vol 1, pp 663–681. https://doi.org/10.1109/aero.2004.1367652
7. Ford DK, Ryan C (1981) Taking technology to market. Harvard Bus Rev 59(2), 117–126
8. Freeman (2013) Measuring technology maturity. J Chem Inf Model. https://doi.org/10.1017/CBO9781107415324.004
9. Gao L et al (2013) Technology life cycle analysis method based on patent documents. Technol Forecast Soc Change 80(3):398–407. https://doi.org/10.1016/J.TECHFORE.2012.10.003
10. Greitemann J et al (2014) Quantitative model of the technology lifecycle for forecasting the maturity of manufacturing technologies. IEEE, pp 290–294
11. Halicka K (2016) Innovative classification of methods of the future-oriented technology analysis. Technol Econ Dev Econ 22(4):574–597. https://doi.org/10.3846/20294913.2016.1197164
12. Héder M (2017) From NASA to EU: the evolution of the TRL scale in public sector innovation. Innov J 22(2):1–23

13. Hicks B, Larsson A, Culley SLT et al (2009) A methodology for evaluating technology readiness during product development. In: International conference on engineering design, ICED'09, pp 157–168
14. Jæger B et al (2016) IoT technological maturity model and assessment of norwegian manufacturing companies. In: 23rd EurOMA conference, June 2016, pp 1–10
15. Lezama-Nicolás R, Bildosola I et al (2018) A bibliometric method for assessing technological maturity: the case of additive manufacturing. Scientometrics 117(3):1425–1452. https://doi.org/10.1007/s11192-018-2941-1
16. Linden A Fenn J (2003) Understanding Gartner's Hype cycles understanding Gartner's hype cycles. Strategic Analysis Report N{\textordmasculine} R-20–1971. Gartner, Inc, May 2003, p 88
17. Malone P et al (2011) The application of TRL metrics to existing cost prediction models: a practitioners guide to applying cost correction factors to technology. In: IEEE aerospace conference proceedings. IEEE. https://doi.org/10.1109/AERO.2011.5747634
18. Mankins JC (2009) Technology readiness assessments: a retrospective. Acta Astronautica Pergamon 65(9–10):1216–1223. https://doi.org/10.1016/J.ACTAASTRO.2009.03.058
19. Modis T (2007) Strengths and weaknesses of S-curves. Technol Forecast Soc Chang 74(6):866–872. https://doi.org/10.1016/j.techfore.2007.04.005
20. Nieto M, Lopéz F, Cruz F (1998) Performance analysis of technology using the S curve model: the case of digital signal processing (DSP) technologies. Technovation 18(6–7):439–457. https://doi.org/10.1016/S0166-4972(98)00021-2
21. Nolte W, Kennedy BC, Dziegiel RJ (2004) Technology readiness calculator. In: 6th annual system engineering conference, pp 1–16. Available at: https://ndiastorage.blob.core.usgovcloudapi.net/ndia/2003/systems/nolte2.pdf
22. Olechowski A, Eppinger SD, Joglekar N (2015) Technology readiness levels at 40: a study of state-of-the-art use, challenges, and opportunities. In: Portland international conference on management of engineering and technology. Portland international conference on management of, Sept 2015, pp 2084–2094. https://doi.org/10.1109/PICMET.2015.7273196
23. Peters W, Doskey S, Moreland J (2017) Technology maturity assessments and confidence intervals. Syst Eng 20(2):188–204. https://doi.org/10.1002/sys.21389
24. Reinhart G, Schindler S (2010) A strategic evaluation approach for defining the maturity of manufacturing technologies. World Acad Sci Eng Technol 4(11):1291–1296
25. Rodriguez-salvador M et al (2019) Lessons learned in assessment of technology maturity. In: Engineering digital transformation, January 2019, pp 120–131. https://doi.org/10.1007/978-3-319-96005-0
26. Sadin SR, Povinelli FP, Rosen R (1989) The NASA technology push towards future space mission systems. Acta Astronaut 20:73–77. https://doi.org/10.1016/0094-5765(89)90054-4
27. Sauser B et al (2006) From TRL to SRL: the concept of systems readiness levels. In: Conference on systems engineering research, Los Angeles, CA, May 2006, pp 1–10. Available at: http://www.boardmansauser.com/downloads/2005SauserRamirezVermaGoveCSER.pdf
28. Sauser BJ et al (2008) A system maturity index for the systems engineering life cycle. Int J Ind Syst Eng 3(6):673–691. https://doi.org/10.1504/IJISE.2008.020680
29. Sauser BJ et al (2008) System maturity indices for decision support in the defense acquisition process. Inst Arch Naval Postgrad Sch. http://hdl.handle.net/10945/33278
30. Schilling MA, Esmundo M (2009) Technology S-curves in renewable energy alternatives: analysis and implications for industry and government. Energy Policy 37(5):1767–1781. https://doi.org/10.1016/j.enpol.2009.01.004
31. Terrile RJ et al (2015) Calibrating the Technology Readiness Level (TRL) scale using NASA mission data. In: 2015 IEEE aerospace conference. IEEE, pp 1–9. https://doi.org/10.1109/AERO.2015.7119313
32. Tillack MS et al (2009) An evaluation of fusion energy R&D gaps using technology readiness levels. Fusion Sci Technol 56(2):949–956. https://doi.org/10.13182/FST09-A9033
33. Vella PC et al (2018) Technology maturity assessment of micro and nano manufacturing processes and process chains. Proc IMechE, Part B: J Eng Manuf 232(8):1362–1383. https://doi.org/10.1177/0954405416668922

34. Wang W, Liu Y, Liu S (2017) Study on assessment model for emerging technology maturity. In: PICMET 2017—Portland international conference on management of engineering and technology: technology management for the interconnected world, proceedings, 2017-Jan(71273030), pp 1–7. https://doi.org/10.23919/PICMET.2017.8125278
35. Watts RJ, Porter AL (1997) Innovation forecasting. Technol Forecast Soc Chang 56(1):25–47. https://doi.org/10.1016/S0040-1625(97)00050-4
36. Whiston TG (1994) Forecasting for technologists and engineers: a practical guide for better decisions: Brian C. Twiss, (Peter Peregrines (on behalf of the Institution of Electrical Engineers), London, 1992) xv, 221 pp., £ 19.00, ISBN 0 86341 285 8', Research Policy, 23(2), pp 231–232. https://doi.org/10.1016/0048-7333(94)90055-8
37. Yoon J, Kim M et al (2014) Tracing evolving trends in printed electronics using patent information. J Nanopart Res 16(7). https://doi.org/10.1007/s11051-014-2471-6

Chapter 23
Evaluation of the Maturity Level of Continuous Improvement Based on Improvement Routines: A Case Study of SMEs of Capital Goods

G. Unzueta, A. Esnaola, and J. A. Eguren

Abstract This study presents an analysis of implementation and assessment of a frame of reference to adapt and execute an evolutionary continuous improvement process (CIP) in a mature small- and medium-sized enterprise (SME) that works in the capital goods sector. For this, the research team developed a continuous improvement model (CIM) to implement improvement routines and develop an organisational culture of continuous improvement (CI) to improve companies' CI maturity level. Case study methodology was used in the investigation, where, in the same context, eight units of the aforementioned company were analysed for two years to have a holistic view of the organisation's evolution. To analyse the implementation of the CIP, we developed an evaluation system based on a questionnaire completed by the company's management and CI leaders. The assessment results explain how the implementation of the CIM through the application of the CIP helps develop improvement routines and increase the organisation's CI maturity level. In this specific case, the application of the assessment system shows that although the assimilation of the routines evolved positively in most cases, not in all routines maturity level 2 was achieved, thus emphasising the needs of the organisation and the future actions to be implemented during the following CIP cycles.

Keywords Continuous improvement process · Continuous improvement routines · Continuous improvement maturity level · Industrial case study

---

G. Unzueta (✉) · A. Esnaola · J. A. Eguren
Industrial Organization and Mechanical Industrial Production Department, Mondragon Unibertsitatea, Mondragon, Spain
e-mail: gunzueta@mondragon.edu

A. Esnaola
e-mail: aesnaola@mondragon.edu

J. A. Eguren
e-mail: jaeguren@mondragon.edu

C. Avilés-Palacios and M. Gutierrez (eds.), *Ensuring Sustainability*, Lecture Notes in Management and Industrial Engineering,
https://doi.org/10.1007/978-3-030-95967-8_23

23.1 Introduction

The outcomes of continuous improvement (CI) implementation are well documented in the literature [5, 13, 15], but the literature also identifies that in many cases, maintaining initial results is difficult and the effectiveness of applied techniques decreases [2]. Many authors point out that to address this problem, it is necessary to adapt the CI system to the needs of each organisation, defining a strategy for the implementation of the CI system to establish a culture of improvement that naturally sustains the implemented system [3, 6, 12, 15, 18]. In order to sustain and increase the CI maturity level of the organisations, taking previous models as references [17], a continuous improvement model (CIM) was designed to deploy and develop an organisational culture of CI. The current paper analyses through a specific assessment system how the implementation of the model positively influenced the development of a CI culture in an industrial small- and medium-sized enterprise (SME), measuring this development through a specific assessment system.

Previous Unzueta et al. work assesses the CI maturity based on five different aspects [17]: (1) management commitment and leadership, (2) training and education programme, (3) improvement teams and promotion of teamwork, (4) participation and involvement in CI activities and (5) generation and assimilation of new CI routines. Based on this previous work, this paper deepens how to assess the CI maturity based on the evaluation of CI routines, aspect 5 of [17]. This paper describes how to assess CI routines and analyses the company as a whole, describing the evolution of a company's CI maturity according to the development of improvement routines. The study is based on data collected over a period of two years (November 2017–July 2019).

The paper is organised as follows. Section 23.2 describes the research methodology. Section 23.3 presents the theoretical framework and explains the CIM, and Sect. 23.4 describes the continuous improvement process (CIP) applied to implement the model. The case study is presented in Sect. 23.5, and finally, Sect. 23.6 presents the results and discussion and Sect. 23.7 the conclusions.

23.2 Methodology

The investigation was based on a case study methodology [19]. To have a holistic view of the organisation, eight units of the company were analysed for two years, as shown in Fig. 23.1, where the deployment of the methodology is presented in detail. First, a literature review was conducted to identify the key elements of CI and establish a basis for the CIM [17]. Subsequently, the CIP was designed, and the units to be analysed were selected. These correspond to the various production areas of the company.

The CI maturity level was evaluated through a questionnaire that measured the extent to which improvement routines were assimilated, based on the constitutive

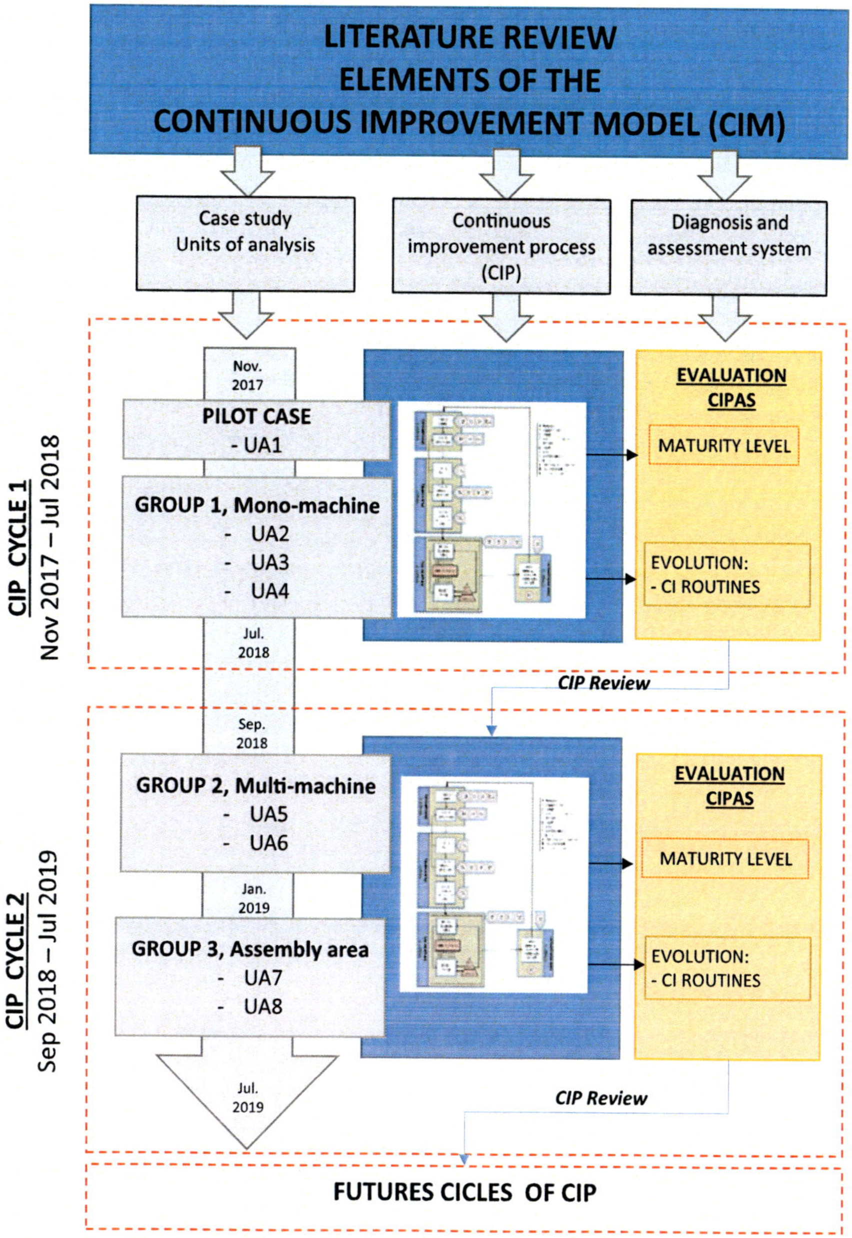

Fig. 23.1 Investigation methodology

routines related to each routine defined by Bessant et al. [4]. Annually, at the end of each CIP cycle, the organisation's management team and CI leaders completed the questionnaire. This study analyses two cycles of the CIP from November 2017–July 2019.

23.3 Theoretical Framework

According to Boer et al. [7], CI is "the planned, organized and systematic process of ongoing, incremental and company-wide change of existing practices aimed at improving company performance". A number of researchers have studied the critical success factors and the elements that influence when a company wants to implement a CIM. Some of them highlighted some specific elements depending on the context and the type of company (e.g. in a large enterprise or an SME, public or private). Considering the literature and the case study company context, the most important elements identified are as follows [17]: management (E1), company culture (E2), strategy (E3), leadership and structure (E4), resources (E5), projects (E6), areas (E7), operating method and improvement tools (E8), training (E9), monitoring and communication (E10), level of involvement (E11) and facilitator or CI leader (E12). Figure 23.2 shows the general structure of the implemented CIM.

According to Bessant et al. [4], CI refers not only to improving results but also to the process by which improvements can be achieved. This process, described in Sect. 23.4, allows the company to evolve its CI maturity level. Bessant et al. [4] proposed an evolutionary CI maturity model consisting of five levels. Based on this model, organisations can identify their level of CI by analysing the uptake of eight improvement routines (see Fig. 23.2). The model proposes that the organisation can advance its level of CI by acquiring the following eight routines [4, 8]: understanding CI (R1), getting the CI habit (R2), leading the way (R3), focusing CI (R4), shared problem-solving (R5), aligning CI (R6), CI of CI (R7) and the learning organisation

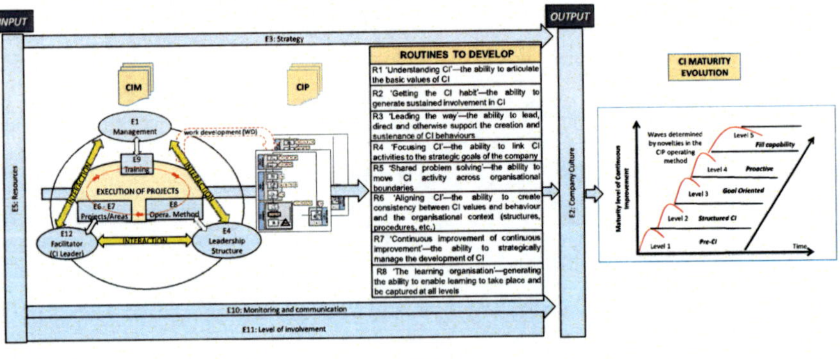

Fig. 23.2 Continuous improvement model. Based on Unzueta et al. [17]

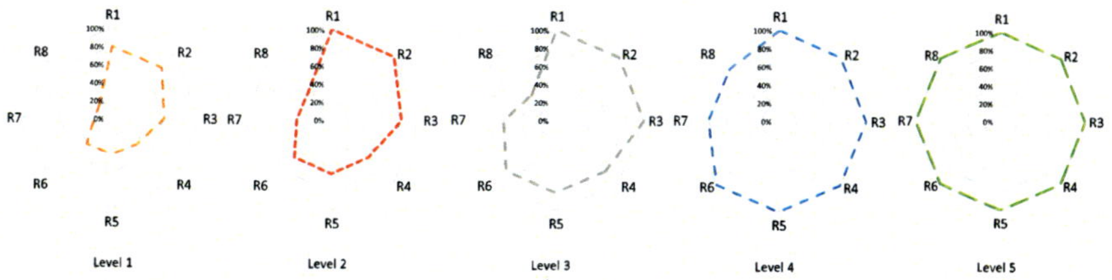

Fig. 23.3 CI routines: complementation objective for each maturity level. Based on Dabhilkar and Bengtsson [8] and Garcia-Sabater et al. [9]

(R8). Considering this evolutionary model analysed [9] how the organisation, in order to increase its level of CI, has to progressively consolidate improvement routines until they are normalised on a day-to-day basis, while taking on new routines that raise its level of CI [8, 9]. As shown in Fig. 23.3, as routines are assimilated, the CI maturity level of the company increases.

23.4 Continuous Improvement Process

To increase the company's CI maturity level, it is necessary to create and assimilate new CI routines, applying systematic improvement methods and tools through a structured process. This structured process is the CIP, the process of implementing the CIM. The CIP presented in Fig. 23.4 identifies four stages, phases of each stage and the significant elements of each phase [17].

Stage 0: Diagnosis. The company should be diagnosed considering the production maturity levels [11] and the improvement tools previously applied. Finally, the appropriate operational method should be extracted for the specific case. Depending on the maturity level, basic lean tools, such as 5S or visual management, or more complex tools based on Six Sigma may be more appropriate. The company's management must also create an organisational structure appropriate to the organisation for the development and implementation of the CIP.

Stage 1: Planning. After having trained the management and the managers of the departments concerned during Stage 0 to ensure their support, the selected operational method is adapted to the organisation and the defined projects. In addition, the management must develop the appropriate channels and activities to communicate the characteristics and benefits of the CIP [16]. Finally, a plan is developed for each project.

Stage 2: The operative stage. In this stage, the execution of the selected projects and the training of participants are done in parallel. The projects are managed and executed following the defined project plan. Training is adapted in accordance with the operating method and to each organisational structure level. Adapting this training is especially important because each organisational level has different activities and responsibilities [1].

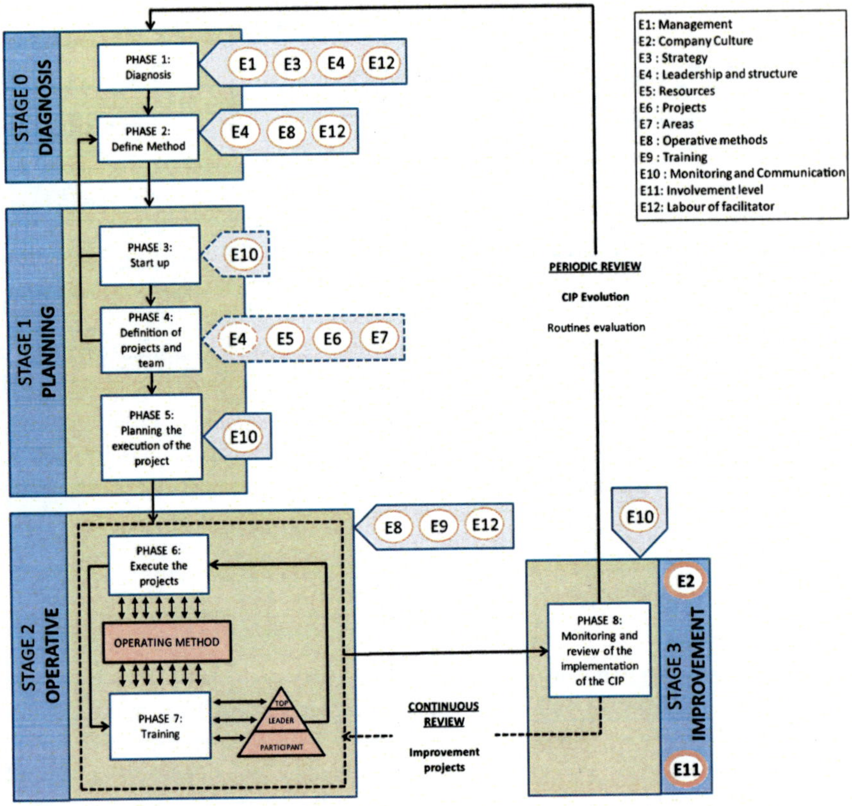

Fig. 23.4 Continuous improvement process. Based on [17]. Based on Unzueta et al. [17]

Stage 3: The improvement stage. This stage has two different reviews: continuous review and periodic review. Regarding the continuous review, Wu and Chen [18] suggested that the metrics used in the evaluation system should be suited to each CI level (continuous review, see Fig. 23.4) to continuously review the progress of the defined projects. On the other hand, considering that self-examination is the most effective way to achieve successful CI [10, 18], organisations must analyse the CIP periodically to understand its weaknesses and implement improvements to increase CI maturity (periodic review, see Fig. 23.4).

In this study, we present how the case study organisation applied the CIP and increased its CI maturity level in accordance with a periodic evaluation of CI routines at the end of each CIP cycle (periodic review, see Fig. 23.4). The assimilation of CI routines was evaluated through a questionnaire based on the constitutive behaviours related to each routine [4]. The company's management team members and CI leaders completed the questionnaire using a four-point Likert scale (see Table 23.1).

Table 23.1 Example of questions posed on the questionnaire based on constitutive behaviours to measure R1 routine

Routines [4]	Constitutive behaviours [4, 8]	Questions [9, 17]	Likert 1–4
R1. Understanding CI The ability to articulate the basic values of CI	People at all levels demonstrate a shared belief in the value of small steps, and everyone can contribute by themselves being actively involved in making and recognising incremental improvements	Do workers have the time and resources to think about, propose and implement small improvements in their daily work?	
	When something goes wrong, the natural reaction of people at all levels is to look for reasons why, etc., rather than to blame individual(s)	Are root cause dynamics implemented? Are these dynamics implemented instead of looking for people to blame? Are there adequate discussion forums to discuss and seek solutions to problems?	
	People make use of some formal problem-finding and problem-solving cycles	Are problem-solving tools (e.g. seven basic quality tools) applied in the discussion forums?	

Questions based on Bessant et al. [4], Dabhilkar and Bengtsson [8], Garcia-Sabater et al. [9] and Unzueta et al. [17]

23.5 Case Study

The CIP and the assessment system were applied in a cooperative model organisation located in Basque Country, Spain. The company had three different businesses, and the projects supported by the CIP were implemented in the power transmission equipment business. Unzueta et al. [17] conducted a previous study on the company, in which the evolution of the implementation of the CIM was analysed by comparing various units of analysis with each other during the first year of the CIP implementation.

In this study, the data obtained in the previous case study [17] and the data achieved during the second year of the CIP implementation were used to analyse the evolution of the company's CIM as a whole and to validate the assessment system based on the assimilation of the improvement routines by the organisation. In the first stage of the CIP (in the first cycle executed between November 2017 and July 2018), it was diagnosed that the company was in the first level of CI, and the management, together with the research team, decided to set a goal to achieve in a period of two years to overcome the second level and establish the basis for the third level of CI maturity. To this end, an organisational structure was defined, identifying the responsibilities related to CI at each level of the structure, and the people involved in the structure were trained in the use of basic improvement tools, starting with middle management and ending with the employees. The training was adapted to each organisational level.

Taking into account that the organisation was at the first level of CI, a basic improvement tool was selected to start laying the foundations of the improvement system. The tool selected was 5S because it is a simple tool to apply, and it facilitates the involvement of staff in a common project for the whole organisation, thus promoting teamwork [14].

23.6 Results and Discussion

The results of the evaluation confirm that by implementing the CIP, the organisation progressively assimilated CI routines and increased its CI maturity level. People involved directly in the CI organisation structure, management, middle management and employees and assimilated several improvement routines at the end of the second cycle. As seen in Fig. 23.5, the first three routines exceed 90% of routine uptake or are close to it. Routines were successfully assimilated, although the objective set for maturity level 2 was not achieved for all routines. It was observed that in routines R6 (aligning CI) and R7 (CI of CI system), which should be assimilated to a greater extent by the management and leaders of CI, the level of assimilation is much higher than the objective defined for the second level of CI. This is evidence of how CIP serves to foster the development of an organisational culture oriented towards CI. The

Routines and constituent behaviours	Routine assimilation	
	CYCLE 1	CYCLE 2
R1 'Understanding CI'—the ability to articulate the basic values of CI	94%	100%
R2 'Getting the CI habit'—the ability to generate sustained involvement in CI	75%	94%
R3 'Leading the way'—the ability to lead, direct and otherwise support the creation and sustenance of CI behaviours	63%	88%
R4 'Focusing CI'—the ability to link CI activities to the strategic goals of the company	38%	56%
R5 'Shared problem-solving'—the ability to move CI activity across organisational boundaries	56%	69%
R6 'Aligning CI'—the ability to create consistency between CI values and behaviour and the organisational context (structures, procedures, etc.)	69%	81%
R7 'Continuous improvement of continuous improvement'—the ability to strategically manage the development of CI	44%	75%
R8 'The learning organisation'—generating the ability to enable learning to take place and be captured at all levels	38%	44%

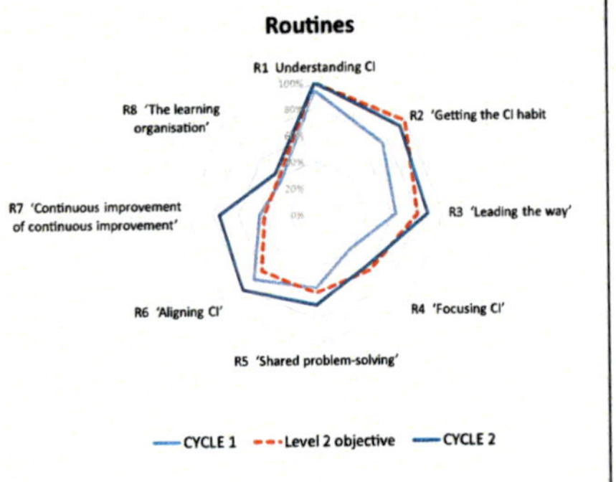

Fig. 23.5 Routine assessment. Based on Bessant et al. [4], Dabhilkar and Bengtsson [8], Garcia-Sabater et al. [9] and Unzueta et al. [17]

main reason for this is the actions implemented in Stage 1 of the CIP, through which the company analyses the evolution of the system on an annual cyclical basis and defines the strategic projects to be developed in the future cycle. On the other hand, the defined organisational structure, taking into account the people, their responsibilities regarding CI and the work dynamics implemented, facilitates the development of the R5 routine (shared problem-solving).

23.7 Conclusions

The questionnaire for evaluating the extent to which CI routines were assimilated was useful in helping the organisation to see the evolution and define its approach to the deployment of each CIP cycle.

Analysing each area individually, the research team observed that the areas where suggestion management systems and manufacturing process measurement were implemented developed more deeply the R2 routine (getting the CI habits). This is because these routines, at the initial levels of CI, must be assimilated to a greater extent by the employees, and the measurement of the manufacturing processes together with the possibility of making suggestions for improvement in a systematic way facilitates the involvement of the employees in the improvement dynamics implemented.

Considering the organisation as a whole, it was observed how the selection of projects has evolved, increasing the difficulty of the projects in the third cycle of the CIP. In the first two cycles, the projects selected were oriented towards the standardisation of workplaces and the establishment of CI dynamics. In the third cycle, projects that are more specific were defined, such as changes in layout and development of quality control through self-monitoring in critical manufacturing processes. These new projects will boost middle management involvement and consolidate routines R4 (focusing CI) and R5 (shared problem-solving).

Compliance with Ethical Standards The results of this work are supported by a questionnaire. Participants of the research study voluntarily agreed and gave informed consent to participate in the questionnaire. All data was anonymised before publication. The "Research Ethics Committee of Mondragon Unibertsitatea" (Ref. IEB-20210924) approved the entire procedure used in the research process.

References

1. Antony J, Gupta S (2019) Top ten reasons for process improvement project failures. Int J Lean Six Sigma 10(1):367–374
2. Bateman N (2005) Sustainability: the elusive element of process improvement. Int J Oper Prod Manage 25(3):261–276
3. Bateman N, David A (2002) Process improvement programmes: a model for assessing sustainability. Int J Oper Prod Manage 22(5):515–526

4. Bessant J, Caffyn S, Gallagher M (2001) An evolutionary model of continuous improvement behaviour. Technovation 21(2):67–77
5. Bhuiyan N, Baghel A (2005) An overview of continuous improvement: from the past to the present. Manage Decis 43(5):761–771
6. Bhuiyan N, Baghel A, Wilson J (2006) A sustainable continuous improvement methodology at an aerospace company. Int J Prod Perform Manage 55(8):671–687
7. Boer H, Berger A, Chapman R, Gertsen F (eds) (2000) CI changes: from suggestion box to organisational learning: continuous Improvement in Europe and Australia. Ashgate Publishing, Aldershot
8. Dabhilkar M, Bengtsson L (2007) Continuous improvement capability in the Swedish engineering industry. Int J Technol Manage 37(3/4):272
9. Garcia-Sabater JJ, Marin-Garcia JA, Perello-Marin MR (2012) Is implementation of continuous improvement possible? An evolutionary model of enablers and inhibitors. Hum Factors Ergo Manuf 22(2):99–112
10. Jorgensen F, Boer H, Laugen BT (2006) CI implementation: an empirical test of the CI maturity model. Creativity Innov Manage 15(4):328–337
11. Kosieradzka A, Ciechańska O (2018) Impact of enterprise maturity on the implementation of six sigma concept. Manage Prod Eng Rev 9(3):59–70
12. Ljungstrom M (2005) Sustainable continuous improvements and work development: important factors for theory and practice. Int J Manage Pract 1(4):330
13. Marin-Garcia JA, Pardo del Val M, Martín TB (2008) Longitudinal study of the results of continuous improvement in an industrial company. Team Perform Manage 14(1/2):56–69
14. Randhawa JS, Ahuja IS (2017) 5S implementation methodologies: literature review and directions. Int J Prod Qual Manage 20(1):48–74
15. Singh J, Singh H (2015) Continuous improvement philosophy: literature review and directions. Benchmarking 22(1):75–119
16. Stankalla R, Koval O, Chromjakova F (2018) A review of critical success factors for the successful implementation of Lean Six Sigma and Six Sigma in manufacturing small and medium sized enterprises. Qual Eng 30(3):453–468
17. Unzueta G, Esnaola A, Eguren JA (2020) Continuous improvement framework to develop cultural change. Case study, capital goods company. TQM J 32(6):1327–1348
18. Wu CW, Chen CL (2006) An integrated structural model toward successful continuous improvement activity. Technovation 26(5–6):697–707
19. Yin RK (2013) Case study research: design and methods, 3rd edn. Sage, London

Part VI
Business Management, Entrepreneurship and Innovation

Chapter 24
Business Model Innovation in SMEs: A Cluster Analysis

D. Ibarra, J. I. Igartua, and J. Ganzarain

Abstract This paper presents an exploratory analysis focused on identifying and characterizing business model innovation (BMI) in small and medium enterprises (SMEs), a phenomenon that has gained increasing attention in management and challenges many companies' competitiveness. Based on a purposive sample of 84 SMEs participating in public-supported BMI projects, we explore different BMI-related elements using a two-step cluster analysis, along with an examination of the predictors' importance and mean differences. The results underline the relevance of BMI management and BMI capabilities in SMEs, as well as stating a degree of importance in the prediction of clusters, suggesting further research opportunities. The research shows two different groups of SMEs that are statistically significant for all clustering variables. The value of this ongoing research lies in its contribution to the quantitative research of BMI in SMEs, as well as in the study of this strategic phenomenon.

Keywords Business model innovation (BMI) · Business model · Business model advantage · SMEs

24.1 Introduction

It is commonly accepted that business models (BMs) describe the business logic of a firm in terms of value creation, delivery and capture [32]. Therefore, in this study, business model innovation (BMI) is defined as the discovery of new and significantly different ways of value creation, delivery and capture within an established business model [34]. BMI can be a source of business opportunities for SMEs, allowing them to respond quickly to market changes, redefine their existing markets or even create new ones by commercializing innovations through new business model configurations [2, 10]. Consequently, BMI can become a source of competitive advantage and superior firm performance [15].

D. Ibarra (✉) · J. I. Igartua · J. Ganzarain
Faculty of Engineering, Mondragon University, Loramendi 4, 20500 Mondragon, Spain
e-mail: dibarra@mondragon.edu

C. Avilés-Palacios and M. Gutierrez (eds.), *Ensuring Sustainability*,
Lecture Notes in Management and Industrial Engineering,
https://doi.org/10.1007/978-3-030-95967-8_24

Despite the potential benefit and relevance of BMI, our understanding of the phenomenon remains limited in relation to SMEs; as is as our knowledge of the factors and processes for its development [15]. Moreover, BMI literature has largely kept a success-driven perspective on large firms, while research on SMEs has only started to gain attention in recent years [4, 8, 12, 24, 29].

SMEs face unique challenges when implementing BMI due to their limited resources [6, 23], manager's influence [4] and environmental contingencies [30]. In addition, organizational inertia and path dependencies can constrain the organizational restructuring and managerial decisions that BMI implies. Nevertheless, the literature suggests that certain drivers related to a firm's behavior [5, 7, 20] and dynamic capabilities [12, 19, 29] could help SMEs overcome those challenges. It is, therefore, essential to understand how SMEs' everyday practices, in the form of capabilities and management, impact on BMI, as well as how they are linked to BMI and its performance [15].

In addition, the research developed to date is based on conceptual works and case studies [3, 35], so scholars are calling for more empirical research, larger samples and replicability of the studies to address these gaps [11, 31, 36].

To succeed in BMI, SMEs need to manage the challenges related to the reconfiguration of their established BM, which demands actions from the top managers and adequate knowledge, capabilities and skills within the company, in order to sense and seize BMI opportunities [25, 33].

24.2 Business Model Innovation in SMEs

Compared to larger companies, SMEs generally have less time, fewer resources and lack a capability-structured approach to innovation [1]. These limitations can represent a challenge for BMI. However, SMEs can compensate for these difficulties by finding ways to develop innovation capabilities and relying on the strengths associated with their size: a more receptive climate, fewer bureaucratic procedures, more flexible structures and greater adaptability [4, 21]. As part of an ongoing project about BMI in SMEs, the following key elements are considered to establish the background of the present research.

Business Model (BM): efers to the internal consistency fit among BM components concerning how value is delivered, created and captured in the SME [26].

Business Model Innovation (BMI): is defined as "designed, novel, non-trivial changes to the key elements of a firm's business model and/or the architecture linking these elements" ([15], p. 17).

Business Model innovation management (BMIM): refers to managerial orientation and an SME's innovative culture. A CEO's individual characteristics and beliefs might influence an SME's capabilities for BMI [4]. An organizational culture that encourages innovation and creativity will lead SMEs' members to take risks, sense new opportunities and pursue new ideas, thereby stimulating BMI [1].

Business Model Innovation Capabilities (BMIC): refers to the set of resources and routines SMEs deploy to (1) sense customer needs, (2) scan technological options, (3) experiment, (4) collaborate and (5) align BMI with their strategy. Rooted in the dynamic capabilities theory [33], these capabilities are considered to drive BMI [4, 15, 25].

Business Model Advantage (BMA): It refers to the business model's predominance toward providing customers with superior benefits than their competitors. This occurs when the BM (1) offers a high value that is perceived as such by customers, (2) is exclusive or provides greater advantages than the competition, (3) allows access to new markets and/or (4) is difficult to imitate [22, 32].

24.3 Objectives and Methodology

The aim of the current study is twofold: (1) to distinguish groups of SMEs according to the BMI elements previously described (BM, BMI, BMIM, BMIC and BMA) and (2) to analyze barriers and drivers for BMI in different groups of SMEs.

The present research, exploratory in nature, aims to explore the phenomenon of BMI in SMEs. The population of interest was SMEs of a Basque Region (Gipuzkoa) that were actively engaged in BMI for at least the last three years. Since the population frame was unknown, purposive sampling was adopted. The sample comprises 267 SMEs that participated (from 2016 to 2018) in some of the Basque Country Regional Government's funding programs for improvement of competitiveness through innovation in value propositions and business models.

An online questionnaire based on a five-point Likert scale was developed to collect data. Variables measuring BM, BMI, BMIM, BMIC and BMA were adopted from previously validated multi-item scales, with slight adaptions to comply with the BMI context. Questions addressing drivers and barriers were developed based on the European Commission Innovation Survey [28] and the Regional Government's strategic concerns.

For data validation, the common method variance was checked using Harman's single-factor test [17]. The factor obtained (14.30%) was below the established limits.

The final sample comprised 84 cases (final valid responses = 31.46%). The survey was mostly completed by senior managers (82.1%). The main participating companies are in the manufacturing industries (59%), followed by companies related to industrial services (18%), ancillary services (7.7%) and ICT industries (10.3%). The sample is predominantly composed of small firms (70.5%), followed by medium-sized firms (25.6%) and micro- (3.8%) according to the EU commission categorization (EU [14]).

24.4 Results

To develop the cluster classification, we used the methodology suggested by previous researchers [27]. We used the two-step cluster analysis [13], with a previous descriptive statistical analysis to test needed conditions. In order to automatically calculate the best number of clusters, both the log-likelihood distance measure and the Schwarz grouping method (BIC) were selected. Afterward, once the number of clusters were fixed, the Euclidean distance was selected for the final membership clustering. Every time, a membership variable was created to perform some of the analyses shown in this paper.

Once the cluster's formation had been validated and identified, the variables with the strongest influence were analyzed (difference between two means). All the analyses presented in this section were carried out using the statistical software SPSS, version 23.

Thus, before proceeding with the cluster analysis, we check for multicollinearity, analyzing the correlation between clustering variables [18].

Correlations (Table 24.1) show a maximum value of 0.719, lower than the limit of 0.90 [18], and it can, therefore, be considered that there are no problems of collinearity between the variables.

The construction of the cluster initially considered four variables (BMIC, BMIM, BM, BMI) and BMA as an evaluation field. Table 24.2 presents the cluster analysis for all cases.

The analysis suggests the creation of two different groups with good quality (silhouette measure of cohesion and separation = 0,54), a value above + 0.5 that lets us assume that the clustering was successful. Note that, the largest number of cases is in the second cluster (transformed SMEs) according to the cluster distribution. Transformed SMEs represent the 65.5% of the SMEs participating in improvement of competitiveness and business transformation public project programs. BMIC is the variable with the greatest impact (Table 24.2). Furthermore, it is worth observing

Table 24.1 Clustering variables correlations

Measures	Means ± SD	BMIC	BMIM	BM	BMI	BMA
Z-value BMI capabilities (BMIC)	3.49 ± 0.51	1	0.719**	0.426**	0.599**	0.570**
Z-value BMI management (BMIM)	3.15 ± 0.62		1	0.390**	0.662**	0.565**
Z-value business model (BM)	3.72 ± 0.51			1	0.394**	0.354**
Z-value business model innovation (BMI)	3.19 ± 0.69				1	0.532**
Z-value business model advantage (BMA)	3.28 ± 0.62					1

**Correlation is significant at the 0.01 level (2-tailed). All Spearman's coefficients

Table 24.2 Frequency and variables position in cluster analysis

Cluster	Cases	BMIC		BMIM		BMI		BM	
		Mean	Std. dev	Mean	Std. dev	Mean	Std. dev	Mean	Std. dev
1	29 (34.5%)	2.9767	35.985	2.5323	39.643	2.5607	56.461	3.3479	49.210
2	55 (65.5%)	3.7665	33.159	3.4775	43.435	3.5265	48.177	3.9122	41.219
Combined	84 (100%)	3.4938	50.785	3.1512	61.654	3.1931	68.700	3.7174	51.485

Table 24.3 Predictor importance in cluster analysis

Variable	Predictor importance	Significant order	
		Cluster 1: ongoing SMEs	Cluster 2: transformed SMEs
BMIC	1	1	2
BMIM	0.96	2	1
BMI	0.76	3	3
BM	0.43	4	4
BMA	0.38	–	

how the order of the variables between BMIC and BMIM changes when analyzing their importance in each of the clusters (Table 24.3).

As outlined in Table 24.4, the mean differences of the main key variables for the configuration of the two clusters (BMIC, BMIM, BMI and BM) and the evaluation field (BMA) were calculated. The differences between means were statistically significant in all cases.

The graphs (Fig. 24.1) show the cluster analysis scatter plot of BMI and the two most significant variables according to the analysis (BMIC and BMIM).

Other elements highlighted in the BMI literature refer to the barriers and drivers for BMI [9]. Thus, based on the two clusters, an analysis of both elements was performed. Although average barrier perception was lower for "transformed SMEs" (2.81 compared to 2.89), no statistical significance was found when analyzing each of the barriers.

As regards the drivers for BMI, although all driver values are higher for "transformed SMEs", only three statistically significant mean differences were found (Table 24.5).

Thus, higher and significant values were identified for drivers regarding diversification of BM (DIV), digital transformation of the BM (DIG) and talent as a BMI driver (TAL).

Table 24.4 Mean differences for the clustering variables

Measures	Cluster 1: ongoing SMEs	Cluster 2: transformed SMEs
	Means ± SD	Means ± SD
BMIC—business model innovation capabilities	2.98 ± 0.36	3.77 ± 0.33
BMIM—business model innovation management	2.53 ± 0.40	3.48 ± 0.43
BMI—business model innovation	2.56 ± 0.56	3.53 ± 0.48
BM—business model	3.35 ± 0.49	3.91 ± 0.41
BMA—business model advantage	2.86 ± 0.59	3.50 ± 0.51

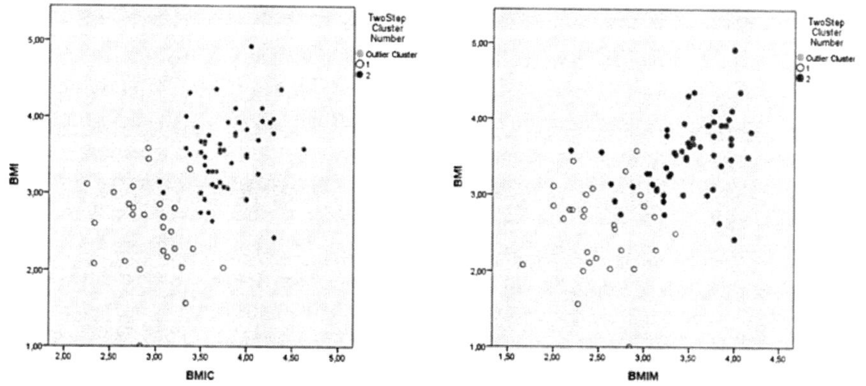

Fig. 24.1 Two-step cluster analysis scatter plot

Table 24.5 Significant drivers for BMI

Measures	Cluster 1: ongoing SMEs	Cluster 2: transformed SMEs
	Means ± SD	Means ± SD
Diversification as BMI driver (DIV)	3.31 ± 0.89	3.85 ± 0.70
Digitalization as BMI driver (DIG)	3.24 ± 0.99	3.95 ± 0.83
Talent management and generational renewal as BMI driver (TAL)	3.83 ± 0.89	4.22 ± 0.71

24.5 Discussion, Conclusions and Further Research

In this study, we extend the BMI research to SMEs, exploring the relationships between different elements using a cluster analysis. We support the classification of SMEs based on variables that in the literature are interconnected in the literature and studies have been considered independently. More precisely, the cluster analyses, based on a sample of convenience of companies already involved in BMI, confirm that, although there might be difficulties in assessing the experiences developed, two clear groups could be considered: (1) on-process SMEs and (2) transformed SMEs.

Based on several variables defined in the literature, we have empirically tested the existence of two groups of SMEs with a good quality estimation. We have also evidenced the order of the dimensions, with the highest significant level for BMIC and BMIM. With regard to those variables, it is important to highlight the changing position of the BMIC and BMIM variables for the two cluster groups.

Transformed SMEs represent the biggest group according to the sample ($n = 55$) with the highest values (higher than 3.48) in all four variables used for the cluster analysis, as well as in the variable used for evaluation. On-process SMEs ($n = 29$) show lower values with values under 3.0 for all variables but one. These statistically significant results (Table 24.4) indicate the effective existence of two different SME groups in relation to the BMI phenomenon.

Our findings highlight the importance of BMIC and BMIM when introducing BMI in companies. Mean differences for the clustering support the role of these two elements when developing BMI. On the contrary, the BM variable with lower mean differences and high values in both clusters (higher than 3.3) indicates that the value creation, value delivery and value capture in both groups of SMEs are consistent. It, therefore, seems that SMEs reconfigured their established BM for the sake of BMI [16]. In the lack of further research, the results suggest that transformed SMEs might have established strategies or developed pilot experiences, which would have allowed them to generate the dynamic capabilities required for BMI. Similarly, ongoing SMEs might not have deployed these capabilities.

The results also explore, based on the cluster configuration, the role of barriers and drivers for SMEs confronting the BMI phenomenon. Although barrier values are higher in the first cluster (on-process SMEs), no statistical significance has been found. On the contrary, the analysis regarding the drivers for BMI suggests the major importance of diversification, digital transformation of the business and talent management. These results are aligned with recent research emphasizing the influence of internal capabilities and managerial actions enabling SMEs to address BMI proactively [4, 15, 25, 32]. Besides, questions over the role of new business trends, such as servitization and circular economy, although analyzed, have not shown statistical significance as BMI drivers. This is turn raises some questions regarding the capability of SMEs to embrace these opportunity streams.

This study characterizes SMEs according to dimensions identified in the literature and based on an analysis of different variables that could lead to further research and the definition of a structured framework that might help to distinguish companies involved in business model innovation.

The limitations of this study are due to time and resources. Further phases will aim to analyze in detail the moderating impact of other context factors, such as management practices, and BMI activities. The value of this ongoing research lies in its contribution to the quantitative research of BMI in SMEs, together with the study of associated elements and their interconnections.

Compliance with Ethical Standards This article does not contain any studies with human participants; all data was gathered from an organizational perspective (position level—1, 2, 3 and years of experience of the respondent), with no personal data in the questionnaire. Companies' data was also anonymized. Participating companies were informed about the process and data management policy though a presentation letter, voluntarily agreeing to participate in a research study by filling in and returning a questionnaire. The authors declare that they have no conflict of interest, nor work for, consult, own shares in or receive funding from any company or organization that would benefit from this article and have disclosed no relevant affiliations beyond their academic appointment. The Research Ethics Committee of Mondragon Unibertsitatea (Ref. IEB-20201201) approved the entire procedure used in the research process.

References

1. Aksoy H (2017) How do innovation culture, marketing innovation and product innovation affect the market performance of small and medium-sized enterprises (SMEs)? Technol Soc 51:133–141. https://doi.org/10.1016/j.techsoc.2017.08.005
2. Amit R, Zott C (2012) Creating value through business model innovation. MIT Sloan Manag Rev 53:41
3. Andreini D, Bettinelli C (2017) Business model innovation: from systematic literature review to future research directions. Springer
4. Arbussa A, Bikfalvi A, Marquès P (2017) Strategic agility-driven business model renewal: the case of an SME. Manag Decis 55:271–293. https://doi.org/10.1108/MD-05-2016-0355
5. Asemokha A, Musona J, Torkkeli L, Saarenketo S (2019) Business model innovation and entrepreneurial orientation relationships in SMEs: implications for international performance. J Int Entrep 17:425–453. https://doi.org/10.1007/s10843-019-00254-3
6. Berends H, Jelinek M, Reymen I, Stultiëns R (2014) Product innovation processes in small firms: combining entrepreneurial effectuation and managerial causation. J Prod Innov Manag 31:616–635. https://doi.org/10.1111/jpim.12117
7. Bock AJ, Opsahl T, George G, Gann DM (2012) The effects of culture and structure on strategic flexibility during business model innovation. J Manag Stud 49:279–305. https://doi.org/10.1111/j.1467-6486.2011.01030.x
8. Bouwman H, Nikou S, de Reuver M (2019) Digitalization, business models, and SMEs: How do business model innovation practices improve performance of digitalizing SMEs? Telecomm Policy. https://doi.org/10.1016/j.telpol.2019.101828
9. Chesbrough H (2010) Business model innovation: opportunities and barriers. Long Range Plann 43:354–363. https://doi.org/10.1016/j.lrp.2009.07.010
10. Chesbrough H, Rosenbloom RS (2002) The role of the business model in capturing value from innovation: evidence from Xerox corporation's technology spin-off companies. Ind Corp Chang 11:529–555. https://doi.org/10.1093/icc/11.3.529
11. Clauss T (2017) Measuring business model innovation: conceptualization, scale development, and proof of performance. R D Manag. 47:385–403. https://doi.org/10.1111/radm.12186
12. Clauss T, Bouncken RB, Laudien S, Kraus S, (2020) Business model reconfiguration and innovation in SMEs: a mixed-method analysis from the electronics industry. Int J Innov Manag 24. https://doi.org/10.1142/S1363919620500152
13. Dietrich T, Rundle-Thiele S, Kubacki K (2016) Segmentation in social marketing: process, methods and application. Segmentation Soc Mark Process Methods Appl 1–214. https://doi.org/10.1007/978-981-10-1835-0
14. EU Commission, 2003. COMMISSION RECOMMENDATION of 6 May 2003 concerning the definition of micro, small and medium-sized enterprises (notified under document number C(2003) 1422) (Text with EEA relevance). Off. J. Eur. Union.
15. Foss NJ, Saebi T (2017) Fifteen years of research on business model innovation. J Manage 43:200–227. https://doi.org/10.1177/0149206316675927
16. Geissdoerfer M, Vladimirova D, Evans S (2018) Sustainable business model innovation: a review. J Clean Prod. https://doi.org/10.1016/j.jclepro.2018.06.240
17. Genc E, Dayan M, Genc OF (2019) The impact of SME internationalization on innovation: The mediating role of market and entrepreneurial orientation. Ind Mark Manag. https://doi.org/10.1016/j.indmarman.2019.01.008
18. Hahs-Vaughn DL (2017) Applied multivariate statistical modeling, 1st Editio. ed. Routledge (Taylor and Francis Group), New York and London. https://doi.org/10.17605/OSF.IO/3X76A
19. Hock-Doepgen M, Clauss T, Kraus S, Cheng CF (2020) Knowledge management capabilities and organizational risk-taking for business model innovation in SMEs. J Bus Res. https://doi.org/10.1016/j.jbusres.2019.12.001
20. Hock M, Clauss T, Schulz E (2016) The impact of organizational culture on a firm's capability to innovate the business model. R D Manag. https://doi.org/10.1111/radm.12153

21. Ibarra D, Bigdeli AZ, Igartua JI, Ganzarain J (2020) Business model innovation in established SMEs: a configurational approach. J. Open Innov Technol Mark Complex 6:76
22. Lecocq X, Demil B, Ventura J (2010) Business models as a research program in strategic management: an appraisal based on lakatos. Management 13:214–225
23. Leithold N, Woschke T, Haase H, Kratzer J (2016) Optimising NPD in SMEs: a best practice approach. Benchmarking. https://doi.org/10.1108/BIJ-05-2015-0054
24. Lopez-Nicolas C, Nikou S, Molina-Castillo FJ, Bouwman H (2020) Gender differences and business model experimentation in European SMEs. J Bus Ind Mark
25. Mezger F (2014) Toward a capability-based conceptualization of business model innovation: insights from an explorative study. R D Manag. 44:429–449. https://doi.org/10.1111/radm.12076
26. Osterwalder A, Pigneur Y (2010) Business model generation: a handbook for visionaries, game changers, and challengers. Wiley
27. Pons M, Bikfalvi A, Llach J (2018) Clustering product innovators: a comparison between conventional and green product innovators. Int J Prod Manag Eng 6:37. https://doi.org/10.4995/ijpme.2018.8762
28. Rammer C (2019) Community innovation survey, in: Handbuch Innovationsforschung. https://doi.org/10.1007/978-3-658-17671-6_66-1
29. Ricciardi F, Zardini A, Rossignoli C (2016) Organizational dynamism and adaptive business model innovation: the triple paradox configuration. J Bus Res 69:5487–5493. https://doi.org/10.1016/j.jbusres.2016.04.154
30. Saebi T (2014) Business model evolution, adaptation or innovation? A contingency framework on business model dynamics, environmental change and dynamic capabilities. Bus Model Innov Organ Dimens. Oxford University Press. https://doi.org/10.1093/acprof
31. Spieth P, Schneckenberg D, Ricart JE (2014) Business model innovation—state of the art and future challenges for the field. R D Manag 44:237–247. https://doi.org/10.1111/radm.12071
32. Teece DJ (2010) Business models, business strategy and innovation. Long Range Plann. https://doi.org/10.1016/j.lrp.2009.07.003
33. Teece DJ (2017) Business models and dynamic capabilities. Long Range Plann 51:40–49. https://doi.org/10.1016/j.lrp.2017.06.007
34. Velu C (2016) Evolutionary or revolutionary business model innovation through coopetition? The role of dominance in network markets. Ind Mark Manag 53:124–135. https://doi.org/10.1016/j.indmarman.2015.11.007
35. Wirtz BW, Daiser P, Göttel V, Daiser P, Pistoia A, Ullrich S, Göttel V, Daiser P, Göttel V, Daiser P (2016) Business model innovation: development, concept and future research directions. J Bus Model 4:1–28. https://doi.org/10.5278/ojs.jbm.v4i1.1621
36. Zott C, Amit R, Massa L (2011) The business model: Recent developments and future research. J Manage 37:1019–1042. https://doi.org/10.1177/0149206311406265

Chapter 25
Crowdsourcing Ecosystem. The Crowd Pillars and Their Implementation Process

H. Castillo, M. Grijalvo, A. Martinez-Corral, and M. Palacios

Abstract Crowdsourcing is a business approach based on collective contributions. It has had successful results in both the private sector and the research field due to the capability to apply the methodology to a variety of disciplines, including technological, operational and managerial ones. Although the topic has been analyzed in a fragmented way throughout the years, this paper has reviewed more than 100 studies in order to define crowdsourcing as a new strategic methodology that provides high-quality solutions and successfully brings companies and users together resulting in a productivity increase and innovation growth. Moreover, this article includes a new component to the pillars theory (Hosseini, M., Phalp, K., Taylor, J. and Ali, R., 2014, May. The four pillars of crowdsourcing: A reference model. In Research Challenges in Information Science (RCIS), 2014 IEEE Eighth International Conference on (pp. 1–12). IEEE.), the evaluation mechanism, which validates and corrects the results obtained on the earlier model that relays on crowd selection, platform efficiency, judgment assertiveness, well-defined tasks and objectives. This also applies to solutions aligned with expected outcomes, which could tip the scale toward an optimal result. Finally, the evolution of crowdsourcing activities at a financial institution is analyzed in all levels of its ecosystem: macro-, meso-, micro- and their interdependence. The case study shows that the evaluation mechanism helps not only to identify the best ideas but to align innovation with business strategy and customer satisfaction.

Keywords Crowdsourcing · Innovation · Pillars · Taxonomy · Evaluation mechanism

---

H. Castillo · A. Martinez-Corral
Universidad Politécnica de Madrid, C/José Gutiérrez Abascal, nº 2, 28006 Madrid, Spain

M. Grijalvo (✉)
Dpto. de Ingeniería de Organización, Administración de Empresas Y Estadística, Universidad Politécnica de Madrid, C/José Gutierrez Abascal, nº 2, 28006 Madrid, Spain
e-mail: mercedes.grijalvo@upm.es

A. Martinez-Corral · M. Palacios
ESCP Business School, Arroyofresno 1, 28035 Madrid, Spain

C. Avilés-Palacios and M. Gutierrez (eds.), *Ensuring Sustainability*, Lecture Notes in Management and Industrial Engineering,
https://doi.org/10.1007/978-3-030-95967-8_25

25.1 Introduction

Nearly fifteen years ago, the world changed radically when users started to collaborate with each other through the Web. This phase was called Web 2.0, which not only allowed growth of people's interaction through the web, but also granted the opportunity to create social trends more rapidly. As a result, companies had to rethink effective strategies in a market that was more susceptible to change in short term and to predict customer demands in the long term.

Throughout this past decade, businesses have been struggling with this issue, but some organizations like P&G, Airbus and LEGO have been able to successfully implement the new business paradigm, crowdsourcing, that involves collective participation to generate solutions, develop ideas and increase productivity [1, 6, 7, 23].

This new approach seeks to give more power to users by allowing them to develop solutions to their own needs. Therefore, the key to success falls into the crowd as it would uncover new trends in society in the long run. [4, 12, 17].

Literature on crowdsourcing is still scarce since it focuses mainly on success stories and has centered on establishing its microsystem level elements: crowdsourcers (source), crowdsourcees (multitude), the task and the channel that connects every node [16], but lacks a clear definition at the meso- and macro-levels.

This study aims to help in filling this gap and moreover other levels of analysis of the innovation ecosystem. Based on that, the main aim of this research is to define the basic characteristics of any crowdsourcing initiative in all levels of its ecosystem: macro-, meso-, microsystem level and their interdependence.

Based on this work, a model is presented and contrasted with cases from the literature and an empirical case of one important entity of the Spanish financial sector. This financial group was selected as a result of implementing and record-keeping crowdsourcing projects for almost a decade in different continents. The results on the financial industry and services indicate a positive impact in society and on its organizational structure.

Findings show that at mesosystem level, the evaluation and its mechanisms build the foundation that holds the four previous elements since they validate, correct and affirm the results obtained in its application, and at the macrosystem level, the legitimacy of crowdsourcing becomes necessary.

This paper begins detailing the methodology used, followed by a theoretical analysis consistent on the literature study and the definition of the model. Later, the case studies are presented, leading to the final conclusions.

25.2 Methodology

The research was designed in three phases. First, a thorough literature review was conducted combining a systematic search on the ISI Web of Science with a careful scan of the reference lists from the relevant articles. The aim was to identify articles that touched upon models and classifications of the crowdsourcing activities. To employ the e-search, we began using the following keywords: crowdsourcing, taxonomy crowdsourcing, crowdsourcing organization, crowdsourcing business, open collaboration, innovative organization, crowdsourcing model y crowdsourcing cases. Later, the search was expanded to include reference lists from the relevant articles found; we selected the journal articles from 2006 and 2016 in English with priority given to magazines focused on businesses and organizations such as those included in databases as: Abi Inform, Springer and the Harvard Business Review. The documents were classified according to the JCR index and Quartile Score. This activity led to the identification of 110 articles relevant to the study.

The second phase consisted of a review of each article to identify the factors, mechanisms and types of crowdsourcing and the organizational models for crowdsourcing activities. This exercise led to a reduction in the number of articles from 110 to a total of 44 documents (27 are referenced in this paper).

The third phase was centered on an evaluation of crowdsourcing activities with cases from the literature and an empirical case from a financial institution. The case was based on an interview to the Events and Activities Manager of the financial institution's Innovation Center. The institution had already accumulated some experience working in this field, and the case was not considered as an analysis of a success case; instead, it sought to provide greater knowledge of the key factors required in the implementation of crowdsourcing initiatives.

25.3 Crowdsourcing Definition and Conceptualization

After Web 2.0, interaction between users paved the way for new conducts that derived in new work techniques driven by new ways of interaction. These concepts were described by [17] when presenting crowdsourcing as "the act of a company or institution taking a function once performed by employees and outsourcing it to an undefined—and generally large—network of people in the form of an open call" ([5], p. 2), distinguishing it from "outsourcing". The theory aims to create a framework where users could be able to collaborate with one another to achieve specific goals throughout an online platform, as the crucial element that unites every participant globally. However, there are several definitions presented by numerous authors in the last two decades who might disagree with Howe's definition. Table 25.1 shows how this investigation lays the first stone to a chronological evolution of the crowdsourcing definition throughout the years according to numerous relevant authors.

The contributions show crowdsourcing as a new strategy that provides high-quality solutions and successfully brings companies and users together within a single core of the business aiming to increase productivity and promote innovation. In the next epigraph, successful implementation of crowdsourcing in businesses will be examined to identify the fundamental factors that ought to be presented in the crowdsourcing structure for its correct implementation, later known as the Pillar

Table 25.1 Crowdsourcing definition according to relevant crowdsourcing researchers

Principal researcher's ideas	Researchers with similar ideas	Contributions
Surowiecki [29]	Leimeister [19] Ford et al. [11] Georgi and Jung [13]	- Business model creates goods and services through collaboration of people - Groups of people tend to provide better responses than the smartest people of that same group
Howe and Robinson [17]	Geiger et al. [12] Chiu et al. [8] Prpic et al. [23]	- Conceptualized crowdsourcing differentiates it from outsourcing - Outsource traditional activities through a network platform - Global collaboration achieves greater results than locally
Brabhaman [4]	Brabham [4] Alonso de Magdaleno and Garcia-Garcia [1] Hossain et al. [14]	- Crowdsourcing as a model to solve problems and provide legitimacy - Possibility to achieve the best solutions through the masses
Estelles-Arolas and Gonzalez-Ladron-de-Guevara [10]	Edgeman et al. [9] Simula and Vuori [27]	- Involves crowdsourcing with innovation - Synergy among individuals and companies as a key factor

Theory [16]. The elements of this Pillar Theory can be identified in the evolution of the definition of crowdsourcing, along with the need for a new element, the evaluation mechanism.

25.4 Crowdsourcing Ecosystem. A Crowd Capital Perspective

Numerous researchers agreed that exist four elements that categorize the fundamentals of a crowdsourcing process. These elements, also called pillars, are: crowdsourcer (source), crowdsourcees (multitude), the task (objective) and the channel (connection) [16]. These pillars influence the way a project is carried out and its possible outcome, since they generate an impact on the sample size, the quality of participants or even the efficiency and effectiveness of the platform. Nonetheless, this investigation has recognized the evaluation and its mechanisms as a new key component, crucial for the pre-implementation analysis due to it validates, corrects and affirms the results obtained (Table 25.2).

Table 25.2 Defining the crowdsourcing pillars

Pilars	Description
Crowdsourcer	- The source that delegates the task to be carried out by a group of people
	- Refers to those in charge of proposing, analyzing and evaluating the task, the candidates, the reward and the means
Crowdsourcees	- Described as "the talent", the private workers, freelancers, experts, young innovation, the crowd, the sample, among others
	- The selection of candidates before starting will define the quality of the solutions provided
Platform	- The channel that connects the crowdsourcers and the crowdsources
	- Allows to develop the ideas and work for a common goal thought the Web
	- Must be reliable, accessible and comply with the basic functionalities required by the tasks; otherwise, it will result in poor execution and, therefore, low performance
Task	- Goal's description and activity design
	- Must be sufficiently simple that can be understood perfectly by all crowdsourcees, otherwise is ineffective. Therefore, the selection process, design and management are crucial to get the best results
Evaluation and mechanisms	- Involves the mechanisms that would ensure reaching the optimum
	- The solutions would be analyzed, so it could fulfill the crowdsourcer's expectations
	- Considers and analyzes the subject's productivity to solve a problem, the means used to achieve the solution, the selection of the evaluation team and the voting mechanisms internal or external

As shown at Table 25.3, a handful of authors emphasized the relevance of this new component: crowd selection, efficient platform, assertive judges, well-defined tasks and objectives and solutions aligned with expected results, which could tip the scale toward an optimal result.

We have noticed that at different levels of the organizations, the academic literature has been centered on success stories of crowdsourcing. Authors have developed models from these success stories at the microsystem level that fall under the pillars. However, when reviewing other levels of the innovation ecosystem, this study has identified other necessary factors for the evolution of the crowdsourcing activities and its success or failure. Thus, at the mesosystem level, we have identified the need of an evaluation system and procedures to govern the relationship among the pillars. In this area, [20, 24] indicated the need for an evaluation mechanism to identify the best ideas, given the limited capacity of the crowdsourcers to incorporate all. Then, [20, 30] included elements like the selection of the evaluating team or the voting mechanisms. This investigation complements the Pillar Theory by adding the evaluation as a new key component that holds the pillars by validating and correcting the results obtained within the earlier model. For instance, the crowd must have previous knowledge of the topic at hand, the results must be independent to avoid biased information, the task must be sufficiently simple to be understood by crowd, the motivation is crucial as well as compensation, and the platform must be accessible and trustworthy. Finally, we observed that at the macrosystem level, the legitimacy of the crowdsourcing activities becomes a required component [22].

25.5 Using the Crowd as a Business Partner. The Case of a Financial Institution

Crowdsourcing operations have been changing this past decade as a result of numerous implementations on the private and public sector. For instance, organizations like Threadless, Innocentive, P&G and LEGO have successfully been reducing operating cost, minimizing technical and economic risks, increasing time to market acceleration and boosting innovation levels by considering and adapting new parameters in their initial crowdsourcing design. The evolution falls back in the evaluation process that has established a change on the process design by focusing on the crowd, its motivation and the final results (quality and feasibility). Another example of the evolution process is the crowdsourcing projects of a Spanish financial institution aiming to provide better financial service experience and to build an online presence in order to evolve its business as fast as market trends. Table 25.4 presents the most famous crowdsourcing platforms of the entity.

The institution started using collaborative models obtaining results aligned with the market tendencies and realizing the importance of working with the crowd. The following years, the institution built its own process improved by trial and error, these changes being correlated to the Pillar Theory and to the evaluation component.

Table 25.3 Pillar classification according to relevant researchers

Reporter	Cr[25.1]	Cs[2]	Pm[3]	Tk[4]	EM[5]
Alonso de Magdaleno and Garcia-Garcia [1]				X	X
Brabhaman [4]		X			
Chanal and Caron-Fasan [6]			X	X	X
Chiu et al. [8]		X	X	X	X
Chesbrough [7]				X	
Bonabeau [2]		X			
Estelles-Arolas and Gonzalez-Ladron-de-Guevara [10]		X		X	X
Saxton and Kishore [25]			X	X	
Geiger et al. [12]		X	X		
Simula and Vuori [27]		X	X		
Simula et al. [28]		X	X		
Hua Ye et al. (2015)					
Howe and Robinson [17]	X	X	X		
Prpie et al. [23]	X	X			X
Boudreau and Lakhani [3]					
Hosseini [15]	X	X	X	X	

(continued)

Table 25.3 (continued)

Reporter	Cr[25.1]	Cs[2]	Pm[3]	Tk[4]	EM[5]
Hossain et al. [14]		X	X		
Palacios et al. [22]	X	X	X	X	
Edgeman et al. [9]	X	X		X	
Nakatsu [21]				X	
Lee et a. [18]			X		
Surowiecki [29]		X			
Zhao and Zhu [30]		X			X
Zheng et al. [31]		X		X	

[25.1] Crowdsourcer, [2] Crowdsourcees, [3] Platform, [4] Task, [5] Evaluation and mechanisms

Table 25.4 Financial institution crowdsourcing projects

Platform/pillars	Crowd	Task	Motivation	Evaluation and comments
Online bank	No crowd—employees	Financial services	–	Internal team Not a crowdsourcing project
Competition of fintech startups	Organization with capital to develop project	Any area of the financial institution	Monetary	External team. An objective delimitation needed, improve the solution quality, hence crowd quality, therefore, upgrade the preselection process and reward
Competition of fintech startups—upgrade	Any company	Any area of the financial institution	Monetary, acknowledge	Internal team. Monetary compensation + crowd acknowledgment = better results. More accurate objective's delimitation to obtain higher quality results. Quality control during the project to take place in the future events
Competition for developers	Developers, independent contractors	Big data	Monetary, acknowledge, company employment	Internal team. Well-delimited subject and objectives (big data applications). Well-delimited motivation parameters: acknowledgment and credit equally important as monetary compensation. Involvement and future integration in the organization. Crowd creation, value capture

(continued)

Table 25.4 (continued)

Platform/pillars	Crowd	Task	Motivation	Evaluation and comments
API market	Organizations and independent contractors	Tools and service development	Monetary, acknowledge, company employment	Tester and internal team. Preselected crowd, quality crowd/experts. Monetary + acknowledge + integration = compensation. Delimited subjects and objectives. Crowd creation, value capture

For instance, selecting the right type of challenge and platform, selecting a qualified crowd and judges for each project and by delimiting the task. Interestingly enough, these same initiatives can be seen on the LEGO case [26] and P&G case [6].

However, when analyzing individuals, it is essential to consider the motivation, the quality of the solutions and compensations sought by each crowd. Nonetheless, the best results were found when control points were set up along the project, the evaluation process corroborated the results on each stage making it possible to tweak the method along the way. This design allowed the institution to create and capture value through crowd implication, feeling of pride and acknowledgment.

25.6 Conclusions

This article reviews the literature on successful implementation of crowdsourcing and offers contributions to both the private sector and the research field. It adds a new element, the evaluation mechanism, to the model provided by scholars at the organization levels [16]. While previous models center on elements at the micro-level, this new element exists at the meso-level. This element sets a more complete framework for the implementation of crowdsourcing. This addition should provide practitioners with further tools to study the evolution of crowdsourcing activities and the success of the methodology.

The existence of this new element is seen in three ways. In the evolution of the definition of crowdsourcing, as the elements of the Pillar Theory appear from one definition to another, including the need for an evaluation mechanism. Then, in the review of the literature of successful crowdsourcing cases. And finally, in the empirical case of a Spanish financial institution.

The current investigation maintains that it is possible to reduce research and development costs, acquire above-average solutions and align with market needs successfully through the addition of the evaluation mechanism as a new component in the model. Consequently, this approach should be able to anticipate or better predict the long-term market need.

Future work with the objective of improving the methodology could include the successful integration of the evaluation process and its mechanisms into the model, the identification of dimensions through which it could be categorized and the evaluation on its impact on business organizations.

Compliance with Ethical Standards The results of this work are supported by an interview to the Events and Activities Manager of the financial institution's Innovation Center, who voluntarily agreed to participate in the study. The interview was semi-structured and included open questions on the implemented initiatives. Company data as well as personal data have been anonymized. Research Ethics Committee of Universidad Politécnica de Madrid approved the research methodology (Ref. DATOS-20210922-MG-Análisis).

References

1. Alonso de Magdaleno MI, Garcia-García J (2014) Crowdsourcing: la descentralización del conocimiento y su impacto en los modelos productivos y de negocio. Cuadernos de Gestión 14(2):33
2. Bonabeau E (2009) Decisions 2.0: The power of collective intelligence. MIT Sloan Manage Rev 50(2):45
3. Boudreau KJ, Lakhani KR (2013) Using the crowd as an innovation partner. Harv Bus Rev 91(4):60–69
4. Brabham DC (2008) Crowdsourcing as a model for problem solving: an introduction and cases. Convergence 14(1):75–90
5. Chanal V, Caron-Fasan ML (2008) How to invent a new business model based on crowdsourcing: the Crowdspirit® case. In Conférence de l'Association Internationale de Management Stratégique, pp 1–27
6. Chanal V, Caron-Fasan ML (2010) The difficulties involved in developing business models open to innovation communities: the case of a crowdsourcing platform. Management 13(4):318–340
7. Chesbrough H (2006) Open innovation: a new paradigm for understanding industrial innovation. Open Innovation: Researching a New Paradigm 400:0–19
8. Chiu CM, Liang TP, Turban E (2014) What can crowdsourcing do for decision support? Decis Support Syst 65:40–49
9. Edgeman R, Engell T, Jensen NG, Vrtik M, Eskildsen J, Tambo T (2015) Structured crowdsourcing: a B2B innovation roadmap. In 10th European conference on innovation and entrepreneurship (ECIE), 17–18 Sep 2015, Univ Genoa, Genoa, ITALY, pp 165–174
10. Estellés-Arolas E, González-Ladrón-De-Guevara F (2012) Towards an integrated crowdsourcing definition. J Inf Sci 38(2):189–200
11. Ford RC, Richard B, Ciuchta MP (2015) Crowdsourcing: a new way of employing non-employees? Bus Horiz 58(4):377–388
12. Geiger D, Seedorf S, Schulze T, Nickerson RC, Schader M (2011) Managing the crowd: towards a taxonomy of crowdsourcing processes. In: AMCIS
13. Georgi S, Jung R (2012) Collective intelligence model: how to describe collective intelligence. In: Advances in collective intelligence 2011. Springer, Berlin, Heidelberg, pp 53–64
14. Hossain M (2015) Crowdsourcing in business and management disciplines: an integrative literature review. J Glob Entrep Res 5(1):21
15. Hosseini M (2014) Crowdsourcing definitions and its features: an academic technical report. In: RCIS 2014 conference
16. Hosseini M, Phalp K, Taylor J, Ali R (2014) The four pillars of crowdsourcing: a reference model. In Research challenges in information science (RCIS), 2014 IEEE Eighth International Conference on. IEEE, pp 1–12
17. Howe J, Robinson M (2006) The rise of crowdsourcing. Wired 14(6):176–183
18. Lee SH, DeWester D, Park SR (2008) Web 2.0 and opportunities for small businesses. Serv Bus 2(4):335–345
19. Leimeister JM (2010) Collective intelligence. Bus Inf Syst Eng 2(4):245–248
20. Martínez-Corral A, Grijalvo M, Palacios M (2019) An organisational framework for analysis of crowdsourcing initiatives. Int J Entrepreneurial Behav Res 25(8):1652–1670
21. Nakatsu RT, Grossman EB, Iacovou CL (2014) A taxonomy of crowdsourcing based on task complexity. J Inf Sci 40(6):823–834
22. Palacios M, Martinez-Corral A, Nisar A, Grijalvo M (2016) Crowdsourcing and organizational forms: emerging trends and research implications. J Bus Res 69(5):1834–1839
23. Prpić J, Shukla PP, Kietzmann JH, McCarthy IP (2015) How to work a crowd: developing crowd capital through crowdsourcing. Bus Horiz 58(1):77–85
24. Riedl C, Blohm I, Leimeister JM, Krcmar H (2010) Rating scales for collective intelligence in innovation communities: why quick and easy decision making does not get it right
25. Saxton GD, Oh O, Kishore R (2013) Rules of crowdsourcing: models, issues, and systems of control. Inf Syst Manag 30(1):2–20

26. Schlagwein D, Bjørn-Andersen N (2014) Organizational learning with crowdsourcing: the revelatory case of LEGO. J Assoc Inf Syst 15(11):754
27. Simula H, Vuori M (2012) Crowdsourcing in business-to-business firms-layers perspective. In: ISPIM Conference proceedings (p. 1). The International Society for Professional Innovation Management (ISPIM)
28. Simula H, Ahola T (2014) A network perspective on idea and innovation crowdsourcing in industrial firms. Ind Mark Manage 43(3):400–408
29. Surowiecki J (2004) The wisdom of the crowd. Doubleday, United States of America
30. Zhao Y, Zhu Q (2014) Evaluation on crowdsourcing research: current status and future direction. Inf Syst Front 16(3):417–434
31. Zheng H, Li D, Hou W (2011) Task design, motivation, and participation in crowdsourcing contests. Int J Electron Commer 15(4):57–88

Chapter 26
Equity-Based Crowdfunding: Pillars and Risks on the New Funding Structure of Collaborative Economy

M. A. Holgado, M. Grijalvo, A. Prada, and E. Ortiz de Lanzagorta

Abstract Crowdfunding is a financing method that seeks to obtain funds from multiple participants to finance a business initiative. This paper aims to study crowdfunding from a risk management perspective, reviewing risks and their connection with stakeholders in a specific crowd model, equity crowdfunding. For that purpose, an extensive review of literature has been carried out to identify the risks involved, and based on the crowdsourcing pillars theory (Hosseini et al in The four pillars of crowdsourcing: a reference model. Proceedings—international conference on research challenges in information Science, 1–12, 2014) as reference of participants, a case study analysis of different entities within the crowd ecosystem was made, proving the pillars/risks connections. As a result of the paper findings, a matrix of risk generators/takers is proposed; an instrument to better detection and mitigation of risks was carried by each pillar. The research showed that this new matrix model approach helps to identify risks not previously detected and simplifies business decision making to enhance each participant's experience, highlighting the role of regulators and technology in risk mitigation and aligning the business strategy with the most risk vulnerable pillar: the funder.

Keywords Crowdfunding · Equity crowdfunding · Risks · Pillars · Collaborative economy · Fintech

26.1 Introduction

The financial crisis of 2008 hindered the access to credit from financial institutions, impacting mainly on the survival of small businesses, start-ups or projects whose business models rely on high disruptive and innovative components [10]. This led

M. A. Holgado · M. Grijalvo (✉) · A. Prada · E. Ortiz de Lanzagorta
Dpto. de Ingeniería de Organización, Administración de Empresas Y Estadística, Universidad Politécnica de Madrid, C/ José Gutierrez Abascal, nº 2, 28006 Madrid, Spain
e-mail: mercedes.grijalvo@upm.es

A. Prada
e-mail: alberto.pradap@alumnos.upm.es

C. Avilés-Palacios and M. Gutierrez (eds.), *Ensuring Sustainability*,
Lecture Notes in Management and Industrial Engineering,
https://doi.org/10.1007/978-3-030-95967-8_26

to the growing popularity of alternative forms of funding [5], which spread globally thanks to the use of the Internet and digital platforms [11]. One of the most important new forms of financing was crowdfunding. As [9] pointed out:

> Crowdfunding involves an open call, essentially through the internet, for the provision of financial resources either in the form of donations (without rewards) or in exchange for some form of reward and/or voting rights in order to support initiatives for specific purposes.

This paper examines equity crowdfunding, the crowd model based on offering private company securities online and with highest volume of capital and participants, as well as the most favorable growth expectations [8, 13]. Some studies interpret risks as drivers in the failure of a campaign [6], focusing on how they could be mitigated [8] or comparing them with the risks of other financial products [14], without considering their relationship with the participants of the campaign. Therefore, an exhaustive study of the existing literature was performed and an analysis of several case studies with different entities of the crowd ecosystem was conducted to extract common elements, main participants or pillars and the risks associated to this financial product. The main objective of the research is to identify intrinsic risks on equity crowdfunding and their potential connection with the stakeholders. Taking crowdsourcing pillars theory [7], due to its similarity as a crowd model, as a base reference to identify participants, the definition of the methodology used is presented; then, the theoretical and practical analyses are described. Afterwards, building upon our literature review and the cases study carried out, the theoretical pillars structure and matrix of risk generators/takers are proposed, highlighting the role of regulators, which will lead to the conclusions.

26.2 Methodology

In order to achieve the objectives stated, the research was designed following a three-phase model of development.

First, a bibliographic search of characteristics and different types of crowdfunding was carried out. The bibliographic search was limited to Q1 on web of science, based on the following keywords: crowdfunding, crowdfunding risks, fintech and equity crowdfunding.

The second phase involved the evaluation, analysis and standardization of the information obtained according to their main subject (Table 26.1). From this, the main participants were identified taking as a base the structure of pillars defined in

Table 26.1 Crowdfunding articles examined for the investigation

Main subject	Researched material	Used material
Crowdfunding	57	31
Fintech	29	11
Total	86	42

the crowdsourcing [7]. Their relations and the risks generated and taken during the process of financing equity crowdfunding projects were recognized.

Third and last phase focused on the cases study as being part of the process that allowed the development of the risk model. They serve to obtain a complete and meaningful view of the characteristics of real events, such as organizational and management processes [15]. The study cases were presented through direct interview data and secondary information, obtained from the different Websites.

A selection of the cases was made (Table 26.2) in order to make them relevant for the study and following the criteria of quality, access to information and relationship with the topic. For this purpose, crowd platforms, collaboratively financed projects (2Gether) and regulatory organizations (AEFI) were considered.

26.3 Equity Crowdfunding Stakeholders

Crowdfunding distinguishes between five major stakeholder groups: investors, researchers, entrepreneurs or owners, crowdfunding platforms and financial institutions (European Crowdfunding Stakeholders [4]. Only three of them participate actively, leaving researchers and financial institutions as regulators. These only include interest groups made up of people, without including the nexus between them, the project.

Although there are slight differences in the way the different authors define those participants, it is possible to identify four common elements (Table 26.3) which mirror the pillars of crowdsourcing.

Crowdfunding literature reflects a process that, without being an additional pillar, is transversal to all of them, verification and evaluation. Carried out by the platforms, its purpose is to guarantee the success of the projects through risk mitigation.

26.4 Equity Crowdfunding: Risk Generator/Taker Approach

Like any investment decision, crowdfunding involves a degree of risk for participants. Lack of collateral and poor risk management can make investors unwilling to give their money [5]. From the analysis of the literature, it has been possible to extract the main risks present in crowdfunding.

The relationship of crowd equity pillars generates a business link with obligations and rights. In addition, from the case studies analyzed (Table 26.2), it was observed that potential non-compliance by a participant gives rise to some of the risks on equity crowdfunding (Table 26.4) that one or more of the pillars will face. Because of this relationship and based on the identified pillars a risk generators/takers matrix in crowdfunding was able to be generated (Table 26.5).

Table 26.2 Equity crowdfunding entities analyzed and main characteristics

	Fundee	Urbanitae	The Crowd Angel	Crowdcube	2Gether	AEFI
Entity Type	Crowd-Platform	Crowd-Platform	Crowd-Platform	Crowd-Platform	Crowdfund Banking Platform	FinTech Association
Characteristics	Renewable energy projects	Real estate projects	Technological or innovative projects	Any project to be funded	Collaborative banking	–
Investors	Any Spanish Resident with 500€	Any Investor from 500€	Any Investor from 3000€	Any Investor from 10€	2Gether investors	–
Entrepreneurs	Companies with renovable energy project	Real estate project promoters	Start-ups Seed/Growth/Pre-MAB	Established companies and Start-ups	2Gether founders	–

Table 26.3 Crowdfunding pillars definition. (Adapted from "Crowdsourcing Pillars Theory" [7])

Crowdsource Pillars[1]	Crowdfunding Pillars	Definition	References
Crowdsourcees	Funder	An individual who, through the Internet, invests or donates a small amount of money to achieve an objective or project	[8] [1] [2] [12] [3]
Crowdsourcer	Receivers	Individual or corporation seeking funding for their projects or objectives through the Internet	
Platform	Platform	An online platform whose objective is to connect small investors with individuals or corporations with financing needs	
Task	Project	Goal or objective to be financed, the receiver seeks to achieve	

Table 26.4 Risks on equity crowdfunding campaigns

Risk	Definition	References
Adverse selection	One of the contracting parties has different or less information available than the other	[3] [6] [8] [14]
Moral hazard	An individual is aware of the consequences of his actions, while others, unaware of their intentions, suffer the consequences of the acts of the former	
Locked in investment	A situation in which the investor is unable to leave an investment position due to regulations, taxes or other reasons	
Risk of loss	Potential loss of the capital investment made	
Lack of information	Limited receipt of information on the development of a project or company	
Non-regulation risk	Absence of laws and market regulator	
Dilution risk	Decrease in the percentage of the project or company by the issuance of shares	
Fraud hazard	Potential deception by one of the participants to make a profit, and with which someone is harmed	
Platform risk	Potential risk of fraud or cessation of activities by the crowdfunding platform	
Non-accounting risk	Lack or inaccurate control of accounting making it difficult to know if money is being used as planned	
Dividend risk	Chances of not collecting the dividend when you own shares in a project or company	
Corporate Governance risk	financial uncertainty faced by an investor who holds securities in a specific firm, not knowing the intentions of the management team	

Table 26.5 Risk generators/takers model in crowdfunding

		Generators			
		Funders	Receivers		Platform
Takers	Funders	–	Item Adverse selection	Dilution risk	Non-regulation risk
			Moral hazard	Fraud hazard	
			Lock in investment	Non-accounting risk	
			Risk of loss	Dividend risk	Platform risk
			Lack of information	Corporate Governance risk	
			Non-regulation risk		
	Receivers		–		Platform risk
	Platform		Non-regulation risk		-
			Platform risk		

(*Source* Authors)

Therefore, these pillars are necessary and play an essential role in the success of the campaign, influencing how the project is carried out.

Emphasis is on improving funder perception since it is the most exposed to risk in terms of quantity. This is consistent with the case of 2Gether (Table 26.2) in which in order to generate trustfulness and credibility blockchain technology was used to ensure security during the crowdfunding campaign for funders, the pillar who venture their capital and who are more averse to participating due to their greater exposure (Table 26.5).

26.5 Conclusions

This paper reviews the equity crowdfunding model both in literature and in real cases in terms of the exposure to risk of different participants.

The Funders, Receivers, Platform and Project have been defined as pillars of equity crowdfunding model, including in turn the role of the regulators as a pillar that verifies and guarantees the compliance and development of the validity of the models.

In terms of risks, the founders have been proved the most exposed part. Correct risk management, mitigation and due diligence make it easier for the potential funders to enter the company, boost the crowd call effect and create a bond of mutual confidence with the participants. As a result of the article, a risk generators/taker matrix was formulated providing to the practitioners an instrument to better detection and mitigation of risks carried based on their position.

However, it is important to emphasize the importance of pillar-to-pillar interaction as well as the identification of risks and contingency plans. One of the keys to the success of a crowdfunding project relies on being able to transmit clearly to the crowd project objectives through an accessible, credible and safe platform.

Future works aiming at improving and testing the methodology would be valuable. They could include verifying the validity of the risk generators/takers matrix on other reward base models or assessing how the lack interaction between pillars impacts on business and campaigns success, and since the scope of this research has been limited to Spanish companies, the expansion of the study to other territories would also be very convenient. Several other points related to capital loss or risks would also deserve some investigation, such as estimating the risks of capital loss for crowd investors or the analysis of different types of crowd campaigns, which would check the risks and the pillars distinctive of each of the existing campaign and crowdfunding models.

References

1. de Andrés P, Monjas M (2018) Investment Crowdfunding: an introduction. Madrid: Universidad Autónoma de Madrid
2. Astrauskaite, I. (2017). Economics of crowdfunding by what-if analysis approach. In: Advances in Economics, Business and Management Research (AEBMR), vol 37
3. Casanova L, Dutta S, Cornelius K (2018) Financing entrepreneurship and innovation in emerging markes. Academic Press, Ithaca
4. European Crowdfunding Stakeholders Forum (2018) Available via. https://ec.europa.eu/transparency/regexpert/index.cfm?do=groupDetail.groupDetail&groupID=3130
5. Hervé F, Schwienbacher A (2018) Crowdfunding and innovation. J Econ Surv 32(5):1514–1530
6. Hornuf L, Schmitt M, Stenzhom E (2018) Equity crowdfunding in Germany and the United Kingdom: follow-up funding and firm failure. Corp Governance Int Rev 26:331–354
7. Hosseini M, Phalp K, Taylor J, Ali R. (2014). The four pillars of crowdsourcing: a reference model. In: Proceedings—international conference on research challenges in information science, pp 1–12
8. Kumar B, Kutsuna K, Hoda T (2018) Crowdfunding. lessons from Japan's approach. Springer, Kobe, Japan
9. Lambert T, Schwienbacher A (2010) An empirical analysis of crowdfunding
10. Lee N, Sameen H, Cowling M (2015) Access to finance for innovative SMEs since the financial crisis. Res Policy 44(2):370–380
11. Megías J (2012) ¿Dónde, cuándo y cómo buscar inversión para mi startup? Available via. https://javiermegias.com
12. Rossi A, Vismara S (2017). What do crowdfunding platforms do? A comparison between investment-based platforms in Europe. In: Eurasian business and economics society. Springer
13. Vismara, S. (2018). Sustainability in equity crowdfunding. Technol Forecast Soc Change
14. Wilson, K. E., & Testoni, M. (2014). Improving the role of equity crowdfunding in Europe's capital markets. Bruegel Policy Contribution.
15. Yin R (2011) Applications of case study research

Chapter 27
Selecting the Best Intergenerational Learning Strategies for a Bakery

R. D. Leon, R. Rodriguez-Rodriguez, and J. J. Alfaro-Saiz

Abstract This research aims to determine which are the most appropriate strategies for fostering IGL among the employees who work in a Romanian bakery. Therefore, a multi-phase methodology that combines documentary analysis with analytic network process is employed. Data are collected from the human resource managers of one of the most important Romanian bakeries and processed using SuperDecisions® software. The results bring forward the fact that intergenerational learning can be fostered through training, mentoring, and communities of practice. Besides, when selecting which intergenerational learning strategy to implement, the managers should take into account employees' commitment, work satisfaction, and also the organizational culture. These findings have both theoretical and practical implications. On the one hand, they extend the literature from the human resource management and knowledge management field by presenting the case of a bakery and on the other hand, they offer the human resource managers a viable tool for deciding how to enhance intergenerational learning.

Keywords Intergenerational learning · Knowledge sharing · Human resource strategies · Bakery · Multi-criteria decision

27.1 Ensuring Company's Sustainability Through Intergenerational Learning

A sustainable organization is defined as "a company capable of surviving, adapting and growing in a turbulent environment by developing short-, medium-, and long-term objectives on economic, social and environmental level. It concentrates on satisfying the needs of all the categories of stakeholders and also on creating an

R. D. Leon (✉) · R. Rodriguez-Rodriguez · J. J. Alfaro-Saiz
CIGIP, Universitat Politècnica de València, Camino de Vera, 46022 Valencia, Spain
e-mail: rleon@cigip.upv.es

C. Avilés-Palacios and M. Gutierrez (eds.), *Ensuring Sustainability*,
Lecture Notes in Management and Industrial Engineering,
https://doi.org/10.1007/978-3-030-95967-8_27

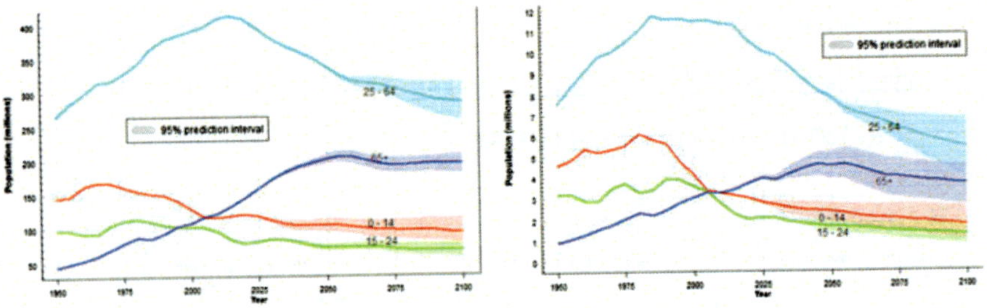

a) Europe b) Romania

Fig. 27.1 Projections regarding population evolution by age groups. Comparative analysis between Europe and Romania [2]

organizational culture that manages to increase stakeholders' awareness to the environmental and social problems, and to encourage and sustain employees' development and empowerment" (Leon [1]: xix). Thus, knowledge is the most important resource and also a source of vulnerability in the context of the contemporary aging society.

According to the United Nations [2], by 2050, one in six people in the world will be over age 65, up from one in 11 in 2019, and Europe is the most affected continent. Since 2000, the evolution of the Romanian population by age group follows the same pattern as the one registered at the European level (Fig. 27.1); thus, the Romanian companies are exposed to an increased risk of knowledge loss due to the mass retirement.

Several scholars [3, 4] claim that in order to avoid corporate amnesia and knowledge loss, intergenerational learning (IGL) strategies should be developed. According to Ropes ([5]: 8), IGL is "an interactive process that takes place between different generations resulting in the acquisition of new knowledge, skills, and values." Therefore, its efficiency depends on both individuals' characteristics (socio-demographic and psychological profile) and organizational coordinates (human capital and structural capital). The former provides the content of the knowledge flows and ensures individuals' availability of sharing and receiving knowledge from those who come from a different generation while the latter defines the context, stimulus, and tools for enhancing intergenerational knowledge sharing.

Despite the complexity of this process, most of the studies developed so far concentrate either on defining the concept [5, 6] or describing several human resource strategies that could enhance it [3, 7]. However, the approaches are mainly based on qualitative analysis and tend to concentrate on the educational industry [8, 9]. An exception is made by Leon [10] who uses the analytic network process as a tool for selecting the most appropriate IGL strategies for the hotel industry, and the results prove that mentoring, on-the-job education, and storytelling are the best choice. Nevertheless, several differences occur between the service and industrial sectors, and these may affect IGL's efficiency. Taking these into account and the fact that, to the best of our knowledge, all of the previously developed studies neglect the impact of the aging society phenomenon on the industrial sector, and the current research aims to determine which are the most appropriate strategies for fostering IGL among the employees who work in a Romanian bakery.

27.2 Intergenerational Learning: Literature Review

According to the generational theory, there is a certain set of values, attitudes, work ethics, and beliefs about personal and professional life that characterize generational cohorts within specific historical periods [11, 12]. These aspects transform each generation into a "community of knowledge" [13] and influence the relationships established among the members of different generations [14]. Thus, they emphasize the knowledge resources owned by the members of a generation and if they are willing to share what they know. Against this backdrop, various studies show that IGL is not a systematic process [15, 16], and it can be enhanced through traditional learning strategies, such as mentoring [12, 17], training [18–20], and on-the-job education [4, 21], and also through modern collaborative learning strategies, like mixed-aged teams [18, 22], storytelling [8, 18], communities of practice [23, 24], and enterprise social networks [3, 4]. Thus, some focus on applying age awareness human resource management practices (such as mentoring, on-the-job education) and exploiting the knowledge owned by the older employees while others concentrate on employing general practices (like training, mixed-aged teams, storytelling, etc.), avoiding the provision of a "special" treatment that may lead to stigmatization [25].

Mentoring is "an intense, dyadic relationship in which a more senior, experienced person provides support and assistance to a more junior, less experienced colleague referred to as a protégé or mentee" ([26], p. 385). Thus, according to the traditional perspective, the older employee is a repository of organizational memory [25] and a role model for the younger co-worker [27]. Within this framework, mentoring becomes a process in which knowledge, such as experience, work ethics, creativity, abilities, skills, and cautiousness, and flows from the older employee to the younger one. However, the intergenerational contact that occurs during mentoring sessions provides opportunities for reciprocal learning [28], and the young employee not only receives the knowledge provided by the older co-worker but he/she can also share some of his/her tacit and/or explicit knowledge. Taking these into account, several researchers [7, 29] bring forward the concept of reverse mentoring which describes a mutually beneficial process where a young person uses his/her knowledge to coach a more senior colleague. Both approaches lead to several individual and organizational outcomes. At the individual level, mentoring is not only developing employees' knowledge and skills, but it also reduces the levels of role ambiguity [30], fosters the development of a feeling of inclusion [31], reduces the negative stereotypes [32], expands the networks [33, 34], and increases job satisfaction and organizational commitment [35]. At the organizational level, it stimulates knowledge creation and transfer [36, 37], and it improves the organizational processes [38]. However, although [27] state that (reverse) mentoring should be implemented in a context-specific way that takes into account the characteristics of the internal and external organizational environment, and none of the previously developed studies take these aspects into consideration. Thus, they treat mentoring as a "successful recipe" and neglect the fact that other IGL strategies may be more appropriate for a certain organization/industry than mentoring.

Training and *on-the-job education* focus on skills acquisition, and they are usually the primary mode of education for staff development Gallo [39]. Thus, they help the members from different generations to communicate more effectively with one another [40] while enhancing knowledge exchange and the development of social capital.

Mixed-aged teams enhance mutual learning due to the fact that "the older workers learn to use their large store of experience and expert knowledge in a way that complements the younger generation's more current knowledge of technological or societal changes" (Ropes [5], p. 8). Hence, they act as a bridge that not only connects the employee from different generation but it also supports a mutual knowledge exchange, and each member has the possibility to be both knowledge receiver and knowledge diffuser. Still, previous studies state that in order for this strategy to be successful, it requires an open organizational culture [24, 41, 42], promoting values like collaboration and trust [42, 43] and developing employees' teamwork skills [44, 45].

Storytelling is a general human resource management practice that fosters IGL and "helps seniors to step forward from isolation" [46]. It puts the employee in the spotlight, and it allows him/her to share what he/she wants quickly, naturally, clearly, truthfully, collaboratively, persuasively, accurately, intuitively, entertainingly, movingly, and interactively [47]. Alfrey et al. [8] state that in order for this strategy to be successful, the narrative should be constructed as an advice-seeking exercise.

Communities of practice describe a group of individuals who have a common interest, solve specific problems, and create new knowledge by practicing together [48]. They provide not only a good environment for tasks performing but they also stimulate mutual adjustment and tasks distribution Rupcic [24]. According to the classical approach, they can be represented by the quality circles. However, in light of the current technological progress and the increased use of social media, *enterprise social networks* provide a new framework for the development of the online communities of practice. The latter is based on five principles, namely (Leon and Damasaru [49], p. 217)

- "knowledge is public—anyone can access individuals and groups memory (the artefacts that the members have already shared) at anytime and from any place in the world; anyone can read and comment previous posts which synthesize what individuals have felt, thought or done in a certain situation;
- the membership is open—anyone can become a member as long as he/she is interested in the subject and is willing to access others' ideas, opinions and experiences and to disseminate what he/she knows or believes within the group;

- the members are volunteers—they become part of the community because they want to and they may leave if they are no longer interested in the subject; nevertheless, they decide to join because they believe they can facilitate the development of future approaches or directions;
- knowledge distribution goes beyond the organizational boundaries—each of the members includes the acquired ideas, thoughts, emotions, beliefs, experiences in his/her mental models and take them with him/her into the real world, fostering their dissemination worldwide;
- interaction is mainly virtual and technology-based—the members meet in a virtual space and communicate with one another without having any type of face-to-face interaction."

Last but not least, this modern approach changes the centers of power; the managers can no longer control who learns what. In the online communities of practice and in the enterprise social networks, knowledge coordination is hierarchical and uncontrollable [3], and members seek information, advice, and quick solutions for what they want not for what managers think they need.

Although several researchers analyze the advantages and disadvantages of each of these IGL strategies, none of them brings forward which is the most appropriate for a certain company or industry except for [10] who argues that mentoring, on-the-job education, and storytelling are the most suitable strategies for enhancing IGL in the hotel industry. Thus, further research is required in order to determine which are the best strategies for the manufacturing area, especially for the bakery sector.

27.3 Research Methodology

This research aims to determine which are the most appropriate strategies for fostering IGL among the employees who work in a Romanian bakery. Therefore, a multi-phase methodology that combines documentary analysis with analytic network process is employed.

Stage 1: Literature review and questionnaire design. It involves a comprehensive analysis of the studies published in SCOPUS and Web of Science, in the last 12 years, regarding IGL and intergenerational knowledge sharing in order to identify the main organizational strategies used for enhancing IGL. The results of the literature review are synthesized in Table 27.1 and used for designing a questionnaire, which involves comparing the criteria and variables, based on Saaty's [50] scale of 1–9.

Stage 2: Data collection. The questionnaire is distributed among the human resource managers of the most important player from the Romanian bakery sector, which will be dubbed further as RBL. RBL is selected due to its position on the market, and its employer brand, and the questionnaire is filled by three out of ten human resource managers, namely the national human resource manager and two regional human resource managers. The research aim, methods, and implications

Table 27.1 IGL strategies and influence factors

Categories	Criteria	Variables	References
Influence factors	Socio-demographic profile	Age	Baily [7]; Bjursell [6]; Ropes and Ypsilanti [51]
		Employment length	Brucknerova and Novotny [9]; Day and Gu [52]; Plant et al. [53]
		Position in the company	Clark and Eastland [54]; Rupčić [24]; Yen et al. [55]
	Psychological profile	Interpersonal trust	Hau et al. [43]; Holste and Fields [56]; Nisula and Metso [42]
		Motivation	Burmeister et al. [57]; Kaše et al. [17]; Plant et al. [53]
		Knowledge self-efficiency	Burmeister et al. [57]; Kaše et al. [17]; Milligan et al. [58]; Van Acker et al. [59]
	Organizational human capital	Commitment	Jeung et al. [60]; Li et al. [61]; Ouakouak and Ouedraogo [62]
		Reward management	Kosir and Soba [20]; Ropes and Ypsilanti [5]; Ropes [51]
		Work satisfaction	Nisula and Metso [42]; Ropes and Ypsilanti [51]
	Organizational structural capital	Company's size	Coetzer et al. [63]; Csillag et al. [64]
		Organizational culture	Kazak and Polat [41]; Nisula and Metso [42]; Rupčić [24]
		Technology	Egloffstein and Ifenthaler [65]; Kaminska and Borzillo [3]; Razmerita et al. [66]
Intergenerational learning strategies	Mentoring		Kaminska and Borzillo [3]; Kaše et al. [17]; Satterly et al. [12]
	Mixed-aged teams		Geeraerts et al. [18]; Kyndt et al. [22]; Ropes [21, 67]

(continued)

Table 27.1 (continued)

Categories	Criteria	Variables	References
	Storytelling		Alfrey et al. [8]; Geeraerts et al. [18]; Harvey [15]
	Training		Geeraerts et al. [18]; Gerpott et al. [19]; Kosir and Soba [20]
	Communities of practice		Geeraerts et al. [23]; Rupčić [24]
	Enterprise social networks		Kaminska and Borzillo [3]; Pauget and Chauvel [4]

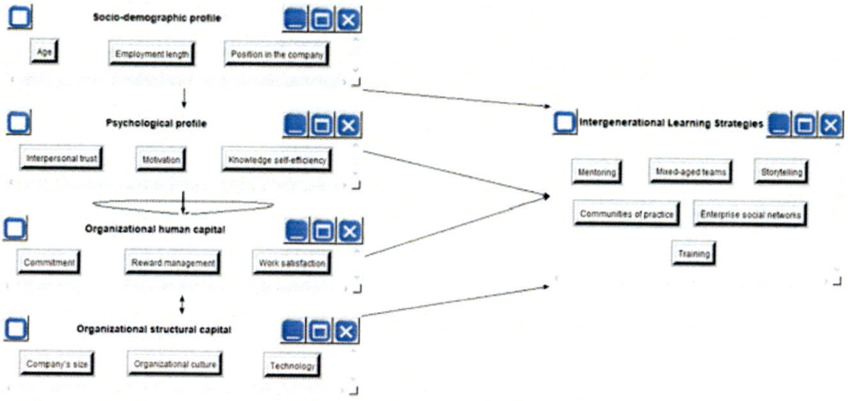

Fig. 27.2 Analytic network process for the IGL strategies used in the Romanian bakery

are presented to the managers, and their participation is voluntary. Thus, an orally informed consent is obtained from the managers.

Stage 3: Data processing and analyzing. The collected data are processed using an analytic network process, which is built with the help of the free software Superdecisions V2.10 (Fig. 27.2).

27.4 Main Results

According to data presented in Table 27.2, the best IGL strategies for RBL are training, mentoring, and communities of practice. These represent the traditional organizational learning strategies and are easily applied for both white-collar and blue-collar employees.

Table 27.2 Overall synthesized priorities for the alternatives

IGL Strategies	Overall priority
Mentoring	*0.197673*
Mixed-aged teams	0.170405
Storytelling	0.116736
Training	*0.269823*
Communities of practice	*0.196442*
Enterprise social networks	0.048920

Table 27.3 Sensitivity to commitment

IGL Strategies	Alpha 0.25	Alpha 0.5	Alpha 0.75
Mentoring	0.203 (2.5252%)	0.198	0.207 (4.5454%)
Mixed-aged teams	0.169 (−0.5882%)	0.170	0.166 (−2.4096%)
Storytelling	0.124 (5.9829%)	0.117	0.105 (−11.4285%)
Training	0.212 (−21.4814%)	0.270	0.332 (22.9629%)
Communities of practice	0.235 (19.8979%)	0.196	0.152 (−22.4489%)
Enterprise social networks	0.056 (14.2857%)	0.049	0.039 (−20.4081%)

Nevertheless, if nodes' sensitivity is considered, several changes occur in alternatives' priority. These most sensitive factors are commitment (Table 27.3), work satisfaction (Table 27.4), and organizational culture (Table 27.5).

Table 27.4 Sensitivity to work satisfaction

IGL Strategies	Alpha 0.25	Alpha 0.5	Alpha 0.75
Mentoring	0.186 (−6.0606%)	0.198	0.254 (28.2828%)
Mixed-aged teams	0.175 (2.9412%)	0.170	0.149 (−12.3529%)
Storytelling	0.120 (2.5641%)	0.117	0.101 (−13.6752%)
Training	0.269 (−0.3704%)	0.270	0.276 (2.2222%)
Communities of practice	0.199 (1.5306%)	0.196	0.181 (−7.6531%)
Enterprise social networks	0.051 (4.0816%)	0.049	0.039 (−20.4082%)

Table 27.5 Sensitivity to organizational culture

IGL Strategies	Alpha 0.25	Alpha 0.5	Alpha 0.75
Mentoring	0.211 (6.5657%)	0.198	0.143 (−27.7778%)
Mixed-aged teams	0.165 (−2.9412%)	0.170	0.193 (13.5294%)
Storytelling	0.108 (−7.6923%)	0.117	0.146 (24.7863%)
Training	0.302 (11.8519%)	0.270	0.168 (−37.7778%)
Communities of practice	0.171 (−12.7551%)	0.196	0.278 (41.8367%)
Enterprise social networks	0.042 (−14.2857%)	0.049	0.072 (46.9388%)

Thus, if the weight of importance of commitment is reduced by 50%, the overall change in the intergenerational learning strategies will sum 20.6221%. On the other hand, if the weight of importance of commitment is increased by 50%, the sum of the percentage changes will be −29.1868%. So, commitment acts as an inhibitor for group intergenerational learning strategies, such as mixed-aged teams, storytelling, communities of practice, and enterprise social networks.

In the case of work satisfaction (Table 27.4), it can be noticed that a decrease of importance of 50% generates a change in the intergenerational learning strategies of 4.6865% while an increase of 50% may cause a decrease of 23.5844%. The most vulnerable strategies to work satisfaction are mentoring and the use of an enterprise social network.

If the organizational culture is enhanced by 50%, the interest in IGL strategies increases by 61.5356% while if its importance is diminished by 50%, this generates a reduction of 19.2567%. Hence, the organizational culture discourages the traditional learning strategies, based on formal knowledge sharing such as mentoring and training, and enhances the online and offline collaborative learning strategies, like mixed-aged teams, storytelling, communities of practice, and enterprise social networks.

27.5 Conclusions and Further Research Directions

An analytic network process is used for selecting the most appropriate strategies for fostering IGL among the employees who work in a Romanian bakery. The results prove that the best strategies are training, mentoring, and communities of practice, and their selection is strongly influenced by employees' commitment, work satisfaction, and organizational culture.

These findings have both theoretical and practical implications. On the one hand, they extend the literature from the human resource management and knowledge management field by emphasizing the most appropriate IGL strategies. Although they are in line with Geeraerts et al. [18] who claim that the classical professional development programs, such as training and mentoring, enhance IGL, it also complements their results by proving that developing communities of practice (a more modern strategy) can also be useful. Furthermore, it complements the findings of Leon [10] by highlighting the IGL strategies that are suitable for a bakery. Unlike the hotel industry where mentoring, on-the-job education, and storytelling can be successfully used for fostering IGL, in a bakery an efficient IGL can be achieved through training, mentoring, and communities of practice.

On the other hand, they offer to the decision-makers a viable tool for selecting the most appropriate strategy for supporting IGL and avoiding corporate amnesia and knowledge loss in the contemporary aging society. Thus, if the policy-makers want to deal with the challenges caused by an aging society, they could focus on mentoring, training, and communities of practice. However, they should take into account the influence of commitment, work satisfaction, and organizational culture on the effectiveness of the aforementioned IGL strategies. If they value commitment, then they should focus on training; if they value work satisfaction, they should concentrate on mentoring; if the organizational culture is considered highly important, then they should encourage the development of communities of practice.

Despite these theoretical and practical implications, the current research is limited by the characteristics of the analyzed case study. It only takes into account the situation of a single bakery and it uses a classical method for multi-criteria decision-making, namely ANP. Based on these, several future research lines can be identified. First of all, the analysis could be expanded to a statistically significant sample of bakeries. Secondly, instead of the ANP, other techniques could be used; for instance, the fuzzy ANP could be applied as it allows the decision-makers to use a range instead of a determined number when comparing the variables. Last but not least, other IGL strategies could be considered in future studies.

Compliance with Ethical Standards The results of this work are supported by responses to a questionnaire filled by three human resource managers. Participants of the research study were informed about the research aim, methods, and implications, and they voluntarily agreed and gave consent to participate in the process. Company data as well as personal data have been anonymized avoiding ethical conflicts.

References

1. Leon RD (2018) Managerial strategies for business sustainability during turbulent times. IGI Global, Hershey, PA
2. United Nations (2019) World population ageing 2019. United Nations, New York
3. Kaminska R, Borzillo S (2018) Challenges to the learning organization in the context of generational diversity and social networks. Learn Organ 25(2):92–101
4. Pauget B, Chauvel D (2018) Intergenerational learning and memory. Learn Organization 25(2):74–80

5. Ropes D (2014) Intergenerational learning in organizations. an effective way to stimulate older employee learning and development. Develop Learn Organizations 28(2):7–9
6. Bjursell C (2015) Organizing for intergenerational learning and knowledge sharing. J Intergenerational Relat 13(4):285–301
7. Baily C (2009) Reverse intergenerational learning: a missed opportunity? AI & Soc 23:111–115
8. Alfrey L, Enright E, Rynne S (2017) Letters from early career academics: the physical education and sport pedagogy field of play. Sport Educ Soc 22(1):5–21
9. Brucknerova K, Novotny P (2017) Intergenerational learning among teachers: overt and covert forms of continuing professional development. Prof Dev Educ 43(3):397–415
10. Leon RD (2020) Fostering intergenerational learning in the hotel industry: a multiple criteria decision-making model. Intergenerational J Hosp Manage 91:102685
11. Parry E, Urwin P (2011) Generational differences in work values: a review of theory and evidence. Int J Manag Rev 13:79–96
12. Satterly BA, Cullen J, Dyson DA (2018) The intergenerational mentoring model: an alternative to traditional and reverse models of mentoring. Mentoring Tutoring Partnership Learn 26(4):441–454
13. Kuyken K (2012) Knowledge communities: Towards a re-thinking of intergenerational knowledge transfer. VINE J Inf Knowl Manage Syst 42(3/4):365–381
14. Howe N, Strauss W (2007) The next 20 years: how customer and workforce attitudes will evolve. Harv Bus Rev 85(7/8):41–52
15. Harvey JF (2012) Managing organizational memory with intergenerational knowledge transfer. J Knowl Manag 16(3):400–417
16. Kuyken K, Ebrahimi M, Saives AL (2009) Intergenerational knowledge transfer in high-technological companies: A comparative study between Germany and Quebec. Paper presented at the Administrative Sciences Association of Canada, Niagara Falls, Ontario, 7 June 2009
17. Kaše R, Saksida T, Mihelič KK (2018) Skill development in reverse mentoring: Motivational processes of mentors and learners. Hum Resour Manage 58(1):57–69
18. Geeraerts K, Tynjala P, Heikkinen HLT (2018) Inter-generational learning of teachers: what and how do teachers learn from older and younger colleagues? Eur J Teach Educ 41(4):479–495
19. Gerpott FH, Lehmann-Willenbrock N, Voelpel SC (2017) A phase model of intergenerational learning in organizations. Acad Manage Learn Educ 16(2):193–216
20. Kosir S, Soba VA (2016) Theoretical and some practical perspectives on age diversity and comparative age management. Int J Innov Learn 20(3):309–327
21. Ropes D (2013) Intergenerational learning in organizations. Eur J Training Develop 37(8):713–727
22. Kyndt E, Gijbels D, Grosemans I, Donche V (2016) Teachers' everyday professional development: mapping informal learning activities, antecedents, and learning outcomes. Rev Educ Res 86(4):1111–1150
23. Geeraerts K, Vanhoof J, Van den Bossche P (2018) Teachers' intergenerational advice and information seeking: content matters! J Prof Capital Community 3(4):256–271
24. Rupčić N (2018) Intergenerational learning and knowledge transfer—challenges and opportunities. Learning Organization 25(2):135–142
25. Angeloni S, Borgonovi E (2016) An ageing world and the challenges for a model of sustainable social change. J Manage Dev 35(4):464–485
26. Hezlett SA, Gibson SK (2007) Linking mentoring and social capital: implications for career and organization development. Adv Dev Hum Resour 9(3):384–412
27. Babnik K, Širca N (2014) Knowledge creation, transfer and retention: the case of intergenerational cooperation. Int J Innov Learn 15(4):349–364
28. Kearney D, Gebert D (2009) Managing diversity and enhancing team outcomes: the promise of transformational leadership. J Appl Psychol 94(1):77–89
29. Tempest S (2003) Intergenerational learning: A reciprocal knowledge development process that challenges the language of learning. Manag Learn 34(2):181–200
30. Lankau MJ, Scandura TA (2002) Mentoring and personal learning: content, antecedents and outcomes. Acad Manag J 45(4):779–790

31. Madan P, Srivastava S (2017) Assessing the impact of demographic variables on mentoring-managerial effectiveness relationship: an empirical study of Indian bank managers. Gend Manage 32(5):369–383
32. Costanzo PR, Hoy MB (2007) Intergenerational relations: themes, prospects, and possibilities. J Soc Issues 63(4):885–902
33. Hackmann DG, Malin JR (2020) From dyad to network: the evolution of a mentoring relationship. Mentoring Tutoring 28(4):498–515
34. Montgomery BL (2017) Mapping a mentoring roadmap and developing a supportive network for strategic career advancement. SAGE Open 7(2):2158244017710288
35. Orpen C (1997) The effects of formal mentoring on employee work motivation, organizational commitment and job performance. Learn Organization 4(2):53–60
36. Dunn MB (2019) Early career developmental networks and professionals' knowledge creation. J Manag 45(4):1343–1371
37. Molitor S, Parker L, Vetter D (2018) Mentoring for all: Building knowledge and community. J Prof Capital Community 3(4):242–255
38. Short TW (2014) Workplace mentoring: an old idea with new meaning (part 1). Develop Learn Organ 28(1):8–11
39. Gallo AM (2011) Beyond the classroom: using technology to meet the educational needs of multigenerational perinatal nurses. J Perinat neonatal nurs 25(2):195–9
40. White R (2006) Four generations learning to work better together. Public Manage 88(10):35
41. Kazak E, Polat S (2018) School administrators' instructional leadership behaviors, intergenerational atmosphere, and intergenerational learning in schools. J Intergenerational Relat 16(4):441–462
42. Nisula AM, Metso S (2019) Factors fostering vocational students' workplace learning success in the real workplace environment. J Educ Work 32(6–7):552–569
43. Hau YS, Kim B, Lee H, Kim YG (2013) The effects of individual motivations and social capital on employees' tacit and explicit knowledge sharing intentions. Int J Inf Manage 33(2):356–366
44. Spender JC (2014) Business Strategy. Managing Uncertainty, Opportunity, and Enterprise. Oxford University Press, Oxford
45. Wok S, Hashim J (2013) Communicating and sharing working relationships with older employees. J Commun Manag 17(2):100–121
46. Osoian A (2014) Memories, a bridge towards intergenerational learning. Procedia Soc Behav Sci 142:499–505
47. Denning S (2000) The springboard: how storytelling ignites action in knowledge-era organizations. Butterworth-Heinemann, Boston MA
48. Lave J, Wenger E (1991) Situated learning: legitimate peripheral participation. Cambridge University Press, Cambridge
49. Leon RD, Damasaru C (2016) TripAdvisor: A virtual community of practice. In: Oprean C, Titu MA (eds) Proceeding of the 2nd international scientific conference SAMRO 2016 "News, challenges and trends in management of knowledge-based organizations." Tehnica, Bucharest, pp 216–223
50. Saaty TL (1980) The analytic hierarchy process. McGraw-Hill, New York
51. Ropes D, Ypsilanti A (2012) Factors influencing intergenerational learning: towards a framework for organisations to ensure successful learning in older employees. In European Centre for the Development of Vocational Training (ed) Working and ageing. The benefits of investing in an ageing workforce. European centre for the development of vocational training, Luxembourg, pp 280–307
52. Day C, Gu Q (2007) Variations in the conditions for teachers' professional learning and development: sustaining commitment and effectiveness over a career. Oxford Rev Educ 33(4):423–443
53. Plant K, Barac K, De Jager H (2017) Developing early career professional auditors at work what are the determinants of success? Meditari Accountancy Res 25(3):368–390
54. Clark KR, Eastland R (2019) Managers' perspectives on generational differences in medical imaging departments. Radiol Technol 90(5):442–449

55. Yen M, Trede F, Patterson C (2016) Learning in the workplace: the role of nurse managers. Aust Health Rev 40(3):286–291
56. Holste JS, Fields D (2010) Trust and tacit knowledge sharing and use. J Knowl Manag 14(1):128–140
57. Burmeister A, van der Heijden B, Yang J, Deller J (2018) Knowledge transfer in age-diverse co-worker dyads in China and Germany: how and when do age-inclusive human resource practices have an effect? Hum Resour Manag J 28:605–620
58. Milligan C, Fontana RP, Littlejohn A, Margaryan A (2015) Self-regulated learning behaviour in the finance industry. J Work Learn 27(5):387–402
59. Van Acker F, Vermeulen M, Kreijns K, Lutgerink J, van Buuren H (2014) The role of knowledge sharing self-efficacy in sharing open educational resources. Comput Hum Behav 39:136–144
60. Jeung CW, Yoon HJ, Choi M (2017) Exploring the affective mechanism linking perceived organizational support and knowledge sharing intention: a moderated mediation model. J Knowl Manag 21(4):946–960
61. Li X, Zhang JX, Zhang SS, Zhou MJ (2017) A multilevel analysis of the role of interactional justice in promoting knowledge-sharing behavior: the mediated role of organizational commitment. Ind Mark Manage 62:226–233
62. Ouakouak ML, Ouedraogo N (2019) Fostering knowledge sharing and knowledge utilization the impact of organizational commitment and trust. Bus Process Manag J 25(4):757–779
63. Coetzer A, Kock H, Wallo A (2017) Distinctive characteristics of small businesses as sites for informal learning. Hum Resour Dev Rev 16(2):111–134
64. Csillag S, Csizmadia P, Hidegh AL, Szaszvari K (2019) What makes small beautiful? Learning and development in small firms. Hum Resour Dev Int 22(5):453–476
65. Egloffstein M, Ifenthaler D (2017) Employee perspectives on MOOCs for workplace learning. TechTrends 61(1):65–70
66. Razmerita L, Kirchner K, Nielsen P (2016) What factors influence knowledge sharing in organizations? A social dilemma perspective of social media communication. J Knowl Manag 20(6):1225–1246
67. Ropes D (2015) Addressing the challenges of an ageing workforce: an intergenerational learning toolkit. Develop Learn Organizations 29(4):14–18

Part VII
Organizational Engineering Education

Chapter 28
Supporting Innovations to Incorporate the SDGs at Universities Through MOOCs

M. Soberón, T. Sánchez-Chaparro, V. Oquendo-Di Cosola, F. Olivieri, and I. Ezquerra

Abstract The 2030 Agenda calls for universities to incorporate the SDGs in their processes, organizations and structures. In the field of teaching, MOOCs have rapidly expanded in higher education institutions and might be a potential ally for incorporating the SDGs in university contexts. However, there is a lack of research in using MOOCs for this purpose. This paper focuses on analyzing how MOOCs can be an effective tool for introducing the SDGs in universities. In order to do so, a case study methodology is applied to an online itinerary developed by the Universidad Politécnica de Madrid. Our findings reveal that MOOCs can be a useful tool to support universities to generate interdisciplinary knowledge on sustainability, break down organizational silos, stimulate collaboration with stakeholders from various sectors and promote action-oriented teaching.

M. Soberón (✉) · T. Sánchez-Chaparro · V. Oquendo-Di Cosola · F. Olivieri · I. Ezquerra
Centro de Innovación en Tecnología Para El Desarrollo Humano, Universidad Politécnica de Madrid (itdUPM), 28040 Madrid, Spain
e-mail: miguel.soberon@upm.es

T. Sánchez-Chaparro
e-mail: teresa.sanchez@upm.es

V. Oquendo-Di Cosola
e-mail: valentina.oquendo@upm.es

F. Olivieri
e-mail: francesca.olivieri@upm.es

I. Ezquerra
e-mail: irene.ezquerra@upm.es

T. Sánchez-Chaparro
Department of Organizational Engineering, Business Administration and Statistics, Escuela Técnica Superior de Ingenieros Industriales, Universidad Politécnica de Madrid, 28006 Madrid, Spain

V. Oquendo-Di Cosola · F. Olivieri
Department of Construction and Technology in Architecture, Escuela Técnica Superior de Arquitectura, Universidad Politécnica de Madrid, 28040 Madrid, Spain

C. Avilés-Palacios and M. Gutierrez (eds.), *Ensuring Sustainability*, Lecture Notes in Management and Industrial Engineering,
https://doi.org/10.1007/978-3-030-95967-8_28

Keywords Sustainable development goals · MOOCs · Universities · Innovation · Multi-stakeholder collaboration

28.1 Introduction

Universities have a significant role in the sustainable transformation demanded by the 2030 Agenda, as they can be key players in the development of interdisciplinary knowledge and multi-actor collaborations [10]. In addition, university campuses can be understood as spaces where thousands of people (teachers, students and researchers) move, consume, produce, generate knowledge, innovate, etc. Hence, university campuses act as laboratories where new sustainability solutions can be tested [7, 8].

Nevertheless, universities often encounter organizational and structural problems to promote collaboration between different academic and non-academic actors [11, 19]. Especially because universities have a long tradition in disciplinary specialization that has led to the fragmentation of their structures [2, 6], this conventional approach can hardly address the dynamic and holistic nature of sustainable development [36]. Indeed, the existence of "silos" and the lack of horizontal collaboration are common problems in organizations with long histories and highly hierarchical structures [27].

In regard to teaching and learning, these challenges are reflected in the need for universities to seek new environments and training experiences that can cover the constant updating of knowledge and the interrelation of the current challenges of our society (represented by the SDGs) [8].

Sustainability centers have been created in recent years to support universities to transition toward sustainability, helping universities get involved in sustainability research, education and co-creation processes [28]. Following [28] definition, sustainability centers are university-based research centers that use sustainability and/or sustainable development concepts as the main framework of their research and outreach activities. Regarding teaching and learning, sustainability centers can support the introduction of interdisciplinary training oriented to the SDGs through academic innovation (assembling curricula that incorporate knowledge from different disciplines and generate new ones), regulatory innovation (creating normative that recognizes new interdisciplinary studies and their academic value) and organizational innovation (opening new spaces to put SDG17 into practice, breaking down internal silos and establishing partnerships that go beyond the institutional framework and integrate actors with the capacity for societal transformations into teaching, so that learning is linked to real contexts).

The digitalization of education programs has made it possible for universities to respond in an efficient way to increasing demands for lifelong training and become more accessible to society, offering wider programs regardless of the student's physical location [5]. In this sense, one of the tools that have received increased attention in the last years is the Massive Open Online Courses (MOOCs). MOOCs have rapidly expanded in higher education institutions, and, after a few years of their creation, they are offered by 950 universities around the world [26]. Currently, universities offer MOOCs from most of the existing disciplines, sustainability among them [12]. The scientific literature on MOOCs is also expanding [21], with studies on the impact of MOOCs on education systems, their integration into academic curricula, the democratization of access and dropout prediction, the role of the Virtual Learning Communities (VLCs) connected to the MOOCs, among others [1, 3, 4, 9, 13, 24]. However, to the knowledge of the authors, no study has been specifically devoted to the MOOC as a tool to promote sustainability in organizations. This research focuses on analyzing how MOOCs can be an effective tool for introducing the SDGs in universities.

To carry out this analysis, a case study methodology has been used, based on the experience of the Universidad Politécnica de Madrid (UPM). This university has a center that can be characterized as a university-based sustainability center [28], the Center for Innovation in Technology for Human Development (itdUPM). Through the facilitation of the project by itdUPM, the UPM has promoted an interdisciplinary action-oriented training itinerary based on MOOCs on sustainable development, named "Strategies and Technologies to Achieve the Sustainable Development Goals in Cities (STA-SDGs in Cities)" [14]. In this initiative, special attention has been paid to the construction of Virtual Learning Communities (VLCs), a space for conversation and information exchange that promotes the interaction between students and the collaborative generation of knowledge [9]. Finally, the UPM campus serves as a space for experimenting innovations and implementing prototypes which could then be transferred to the city of Madrid. At this point, it is important to highlight that the university has a multi-stakeholder partnership between Madrid City Council and other private and social actors in the city [29]. This partnership aims to develop initiatives to promote city transformation toward climate neutrality.

28.2 Objectives

This research intends to shed light on the way universities can benefit from MOOCs to introduce sustainability in their core activities. Specifically, we will focus on the capacity MOOCs have as a transformative tool for sustainability centers or other structures within the higher education institutions responsible for leading the change toward the SDGs. Therefore, the research question is:

How can MOOCs support sustainability-related innovations to implement the SDGs in a university context?

28.3 Methodology

In this research, a case study methodology is applied referring to the online itinerary "STA-SDGs in Cities". Case studies have been extensively used to investigate contemporary phenomena in depth and within its real-world context in multiple fields [37], including sustainability and the SDGs [22, 23, 27]. The case study has been built using the following sources of information: documentary review of the process (itinerary outputs, meeting minutes), analysis of data generated by MOOCs (students dropout, students surveys) and analysis of the organizational structure and regulations of the university.

Additionally, as this online itinerary was facilitated by itdUPM, it is worth highlighting the research-action framework itdUPM applies to co-design and transfer sustainability solutions to identify needs through different levels [19]: campus, ecosystem and society (Fig. 28.1). The first level, the campus, is the locus of interdisciplinarity, with professors, students and university staff conceptualizing potential actions and solutions toward the identified need. In the second level, the ecosystem, the multi-actor collaboration takes place, with people and organizations from the itdUPM "ecosystem" being involved in a multi-stakeholder conversation and the co-designing of the solution. The ecosystem is formed by actors of different nature with whom itdUPM has a relationship of some kind, from stable collaborative relationships—the so-called partners—to more sporadic relationships—so-called peripheral

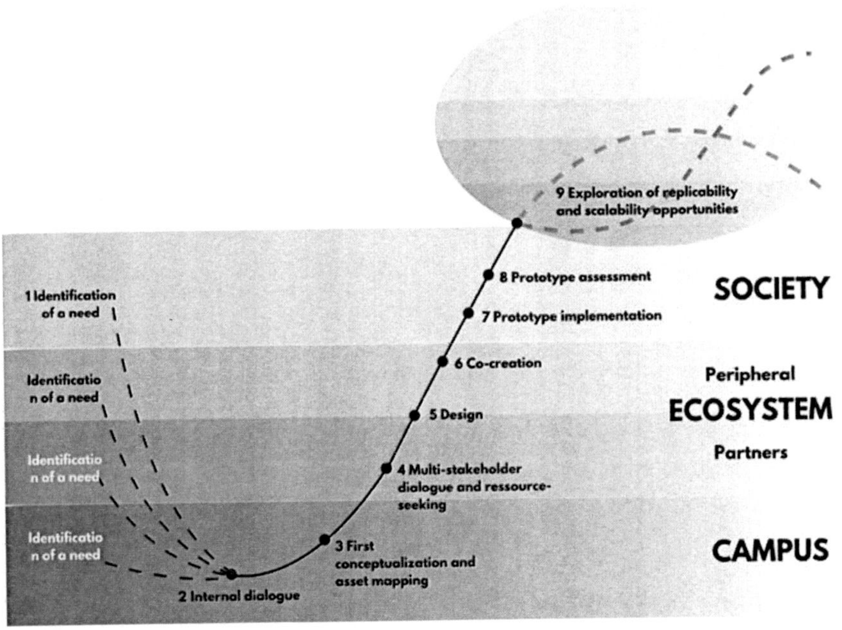

Fig. 28.1 Action-research framework of itdUPM

actors. Finally, in the third level, society, the first prototypes of the solution can be tested and scaled up in case of successful response to identified needs.

28.4 Case Study

The UPM is a medium-size technological public university based in Madrid, divided into four campuses, 17 colleges and 23 research centers. Its academic community brings together approximately 40,000 students from bachelor, master and Ph.D. levels, 3000 lecturers and 2000 staff members [30].

The itdUPM is an innovation center for sustainability created in 2012 by the UPM to promote interdisciplinary and multi-actor collaboration in action-research [19, 20]. These two elements are the basis of all the innovation and knowledge generated in the itdUPM, both in the field of research, teaching and transfer to society. From this approach, and consistently with UPM's sustainability strategy [31], the itdUPM has oriented its activities to:

- Support university transformation toward the SDGs [19, 20, 25, 32].
- Generate knowledge, teaching future sustainability professionals and support organizations on their path toward the internalization of the SDGs [15, 16, 27].
- Bring actors together and build partnerships to catalyze the SDGs through actions at scale [22, 23].

In this context, in 2017 the itdUPM identified different needs in the area of teaching and learning coming from the three levels explained in the methodology. At the campus level, there was a need to integrate sustainability subjects inside university programs under an interdisciplinary perspective. At the ecosystem level, there was a growing need for developing lifelong learning offers to support organizations in incorporating the SDGs into their strategies. At the society level, there was a need for connecting the knowledge created in the university context with action toward the achievement of SDGs, since the SDGs only will be achieved through collective action at global scale. To address these needs, the itdUPM, together with the vice-rectorate of academic strategy and internationalization and department of innovative education of UPM, launched an initiative based on the creation of a training itinerary hosted on MiríadaX, an online teaching platform, composed of three MOOCs:

- "SDGs: An Ineludible Transformation", on how to stimulate sustainable transformation in organizations through creativity and the promotion of a new organizational context.
- "Green Cities: Urban Naturation and Agriculture", on the application of nature-based strategies to combat climate change in cities.
- "Energy and the City: Towards the Energy Transition", on new energy models to promote sustainable transition in cities.

The objectives with which this initiative was proposed aligned with the objectives of the university were

- Expand the educational offer of the UPM in the field of sustainability, internationally. Thus, attract new students and professionals, taking advantage of the reach of online courses.
- Promote collaboration within the university as well as with other organizations to generate interdisciplinary knowledge based on a multi-actor approach around sustainability and the SDGs.
- Be the front-runner of a new line of online training at UPM and generate guidelines to facilitate the creation of subsequent courses.

The itinerary offered an additional VLC outside the platform which aimed to: (i) create conversations about the contents of the course and interactions with the members of the community (on Facebook) (ii) give access to new content streaming and networking opportunities with external actors (through conferences, workshops, walks, etc.) and (iii) promote collective intelligence to transfer the knowledge generated in the community to real-life contexts (on CoLab.UPM). CoLab.UPM is a digital platform that allows to collaboratively imagine solutions around a predetermined problem [17].

28.5 Results

28.5.1 *Academic Innovation*

The academic innovation was implemented through three core elements (Fig. 28.2): collaborative design, VLC and MOOC format.

The collaborative design of the itinerary was built through two fundamental drivers: interdisciplinarity and multi-actor approach. The former was achieved by the coordinated contribution of 34 professors and five researchers from six Higher Technical Schools of Engineering and Architecture of the UPM and 15 different research departments. The latter involved the collaboration in the design phase of four professors from other universities, 13 workers from the private sector, five employees from the public administration and six from non-governmental organizations.

The itinerary incorporated a VLC that welcomed a wide variety of professionals, with varying degrees of experience and knowledge in different disciplines. Thus, the itinerary generated a space parallel to the formal transfer of knowledge, where the VLC functioned as a great collective intelligence. The VLC of the itinerary was composed of 3.243 people, with diverse levels of involvement regarding the type of actor in the different spaces created (events, workshops, CoLab.UPM). In some cases, participants were asked to share information and lessons that they had acquired related to the subject of the courses, and in other cases, they were asked to

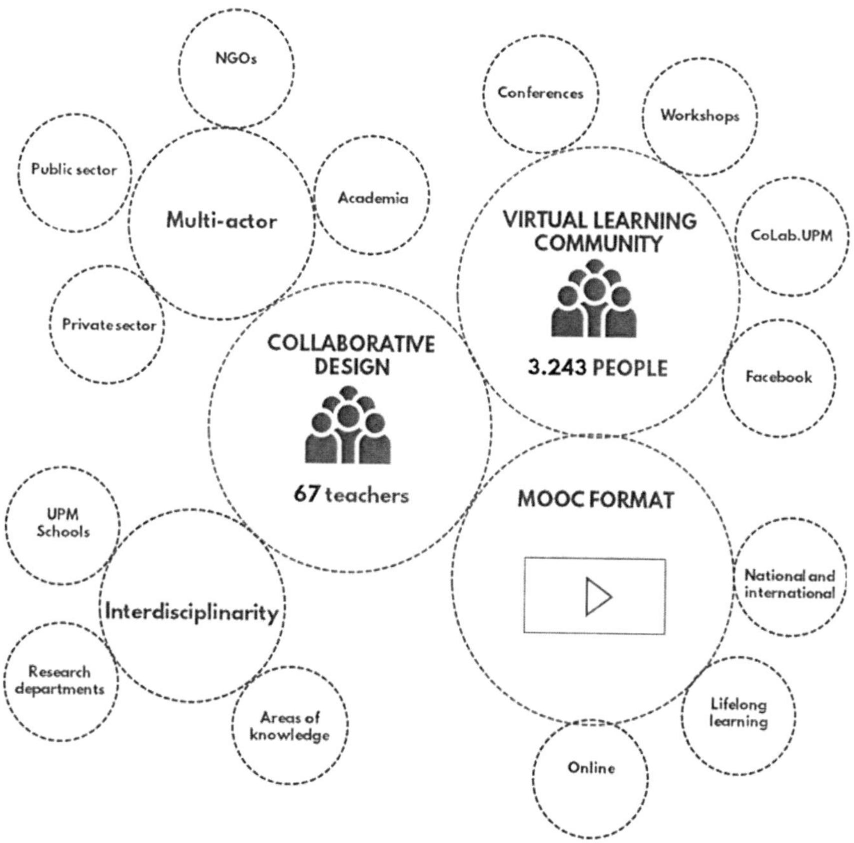

Fig. 28.2 Core elements for the academic innovation

contribute with their knowledge to a co-creation process in a real project, following a project-based learning methodology.

The flexibility offered by the MOOC format allowed the sustainability topic to be accessible to both on campus students and off-campus students, having facilitated access to content that the UPM offers to anyone, anywhere. Thus, the itinerary reached national and international students that were looking for lifelong learning programs. The first edition of the three MOOCs had a total of 7.058 enrolments from 30 different countries. Only 19.3% of the enrollments reported being students as their main occupation, 58.96% being working, 8.85% being unemployed, 1.15% being retired and 12.74% responding “other”. A completion rate of 25.64% from those enrolled and 35.21% from those who started was registered. In addition, 1530 people answered a final satisfaction question measured by a Likert scale (1–5), resulting in a mean satisfaction of the three courses of 4.31 out of 5.

Finally, in addition to the results mentioned above, thanks to these academic innovations, different spin-offs of the course were generated. The most salient one was the adaptation of one of the MOOCs to be taught in a specific organization, the Spanish Agency for International Development Cooperation (AECID), becoming a Small Private Online Course (SPOC). The SPOC was implemented in the course "The 2030 Agenda: Implementation of the Plan of Action in International Cooperation for Development", and its purpose was to generate a reflection among the employees of the AECID about their role in the context of the 2030 Agenda while specific knowledge of the SDGs was acquired, replicating the VLC model used in the itinerary. The course counted on the participation of 197 employees located in 34 different countries and at different hierarchical levels of the organization, eight streaming conversations were generated, and more than 500 comments were produced in the VLC through the activities. Through the course, the following areas were identified as the most relevant to integrate the SDGs in the organization: internal organizational model, systemic approach, multi-stakeholder alliances and social innovation.

28.5.2 (Intra- and Inter-) Organizational Innovation

The process of creation and institutionalization of the online itinerary required the collaboration between the different structures of the university (Fig. 28.3): the vice-rectorate of academic strategy and internationalization drew up the regulations that recognized this type of training; the department of educational innovation dependent of the vice-rectorate of Management and Research Teaching Staff mobilized resources and created a new call to promote online itineraries based on MOOCs; the vice-rectorate for technological services accompanied the entire process of recording and audiovisual support of the itinerary; the vice-rectorate of quality and efficiency provided through the university campus context challenges for the action-oriented learning; UPM schools contributed with their interdisciplinary knowledge to the design of the itinerary (see Sect. 5.1); and the itdUPM facilitated the project encouraging the engagement of the parties involved in its creation and coordinating their contributions. Finally, by having systematized the process and generated guidelines for the design and development of MOOCs, the itinerary sought to facilitate the continuity in the collaborations generated for the more efficient development of new itineraries or collective institutional actions toward the SDGs.

Regarding the inter-organizational innovation, the itinerary generated a space for collaboration with other organizations, not only in the design of the itinerary (see Sect. 5.1) but also during the teaching, to collaboratively look for SDG-oriented solutions to university challenges. Specifically, the commitment that the UPM assumed to achieve climate neutrality by 2040 [33] was used as a context in which to apply the learning outcomes of the course. Thus, the challenge "Achieve a zero-carbon emission campus" was proposed for the VLC to demand solutions in a collaborative way [18]. In order for the solutions to be applicable, a description of the context was prepared with all the data available from the university, 31 experts were involved to recommend improvements and evaluate solutions, and there was support from organizations with the ability to put into practice, at scale, those solutions that were best

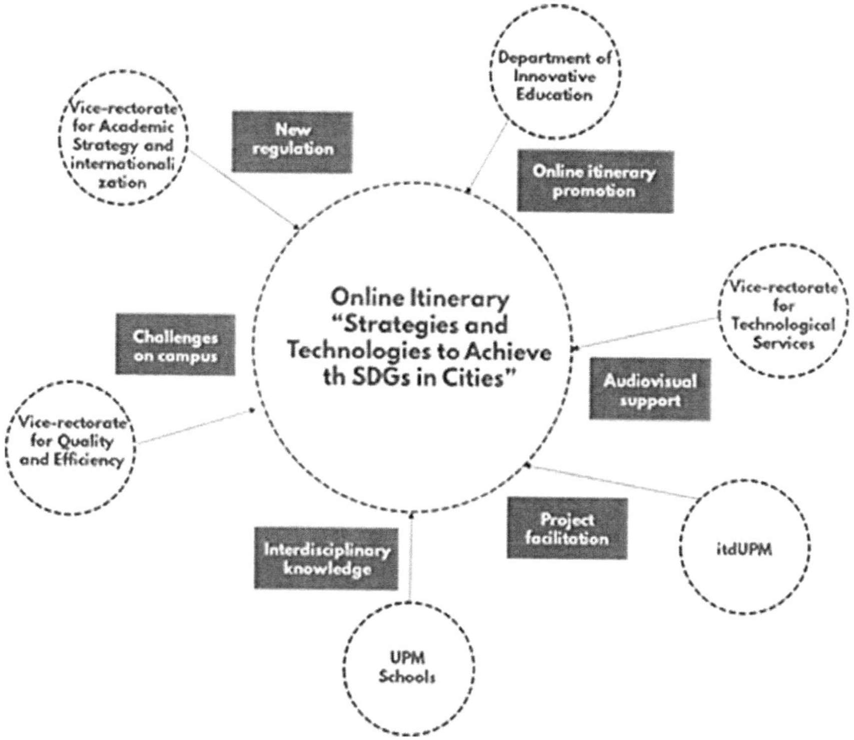

Fig. 28.3 Contributions to the creation and institutionalization of the online itinerary

valued. In addition, students of the engineering management degree at UPM were incorporated into the process of searching for solutions so they could contribute with proposals as campus users. Figure 28.4 shows the different actors involved in the collaborative search for solutions to the UPM challenge.

All proposals were managed through the CoLab.UPM platform, following a co-creation process facilitated by itdUPM through interaction in the platform and a proposal connection workshop. In total, 84 solutions were proposed, related to self-consumption, digitalization, measurement, awareness, mobility and electrification of heat, among others [17]. Since both the MOOCs and the CoLab had organizations and experts with the capacity to implement the solutions, and the UPM has a partnership with the Madrid City Council to develop initiatives to promote city transformation to climate neutrality, it is expected that if these solutions prove successful when implemented on campus, they can be scalable to the city.

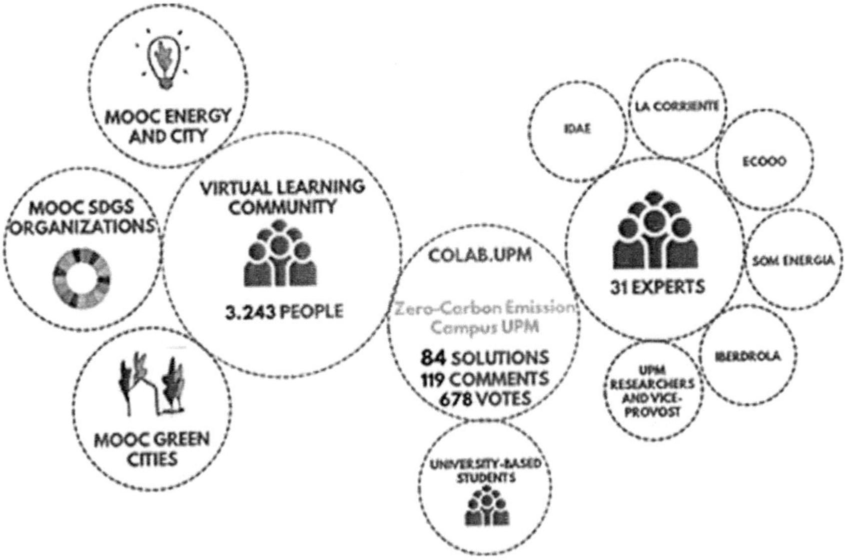

Fig. 28.4 Actors involved in the search for collaborative SDG solutions to "Achieve a zero-carbon emission campus"

28.5.3 Regulatory Innovation

For this itinerary to achieve academic recognition, apart from fitting with university standards, a new regulation was needed. In October 25, 2018, the UPM Governing Council approved a new regulation on UPM degrees that recognizes that:

MOOCs and other varieties of open education of the UPM may be structured in such a way that they constitute a specific training program that articulates training itineraries of at least three courses of this type with a minimum of six ECTS. Its completion will lead, where appropriate, to the obtaining of the corresponding Specific Training Certificate [30, 31].

Under this regulation, the complete itinerary was recognized by the UPM as part of its own academic offer, being the first itinerary of the UPM formed by MOOCs that grants official university credits with an academic load of 7.5 ECTS [34]. However, to connect the itinerary with the official degrees, different pathways have been followed up until now (Fig. 28.5). At the master level, one of the MOOCs of the online itinerary became part of the official master's degree in Strategies and Technologies for Development in 2019 [35], as an optional subject of 2.5 ECTS. At the undergraduate level, the adaptation of one of the MOOCs to a reduced version became part of the Catalog of Creditable Activities of the UPM in 2019, with one ECTS, which can be taken by any student of undergraduate degrees implanted at UPM.

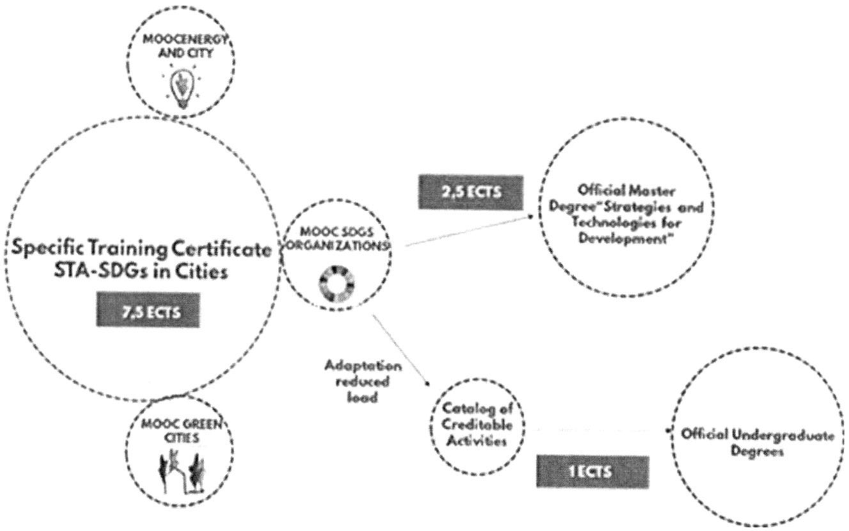

Fig. 28.5 Different educational offers generated by the online itinerary

28.6 Discussion

Following the itdUPM's action-research framework, the online itinerary based on MOOCs has contributed to the university response to the needs identified (interdisciplinarity, action-oriented learning, lifelong learning) in three ways.

In the first place, the academic innovation presented by the collaborative design, together with the intra-organizational innovations that fostered collaboration between different areas and departments of the university (breaking down silos), has allowed the university to create and institutionalize a new teaching and learning itinerary on sustainability based on interdisciplinary knowledge. This itinerary can be taken as a prototype that has opened the organizational space to scale this type of education, developing further MOOC-based programs with more academic load in the field of sustainability.

Second, through the VLC, the itinerary has enabled the creation of the appropriate context for multi-actor collaboration, the use of collective intelligence and the generation of applicable solutions to real contexts. The VLC was an element of the academic innovation that enabled an inter-organizational innovation for the university, bringing MOOC students, university students, experts and other organizations together to develop sustainability solutions on spaces like CoLab.UPM. Understanding university campuses as spaces for the mobilization of citizens and laboratories for new ideas enables the interaction of very diverse agents, who find a space to deliberate and co-create in an environment of symmetry and trust. This context offers an opportunity for university programs to connect teaching with action, allowing students to apply what they are learning into real-life challenges and take part in the collective search of the solutions. The VLC was a context prototype created at university campus level that can be scaled to the city context.

Thirdly, the MOOC format, together with the incorporation of actors from the public, private and social sectors in the design phase from the very beginning, helped the adaptation of the MOOCs for different purposes, being useful for UPM subjects as well as for specific lifelong learning needs reported by students in the survey (80.7% of enrolments were not studying as main occupation) and organizations (i.e., Spanish Agency for International Development Cooperation). Besides, for this to have an academic value, a regulatory innovation was developed to establish mechanisms that enabled MOOCs to be recognized in different modalities. This experience gave rise to two prototypes, one related to the recognition of MOOCs with ECTs that can be escalated to other degrees of the UPM and the other one related to the SPOC for a specific organization that can be escalated to many more.

MOOCs are by themselves a technological innovation that has allowed university knowledge to spread beyond the physical limits of the campus. This tool has demonstrated the ability to bring together numerous groups of people, which is foreshadowed as a great opportunity to promote processes of change to address the challenges of the SDGs. However, to do so, MOOCs need to incorporate a social approach, such as that proposed in the case study, to take advantage of the diversity and heterogeneity of participants, not only of students, but also of those who are involved in the design process. This case study also demonstrates the ability of VLCs to create contexts of knowledge exchange and collaborative design that, by fostering the interaction of multiple actors, link the teaching and learning process with a real context where learning can be applied.

28.7 Concluding Remarks

MOOCs can be a powerful ally for universities to introduce innovations that facilitate the incorporation of the SDGs. In addition to the benefits of online education, when MOOCs are designed with a social approach, they can become the source of interdisciplinary knowledge, multi-stakeholder collaborations and action-oriented processes of teaching and learning. University campuses are spaces for the development and testing of collaborative SDG solutions that can be scaled to the context of cities.

References

1. Alario-Hoyos C, Pérez-Sanagustín M, Delgado-Kloos C et al (2013) Analys-ing the impact of built-in and external social tools in a MOOC on educa-tional technologies. In: Hernández-Leo D, Ley T, Klamma R, Harrer A (eds) Scaling up learning for sustained impact. Springer, Berlin, Heidelberg, pp 5–18
2. Becher T (2001) Tribus y Territorios Académicos: la Indagación Intelectual y la Culturas de las Disciplinas. Gedisa Barcelona, Barcelona

3. Borrás-Gené O (2019) Empowering MOOC participants: dynamic content adaptation through external tools. In: Calise M, Delgado Kloos C, Reich J et al (eds) Digital education: at the MOOC crossroads where the interests of academia and business converge. Springer International Publishing, Cham, pp 121–130
4. Borrella I, Caballero-Caballero S, Ponce-Cueto E (2019) Predict and inter-vene: addressing the dropout problem in a MOOC-based program. In: Proceedings of the sixth (2019) ACM Conference on learning @ Scale. ACM, New York, NY, USA, pp 24:1–24:9
5. Cendon E (2018) Lifelong learning at universities: future perspectives for teaching and learning
6. Clark BR (1991) El sistema de educación superior. Una visión comparativa de la organización académica, Editorial Nueva Imagen. México
7. Evans J, Jones R, Karvonen A et al (2015) Living labs and co-production: university campuses as platforms for sustainability science. Curr Opin Environ Sustain 16:1–6. https://doi.org/10.1016/j.cosust.2015.06.005
8. Favaloro T, Ball T, Lipschutz RD (2019) Mind the gap! developing the cam-pus as a living lab for student experiential learning in sustainability. In: Leal Filho W, Bardi U (eds) Sustainability on university campuses: learning, skills building and best practices. Springer, Cham, pp 91–113
9. Fidalgo-Blanco Á, Sein-Echaluce ML, Borrás-Gené O, García Peñalvo FJ (2014) Educación en abierto: Integración de un MOOC con una asignatura académica
10. Filho WL, Azeiteiro U, Alves F, Molthan-Hill P (eds) (2017) Handbook of theory and practice of sustainable development in higher education, vol 4. Springer
11. Genus A, Theobald K (2015) Roles for university researchers in urban sus-tainability initiatives: the UK Newcastle low carbon neighbourhoods pro-ject. J Clean Prod 106:119–126. https://doi.org/10.1016/j.jclepro.2014.08.063
12. Hajdukiewicz A, Pera B (2020) Education for Sustainable development—the case of massive open online courses. Sustainability 12:8542. https://doi.org/10.3390/su12208542
13. Hakami M (2019) The impact of integrating a MOOC platform on the teach-ing of computer science course: a case study. J Educ Issues 5:44–53
14. itdUPM (2015) Máster en Estrategias y Tecnologías para el Desarrollo - For-mando a los profesionales del futuro en el ámbito de la cooperación, el desar-rollo y la sostenibilidad. In: Máster en Estrategias y Tecnologías para el De-sarrollo. https://www.upm.es/master/etd/. Accessed 13 Dec 2020
15. itdUPM (2017) Cátedra IBERDROLA para los Objetivos de Desarrollo Sos-tenible. In: Centro de Innovación en Tecnología para el Desarrollo Humano. http://www.itd.upm.es/catedraods/. Accessed 31 Dec 2020
16. itdUPM (2019) Cátedra IABS para los ODS y la agricultura baja en carbono. In: Centro de Innovación en Tecnología para el Desarrollo Humano. http://www.itd.upm.es/catedra-iabs-para-los-ods-y-la-agricultura-baja-en-carbono/. Accessed 13 Dec 2020
17. itdUPM (2018) CoLab UPM. In: CoLab UPM. https://madrid-colab.xcolab.org/. Accessed 14 Jan 2021
18. itdUPM (2019) Lograr un campus carbono cero - CoLab UPM. https://madrid-colab.xcolab.org/resources/2019/lograr-un-campus-carbono-cero. Accessed 31 Dec 2020
19. Mataix C, Romero S, Mazorra J et al (2017) Working for Sustainability Transformation in an Academic Environment: The Case of itdUPM. In: Leal Filho W, Mifsud M, Shiel C, Pretorius R (eds) Handbook of theory and practice of sustainable development in higher education. Springer, Cham, pp 391–408
20. Mataix C, Romero S, Moreno-Serna, et al (2017) Regeneración de campus para la creación de un laboratorio vivo de sostenibilidad ("living lab") en el Campus de Excelencia Internacional de Moncloa
21. Meneses EL, Cano EV, Graván PR (2015) Analysis and implications of the impact of MOOC movement in the scientific community: JCR and scopus (2010–13). Comunicar Media Educ Res J 23.https://doi.org/10.3916/C44-2015-08
22. Moreno-Serna J, Purcell WM, Sánchez-Chaparro T et al (2020) Catalyzing transformational partnerships for the SDGs: Effectiveness and impact of the multi-stakeholder initiative El día después. Sustainability 12:7189. https://doi.org/10.3390/su12177189

23. Moreno-Serna J, Sánchez-Chaparro T, Mazorra J et al (2020) Transforma-tional collaboration for the SDGs: the Alianza Shire's work to provide ener-gy access in refugee camps and host communities. Sustainability 12:539. https://doi.org/10.3390/su12020539
24. Pant T (2018) The impact of MOOC on education system: the current sce-nario and the future. In: Library and information science in the age of MOOCs. www.igi-global.com/chapter/the-impact-of-mooc-on-education-system/204192. Accessed 31 Dec 2020
25. SDSN (2020) Accelerating education for the SDGs in Universities
26. Shah D (2020) By the numbers: MOOCs in 2020—class central. In: The report by class central. https://www.classcentral.com/report/mooc-stats-2020/. Accessed 30 Dec 2020
27. Soberón M, Sánchez-Chaparro T, Urquijo J, Pereira D (2020) Introducing an organizational perspective in sdg implementation in the public sector in Spain: the case of the former ministry of agriculture, fisheries food environment. Sustainability 12:9959. https://doi.org/10.3390/su12239959
28. Soini K, Jurgilevich A, Pietikäinen J, Korhonen-Kurki K (2018) Universities responding to the call for sustainability: a typology of sustainability centres. J Clean Prod 170:1423–1432. https://doi.org/10.1016/j.jclepro.2017.08.228
29. Universidad Politécnica de Madrid (2020) Madrid firma oficialmente su ad-hesión al Deep Demonstration de clean & healthy cities. In: Spain. http://spain.climate-kic.org/news/madrid-firma-oficialmente-su-adhesion-al-deep-demonstration-de-clean-healthy-cities/. Accessed 14 Dec 2020
30. Universidad Politécnica de Madrid (2018) Memoria Estadistica 2018
31. Universidad Politécnica de Madrid (2018) Plan de Sostenibilidad Ambiental UPM
32. Universidad Politécnica de Madrid (2019) Seminarios UPM: Tecnología e In-novación para los Objetivos de Desarrollo Sostenible. Universidad Politéc-nica de Madrid. https://www.upm.es/Investigacion/difusion/SeminariosUPM. Accessed 31 Dec 2020
33. Universidad Politécnica de Madrid (2019) Declaración de Compromiso con la Lucha contra el Cambio Climático
34. Universidad Politécnica de Madrid (2018) Normativa titulos propios UPM
35. Universidad Politécnica de Madrid, Universidad Complutense de Madrid (2019) Plan de Estudios Máster Universitario en Estrategias y Tecnologías pa-ra el Desarrollo
36. Waas T, Verbruggen A, Wright T (2010) University research for sustainable development: definition and characteristics explored. J Cleaner Prod 18:629–636. https://doi.org/10.1016/j.jclepro.2009.09.017
37. Yin RK (1981) The case study as a serious research strategy. Knowledge 3:97–114. https://doi.org/10.1177/107554708100300106

Chapter 29
On-Premise Free Data Visualization Tools Within the Design of a Business Intelligence (BI) Learning Activity

M. Fernández, A. Duran-Heras, G. Castilla-Alcala, and S. Ramos

Abstract Business intelligence (BI) platforms are increasingly being implemented in companies in their search for competitive advantages. Nevertheless, their complexity and cost are an obstacle to their inclusion both in companies (especially small and medium-sized enterprises—SMEs) and in university syllabi. This paper analyzes how to overcome these obstacles by comparing, in the context of developing BI learning activities within engineering management programs, three free and easy-to-use software applications (Pentaho Report Designer, Saiku with Pentaho and Power BI) that could be a first step in the introduction to the field of business intelligence through data visualization.

Keywords Data visualization · Open source · Free software · Business intelligence

29.1 Introduction and Objectives

The term business intelligence (BI) encompasses a set of strategies, applications, technologies, data and technical architecture to achieve its goal, namely the harnessing of information to help in decision-making. To obtain this useful information, it is necessary to transform the stored data into information, which, in turn, must be deployed in support of a strategy plan. Within the broad field encompassed by business intelligence, the visualization of data is one of its fundamental components.

M. Fernández · A. Duran-Heras (✉) · G. Castilla-Alcala · S. Ramos
Dpto. Ing. Mecánica, Área Ing. Organización, Escuela Politécnica Superior, Univ. Carlos III de Madrid, Avda Universidad 30, 28911 Leganés, Madrid, Spain
e-mail: duran@ing.uc3m.es

M. Fernández
e-mail: mirifern@ing.uc3m.es

G. Castilla-Alcala
e-mail: castilla@ing.uc3m.es

S. Ramos
e-mail: saramosg@ing.uc3m.es

C. Avilés-Palacios and M. Gutierrez (eds.), *Ensuring Sustainability*,
Lecture Notes in Management and Industrial Engineering,
https://doi.org/10.1007/978-3-030-95967-8_29

Data visualization consists of a process focused on the search, interpretation, comparison and presentation—in illustrated or graphic format—of data. This process allows an in-depth knowledge of these data to be achieved, so that they become understandable information for the user. Likewise, it also helps in the task of decision-making, by visually presenting the appropriate information. In this way, the comprehension of difficult concepts is facilitated, as well as the identification of new patterns.

Thus, full-fledged BI platforms are currently a potentially very valuable strategic tool for companies. Consequently, it would be highly desirable to equip our engineering management students, through appropriately designed hands-on practice based on real-life platforms, with the expertise and skill set required to productively exploit such tools. This might both improve their career prospects (since this skill set in increasingly being sought for by the market) and complement and reinforce their learning in related areas.

However, the monetary and non-monetary cost, the cost of implementation and learning of commercial full-fledged BI platforms are a significant obstacle for their universal, efficient and optimal deployment, particularly among Small and Medium Enterprises (SMEs). It also hinders the otherwise highly desirable inclusion of real-life hands-on BI learning activities in university syllabi. Two major stumbling blocks hamper specifically the implementation of these hands-on BI learning activities: (a) their overall cost, encompassing both infrastructure costs (Hw, Sw licenses…) and the effort involved in their setup and operation, and (b) student time required: particularly, the "entrance barrier" posed by complex software requiring long "familiarization" time by students before they can tackle meaningful tasks.

Therefore, the objective of this paper is to analyze how to overcome these hindrances: infrastructure, setup and operational costs, as well as student's entry barriers. Even though its immediate application will be the design and implementation of BI learning activities within university syllabi, a complementary goal is to help companies, particularly SMEs, to identify a viable approach to the gradual implementation of BI within their own resource constraints. The project is centered in the analysis of software platforms that do not require purchasing licenses; thus, it involves the comparative analysis of three easy-to-use and free data visualization software applications.

29.2 BI Infrastructure

The infrastructure needed for a BI platform is basically software, since hardware is relatively inexpensive, particularly if the quantity of data and the workload are not critical.

To avoid this main cost, it is possible to utilize software whose use does not require the payment of licenses. There are open-source software and commercial software without license costs. In commercial software, the source code is not available, and owners impose certain restrictions on use.

Nowadays, it is increasingly common to find companies and universities that use free open-source software and free commercial software.

Some examples of free "commercial" proprietary software are Google (Chrome, Drive, Google Docs, etc.) or Microsoft software (Microsoft Office online, Power BI Desktop, etc.). These free versions usually imply limitations versus the "full" commercial editions with monetary cost.

Hence, this paper explores on premise open-source software and free commercial software.

29.3 Business Intelligence

The first step to analyze and visualize data is to obtain raw data and convert it to formats compatible with business intelligence tools.

Hence, to be able to carry out the visualization and analysis of data, it is necessary to follow a series of procedures (extraction, transformation and load, or ETL), such as accessing or extracting the data and the preparation of the same for subsequent analysis and/or visualization. These processes will now be reviewed.

29.3.1 Data Extraction

The data extraction process must be accomplished by applying specific tools. Data might come from disparate sources: paper, web page, file, database (relational or otherwise), information system, etc. Specific technologies are required to collect and extract them: Optical character recognition (OCR), web scrapping, database drivers, etc. Some data are structured, but others are unstructured [2]. Through numerous electronic components, such as sensors and intelligent actuators, data can be collected and stored. Most of the data storing projects combine data from different sources.

A very important source for data extraction in companies is transactional systems, from which periodic information is obtained and stored separately (data warehouse).

Thanks to the decrease in the cost of extraction, collection, storage and administration of data, companies are capable of carrying out these tasks, since they need great quantities of data to get results [4].

29.3.2 Data Preparation

Data preparation can occasionally be a complex task depending on the type of data extracted.

The data obtained, both from sensors and that downloaded directly from web pages on the Internet, are usually unstructured, so it must be prepared before it can be analyzed. This process is crucial, and its importance lies in the fact that, even when starting from the same data set, the result can vary greatly depending on the decisions made in this step.

The main data preparation activities can be validation, cleansing, creation of calculated or derived data, organization of data and aggregation.

29.3.3 Data Analysis and Visualization

Once the data in the appropriate format are available, the tools that facilitate its analysis through different algorithms come into play. The range of tools is very wide, the simplest being those of descriptive data analysis, where the analysis tasks consist of representing graphs, creating tables and reports (Microsoft Excel, Calc Libre Office, etc.).

Other tools allow more complete analysis with small volumes of data and without the need for customized models. Examples of this are commercial tools such as SPSS, Rapidminer, PowerBI, Tableau and Qlik, as well as non-commercial tools such as Pentaho, Knowage Community Edition and Weka. These applications can implement traditional algorithms, such as regression and classification techniques, or more advanced algorithms such as machine learning techniques (neural networks, deep learning techniques…).

For large volumes of data and advanced analysis models, more flexible tools involving programming and customization of algorithms are used. SAS and MATLAB are two of the most used commercial tools; however, R, Python or Octave (the open-source alternative to MATLAB) are often used at the free software level.

This article focuses on data visualization tools that constitute a first step in the analysis. All the companies are nowadays aware, to a lesser or greater extent, of the competitive advantage that taking better or at least better-informed decisions may provide [3]. That is why, under this premise, data visualization has become a prominent element within companies that want to be leaders in the development of BI.

Through the visualization tools, the information obtained from the extracted data can be represented more intuitively. Consequently, the communication flows more clearly, simply and effectively, and hence, the decisions derived from it will predictably be of a higher quality.

The presentation of the data can be done through different elements, such as graphs, maps and scatter diagrams[1]. However, there is usually a type of representation that is more efficient and optimal in each situation.

Likewise, interactive visualization allows users to analyze the information in detail. For this, it is necessary to use technology that allows a greater in-depth study of diagrams and graphs.

In short, companies need to obtain information, not data, to help in decision-making. In order to get this useful information, it is essential to transform the data stored into information and this, in turn, into knowledge directed to a strategy plan.

29.4 Visualization Tools

The simplest starting point is to try to answer a business question through the visualization of data in a clear format that can be easily comprehended.

However, many small companies, due to the complexity of the implementation of these tools, are constrained to the use of Excel to present information.

Nevertheless, it is possible to go further and enter the world of advanced graphics. For this, it is not necessary to make a large monetary investment or to learn technical knowledge, since there are tools that are easy to use and do not require an economic investment. These tools cover a wide range, from programming environments to specific packages.

As mentioned above, besides open-source tools, there are also tools for commercial data visualization, whose owner allows free use with certain limitations. An example of this is Microsoft Power BI software.

Nonetheless, there are big differences between these two families of tools (programming environments vs. specific packages). In the case of a programming environment, a greater technical knowledge is required to be able to extract data and its subsequent visualization. On the other hand, using Power BI, the user already has the software created, in this case by Microsoft, with free basic characteristics and with the possibility of increasing and improving the benefits by paying for them. Besides, the technical knowledge required for its use is reduced.

Thus, this article analyzes some scenarios of BI platforms for data visualization. Specifically, this study focuses on easy-to-use free or open-source software: Pentaho Report Designer, Saiku with Pentaho and Power BI.

29.4.1 Pentaho Report Designer

Pentaho is a leader in open-source business intelligence software, with a community version, as well as an enterprise version. This tool encompasses a set of applications that allow not only data visualizations, but also ETL processes and advanced analysis.

Pentaho Report Designer (PRD) is one of the open-source reporting components that can be implemented as a stand-alone tool or as part of the Pentaho Business Analytics distribution.

The reports generated through this tool can be highly detailed; as for the data, from which these reports are created, it can come from virtually any source.

Through this software, the generated documents are divided into different sections of data, in which the different components of each report can be located. That is why this way of generating documents has certain limitations; however, they can be overcome by using sub-reports.

In addition to allowing the user to work with data from various sources, such as Olap4J, Pentaho Analysis, Pentaho Data Integration or XML, the reports generated

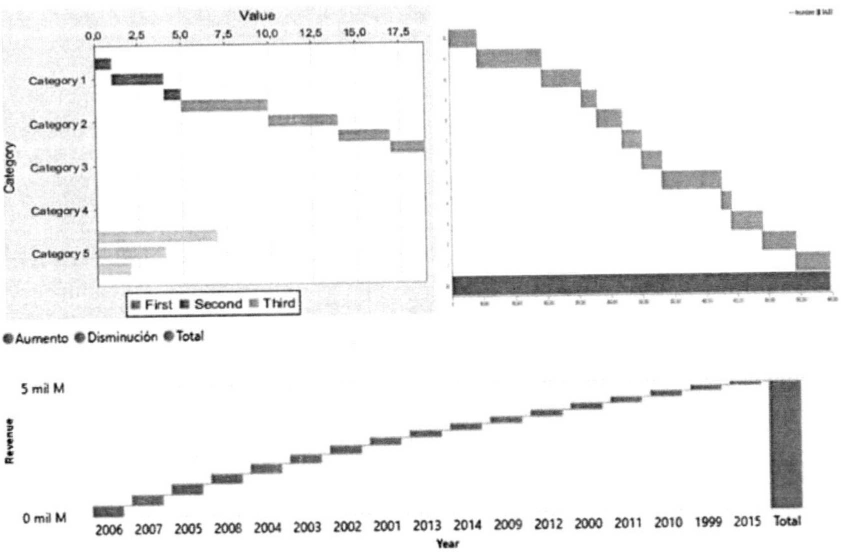

Fig. 29.1 Example of waterfall chart in (a) PRD, (b) Saiku with Pentaho and (c) Power BI

through this tool can also be viewed in different formats, such as PDF, HTML, XLS, RTF and CSV.

This software offers many types of graphs for data visualization: for trend analysis (e.g., line, area, *XY* StepArea, *XY* Step, *XY* Line, sparkline), for comparisons (pie or multi-pie, ring, bar, line, area, radar, *xy* difference, scatter plot, bubble) and to show relationships (bar line combination, waterfall (Fig. 29.1a)).

In the opinion of the authors of this paper, the PRD is not easy to use and requires a certain level of technical knowledge. Pentaho Enterprise components allow interactive reports as opposed to PRD.

29.4.2 *Saiku with Pentaho*

Saiku is an online analytical processing (OLAP) viewer, which is why it provides the user with software to perform data visualization, among other processes, in an easy, simple and intuitive way. But it should be considered that you need to previously create a data cube with Pentaho Mondrian (Workbench schema).

One of the main characteristics that defines Saiku is that it is a flexible and versatile open-source project thanks to its technical architecture. There is a community (CE) version, as well as an enterprise (EE) version that provides additional features.

This application allows you to perform OLAP studies on a separate server, or alternatively to easily incorporate it into a Pentaho server as a plug-in, in order to expand the functionality of this program.

Furthermore, Saiku has a straightforward, simple and intuitive interface, which reduces the barrier for the use of this software. Thanks to this interface, users can design and create their own tables, with the possibility of editing cells, simply "dragging" and "dropping" the fields to be included.

This program also allows you to present the data in different formats, such as in a pivot table or in graphical mode. In fact, Saiku CE includes a wide range of chart types: waterfall (Fig. 29.1b), line, area, sunburst, bubbles grid chart, time wheel, scatter chart, tree map and radar. It also supports the inclusion of mini graphics in the created tables. Additionally, it is possible to install a plug-in for advanced graphics (Saiku Chart Plus) from the Pentaho Marketplace.

This tool is easy to implement and more intuitive than the previous one; in addition, the user experiences a fast learning curve.

29.4.3 Power BI

Power BI Desktop is a free commercial software from Microsoft used to compile interactive reports and visualize data on a Windows personal computer. There are versions that involve a cost (Pro and Premium) and that increase the functionality and make it possible to publish and share the visualizations with other users through the Power BI services in the cloud.

This tool has an easy and simple interface, so that users can create their own reports, maps, graphs and panels quickly and intuitively. One of the main distinguishing features of Power BI is the ability to load customized visualizations.

Likewise, not only does this software allow creating a great variety of types of graphs (Fig. 29.1c), but it also provides the user with data access capabilities from multiple sources, extraction, transformation, data exploration and design of dynamic graphics and interactive panels.

Power BI includes advanced graphics, among which are funnel, donut, tree map, map, gauge, card/multi-card and slicer. Additionally, upon free registration in a Power BI account, an extra repository of graphic objects (AppSource) developed by the community and by Microsoft (similar to the Saiku Marketplace) is available. This repository contains powerful visualizations tools and data mining algorithms.

Due to all of this, out of the three applications analyzed, Power BI is, undoubtedly, the one that allows the greatest variety of graphic visualizations and the easiest to implement. It also presents a rapid learning curve, like Saiku.

29.5 Conclusions

Business intelligence (BI) helps companies to gain competitive advantage. This is due to its ability to generate useful information from the stored data in business management systems. Data visualization is one of the key techniques encompassed by business intelligence.

However, the cost, both of implementation and learning, of full fledge BI platforms makes their implementation in companies, as well as their inclusion in university syllabi, costly and complex. Therefore, this paper presents a project in which three easy-to-use tools without licensing costs have been compared: Pentaho Report Designer, Saiku with Pentaho and Power BI. This comparative analysis suggests that by properly deploying these tools, both SMEs and universities can overcome the main hindrances to BI implementation, namely high infrastructure, setup and operational costs, as well as student's entry barriers.

Acknowledgments This work has been supported by the Madrid Government (Comunidad de Madrid-Spain) under the Multiannual Agreement with UC3M in the line of Excellence of University Professors (EPUC3M20), and in the context of the V PRICIT (Regional Programme of Research and Technological Innovation).

References

1. Anon (2014) Data visualization. educational measurement: issues and practice 33(2):3–4
2. Horton J et al (2010) Systematic review data extraction: cross-sectional study showed that experience did not increase accuracy. J Clin Epidemiol 63(3):289–298
3. Martínez D (2018) El petróleo del siglo XXI. Big Data. Actualidad Económica, p 27
4. Wells S, Whittington A, Talwar R (2017) The fourth industrial revolution. Training J 30–32

Chapter 30
Applying a Cloud-Based Open Source ERP to Industrial Organization Learning Through the Materials Requirements Planning Module

S. Ramos, A. Duran-Heras, G. Castilla-Alcala, M. Fernández, and J. I. Ortiz-Gonzalez

Abstract This paper presents a project aimed at equipping Engineering Management students with the competences and skills required for the forthcoming Connected Industry 4.0. Specifically, it is aimed at applying an open source, cloud-based ERP for teaching Materials Requirement Planning (MRP) at a core operations course taught in several undergraduate programs at the Universidad Carlos III de Madrid (UC3M).

Keywords Engineering education · MRP · Odoo · ERP

30.1 Introduction and Objectives

The industrial sector evolves toward a new paradigm with Connected Industry 4.0 (CI4.0). This new scheme, where traditional production combines with digitalization and Internet, has led many experts to consider it as the Fourth Industrial Revolution [5]. The large volume of data that must be analyzed at high speed is a defining feature of CI4.0.

S. Ramos · A. Duran-Heras (✉) · G. Castilla-Alcala · M. Fernández · J. I. Ortiz-Gonzalez
Dpto. Ing. Mecánica, Área Ing. Organización. Escuela Politécnica Superior, Univ. Carlos III de Madrid, Leganés, España
e-mail: duran@ing.uc3m.es

S. Ramos
e-mail: saramosg@ing.uc3m.es

G. Castilla-Alcala
e-mail: castilla@ing.uc3m.es

M. Fernández
e-mail: mirifern@ing.uc3m.es

J. I. Ortiz-Gonzalez
e-mail: joortizg@ing.uc3m.es

C. Avilés-Palacios and M. Gutierrez (eds.), *Ensuring Sustainability*,
Lecture Notes in Management and Industrial Engineering,
https://doi.org/10.1007/978-3-030-95967-8_30

Nevertheless, this change does not only affect the business ecosystem, but also the educational sector. Higher education must reshape its structure to prepare professionals to develop their work in a dynamic environment. Therefore, it must be able to continuously update their training, introducing all these changes in its educational programs. The academic group behind this paper is thus actually reshaping its educational programs, both incorporating specialized CI4.0 subjects at the master level and adapting the way previous subjects are taught to pave the way for such advanced content.

Effectively handling the data volumes involved in Connected Industry 4.0 and extracting their potential value not only require business processes to be fully supported by Information Systems (IS), but also require a fully integrated approach, since the value of data is greatly enhanced when various data streams and data sets are exploited in an integrated, as opposed to disjoint, manner.

This is one of the major reasons behind the gradual adoption by most companies (starting with the large ones, but increasingly including SMEs) of integrated Enterprise Resource Planning (ERP) systems duc to the fact that all functional modules share the same database where enterprise data are unique and not duplicated. With CI4.0, smart factories must have the capacity to collect a lot of raw data from sensors that monitor business processes and software applications and be able to analyze them efficiently.

Furthermore, the interconnectivity implicit in CI4.0 is an additional incentive, along with potential infrastructure and cost savings, for companies to move their Information Systems to the cloud.

Thus, upon graduation, our current students will most likely face an Information Systems-based working environment orchestrated by integrated, ERP-like systems, often residing in the cloud [4].

On the other hand, we still need to teach our engineering students core techniques, such as Materials Requirement Planning (MRP). This extensively used production planning technique allows anticipating the launch of production and purchase orders, while considering the lead times, in order to meet the demand.

Traditional purely paper-based learning approaches might fail to appropriately prepare our future graduates for the IS-based environment in which they will actually work. Even courses employing stand-alone MRP systems might forego the opportunity to immerse students, at early stages in their learning process, in the type of integrated, IS-enabled, network-enabled settings envisioned in IC4.0, and thus to develop the skill set required by the industry [1].

Thus, the academic group behind this paper has embarked in a multiyear project to reorient Engineering Management education along these lines. For various reasons, including but not limited to budget limitations, open source or at least free software was favored.

Initially, these efforts were aimed at advanced (Master level) and/or specialized courses. This paper presents a project aimed at extending this approach "upstream", to earlier and more basic courses. Specifically, aimed at applying an open source cloud-based ERP for teaching production planning at the University Carlos III de Madrid (UC3M).

In this university, there is a compulsory course called "Industrial Organization" in the curriculum of all the undergraduate degrees that belong to the "Industrial Engineering" family (Industrial Technologies, Mechanical, Electrical...). In this course, they are exposed to an introductory approach to this field of engineering. It is within this context that the design of a learning activity to experience with an MRP software takes place.

Thus, this paper presents a project whose objective is to extend that educational approach "upstream", to earlier and more basic courses. More specifically, it is aimed at applying, in this core operations course, an open source, cloud-based ERP for teaching Materials Requirement Planning (MRP), in a way that, while preserving the traditional learning outcomes (understanding of the key operations concepts and techniques) is also conducive to the gradual development of the abovementioned competences and skills, increasingly required by the job market.

30.2 Literature Review

Among the key learning outcomes to be achieved within the field of industrial engineering in higher education, are those related to logistics and production.

The trend toward CI4.0 implies that conventional techniques and skills must be adapted to the technological and organizational changes involved in the connection of the physical world with the virtual world. This requires communication between the Internet of Things (IoT) elements and information systems, which allow the collection and analysis of very large amounts of data. Besides, nowadays, these processes usually take place in the cloud.

30.2.1 MRP

Material Requirements Planning (MRP) is a method that allows obtaining the manufacturing and purchase orders for products with dependent demand by forecasting the independent demand for final products.

Nowadays, many firms utilize this production planning method, not least because of its simplicity. Through its inputs (Bill of Materials or BOM, Material Production Schedule or MPS and an inventory card), outputs (production orders, purchase orders and possible errors) are computed by using only simple arithmetic operations: addition, subtraction and multiplication. This is one of the reasons for its success, since the logic and the result can be easily understood. This is an advantage over other methods that look like "black boxes" where the logic for arriving at output results is not evident [2]. It also has some disadvantages, since it assumes lead times to be fixed and invariable. Besides, simple MPR system does not take into account capacity constraints (that requires MRP II systems, or, in more sophisticated environments, advanced planning and scheduling (APS) systems), thus the output may be not feasible [3].

In any case, MRP provides at least a preliminary production planning by ensuring the delivery to customers in the shortest possible time. Its main functions can be summarized in three points:

- Ensure that products from the suppliers are available on time. This requires considering the different batch sizing techniques and lead time for each supplier.
- Production must be organized. The MRP considers both the manufacturing lead time and the relationships implied in the Bill of Materials.
- In addition, MRP must comply with the inventory management policies of the firm. That is, stock is kept as low as possible while maintaining the safety stock.

MRP calculations are based on different tables, one for each product, linked through the BOM. The whole process is sufficiently straightforward to be done by hand (at least in educational settings), as the calculations are simple. However, the process becomes tedious when there are many products and different batch sizing techniques.

Therefore, there are multiple MRP packages that automate the operation of an MRP system and perform all the calculations automatically. This way, the operator just has to launch the production and purchase orders that come out as a result of the MRP. Therefore, the different processes will be triggered to obtain the necessary products at the right time.

30.2.2 Software for the Practical Session

The first decision for choosing an application software for the learning activity comes from analyzing two main alternatives. On the one hand, a stand-alone MRP software designed specifically for operations management can be used. On the other hand, it is possible to utilize the MRP module of an integrated ERP. The second alternative is chosen for the reasons outlined in the Introduction section. As discussed there, introducing this software in a basic, undergraduate course has a greater learning advantage since it gives students a first contact with the integrated environment that will be analyzed in further depth at later stages.

Next, it is necessary to analyze the different solutions based on ERP applications. The UC3M Engineering Management area carried out an evaluation process of ERP applications to implement learning activities in the field of Management Information Systems (MIS). The open source Odoo ERP was selected, and several Odoo-based hands-on learning activities have already been implemented in MIS-related courses. The main advantage of this ERP program is that it offers a free educational version with a 10 months duration. After the successful implementation of a learning activity for the subject "Management and Information Systems" in the first year of the Master's Degree in Industrial Engineering, it was decided to extend the use of this software. Therefore, the learning activity for the previously mentioned degree course "Industrial Organization" uses the Odoo manufacturing functional module.

30.2.2.1 Odoo Manufacturing Module

Odoo is an open source ERP, so anyone can access its source code and modify it. In addition, it is backed by a large community of programmers and developers. As a matter of fact, they release a new version almost every year, thus leading to an ever-improving, flexible product.

Although its code is available on the Internet, there is a free Community Edition license and an Enterprise Edition license that the Odoo company sells and for which it offers technical support, both for local and for cloud installation. The price of the license depends on the number of modules, and in the case of the four logistic modules (manufacturing, inventory, sales, purchase), it is less than 90 euros per user and month. Therefore, it is more affordable than other ERP licenses from the leading software companies.

Odoo is an ERP for small and medium-sized enterprises (SMEs) since it offers versatility at a low price. It includes the most traditional ERP modules (accounting, human resources, manufacturing, purchasing, sales…) as well as other more specific modules such as a point of sale, e-commerce or a barcode reader. In addition, it allows operators to make changes in real time from a tablet, as it is compatible with any device. This allows potential users to reduce infrastructure costs and therefore the general costs.

Anyhow, Odoo offers a free educational program for schools and universities. In this program, it allows the user to choose the desired modules and to create a database instance where many practical activities can be accomplished.

This database instance has a 10-month time limit. Once it is activated by e-mail, it can be shared in the "Odoo Educational Program" available on its website. After the publication, each student, in turn, creates an individual instance, which is a copy of the original database. This copy is valid for 4 h, or for 10 months if the student activates it by e-mail. Therefore, there is a double advantage: On the one hand, the students can introduce as many transactions as they need in their instance without affecting either the original database or fellow pupils. On the other hand, the whole process is online, since Odoo's educational offering is based on its cloud.

The Odoo MRP tables are in the manufacturing module, under the MPS section. These tables are related to the other modules, since entering the demand forecast for the chosen time period suffices for obtaining the desired outputs. Thus, the corresponding purchase orders are launched from the purchase module, and the manufacturing orders are in the manufacturing module.

It is important to highlight that the main advantage of an ERP system lies in this interconnection: Modules share a common database. Therefore, any change (transaction) entered through one module will be directly reflected in the database and thus will be instantly available for the other modules.

30.3 Case Study

A learning activity has been designed to allow students to apply the MRP techniques they have learned in a realistic business and IS settings. Students of the "Industrial Organization" course in any undergraduate engineering degree within the "Industrial" family will thus be able to verify the functioning of an MRP software and to observe the calculations for a demand forecast of a fictitious company. Each student accesses the database shared in Odoo educational program and follows the steps to create an instance. It is important to note that each instance is totally independent. Besides, each student can create more than one instance if they need to reset the database. This increases flexibility, since students can change elements and see how they affect the different modules.

In order for everyone to start from the same point, some inputs have been introduced in the original database:

- Inventory modulc: Thc different products and corresponding BOM have been created. In addition, the initial inventory level for each product and its lead time have been introduced.
- Manufacturing module: Some parameters are included in the MRP tables such as the desired safety stock or the maximum or minimum quantity of each order. In addition, it is necessary to activate the MPS as it is not installed by default.

After the students access the database, they are instructed to carry out some tasks, in order to get a better grasp of how the MRP works. First, they must access each of the modules that are installed on the database (manufacturing, purchasing, sales and inventory). This way, students see the different products and the Bill of Materials that relate them.

They then go to the manufacturing module, where the MRP tables can be visualized, and they fill them with a forecast of the demand that comes in the script. From this point on, the MRP automatically calculates both the purchase and manufacturing orders according to their lead times. In addition, it considers the sizing techniques: maximum and minimum quantity that can be carried by the trucks at each order. It is important to mention that currently the "free" Odoo does not support lot size ordering, since this requires the installation of an Odoo app which involves payment.

Finally, supply orders for that day are launched. In this way, both purchase and manufacturing orders are automatically generated. This is the major advantage of an ERP compared to a stand-alone MRP software: The process is fully integrated since all the modules share a common database.

In addition, students can follow the logical sequence of an order fulfillment process. That is, production orders cannot be accepted until the materials are available in the warehouse. Therefore, purchase orders must be accepted first, and it must be assumed that the products have already been received.

Finally, there are some questions that students have to answer using the ERP software. This way, all students can investigate the use of Odoo on their own, so practical learning is encouraged. Figure 30.1 shows the MPS tables for an independent demand

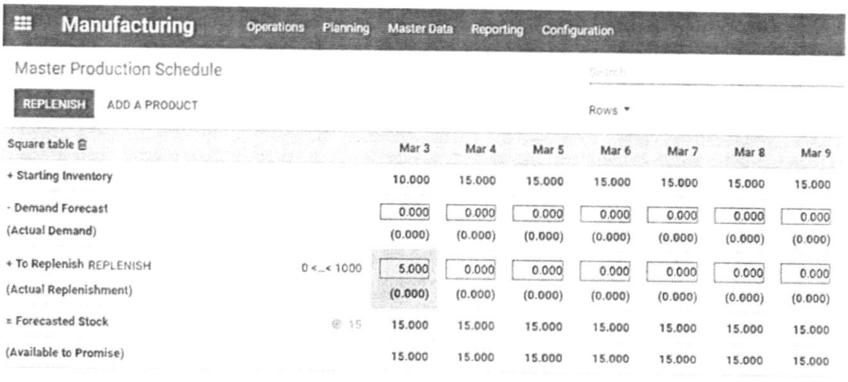

Fig. 30.1 MRP table in Odoo

product in the Odoo manufacturing module.

30.4 Conclusions

In the context of the forthcoming Connected Industry 4.0, there is both a pressing need and a significant opportunity for improving the learning methods. Therefore, higher educations should adapt the teaching of engineering core techniques to prepare students for the Information Systems-based working environment they will experience in industry. The objective of this paper was to extend this approach "upstream" to earlier courses using a learning activity, to be implemented in an operations core course.

More specifically, since most companies use integrated ERP software to manage their business processes, engineering students should graduate with at least a basic knowledge of how they work. That is why it was decided to use a cloud-based, open source ERP to consolidate the concepts they learn in theory about the MRP method. Using the Odoo manufacturing module, students can better understand both the inputs introduced and the outputs obtained in an MRP system. Besides, they can make changes and see the results automatically, without needing to operate by hand. Therefore, with this learning activity, students of the course "Industrial Organization" will not only have a first contact with an ERP and thus with the integrated, cloud-based working environment they will soon face, but will also consolidate the theoretical underpinning of the core techniques taught in the lectures.

Acknowledgments This work has been supported by the Madrid Government (Comunidad de Madrid-Spain) under the Multiannual Agreement with UC3M in the line of Excellence of University Professors (EPUC3M20), and in the context of the V PRICIT (Regional Programme of Research and Technological Innovation).

References

1. Cvetić B, Ilić OR (2015) Using educational software to enhance the teaching and learning in computer integrated manufacturing: material requirements planning module. Comput Appl Eng Educ 23(5):682–693
2. Milne RJ, Mahapatra S, Wang CT (2015) Optimizing planned lead times for enhancing performance of MRP systems. Int J Prod Econ 167(C):220–231
3. Rossi T et al (2017) Improving production planning through finite-capacity MRP. Int J Prod Res 55(2):377–391
4. Schwade F, Schubert P (2016) The ERP challenge: an integrated E-learning platform for the teaching of practical ERP skills in universities. Procedia Comput Sci 100:147–155
5. Wells S, Whittington A, Talwar R (2017) The fourth industrial revolution. Training J 7:30–32

Index

C. Avilés-Palacios and M. Gutierrez (eds.), *Ensuring Sustainability*, Lecture Notes in Management and Industrial Engineering, https://doi.org/10.1007/978-3-030-95967-8